Proceedings of the Conference on

ASSESSING THE EFFECTS OF POWER-PLANT-INDUCED MORTALITY ON FISH POPULATIONS

Proceedings of the Conference on

ASSESSING THE EFFECTS OF POWER-PLANT-INDUCED MORTALITY ON FISH POPULATIONS

Edited by

Webster Van Winkle
Environmental Sciences Division
Oak Ridge National Laboratory
Oak Ridge, Tennessee 37830

Riverside Motor Lodge
Gatlinburg, Tennessee
May 3–6, 1977

Sponsored by

- OAK RIDGE NATIONAL LABORATORY
  Operated by UNION CARBIDE CORPORATION
- ENERGY RESEARCH AND DEVELOPMENT ADMINISTRATION
- ELECTRIC POWER RESEARCH INSTITUTE

PERGAMON PRESS
New York/Toronto/Oxford/Sydney/Frankfurt/Paris

Pergamon Press Offices:

U.S.A.	Pergamon Press Inc., Maxwell House, Fairview Park, Elmsford, New York 10523, U.S.A.
U.K.	Pergamon Press Ltd., Headington Hill Hall, Oxford OX3, OBW, England
CANADA	Pergamon of Canada, Ltd., 207 Queen's Quay West, Toronto 1, Canada
AUSTRALIA	Pergamon Press (Aust) Pty. Ltd., 19a Boundary Street, Rushcutters Bay, N.S.W. 2011, Australia
FRANCE	Pergamon Press SARL, 24 rue des Ecoles, 75240 Paris, Cedex 05, France
WEST GERMANY	Pergamon Press GmbH, 6242 Kronberg/Taunus, Frankfurt-am-Main, West Germany

---

Library of Congress Catalogue Card No. 77-81956

Contract No. W-7405-eng-26

Printed in the United States of America

ISBN 0-08-021950-0

CONTENTS

LIST OF CONTRIBUTORS ix
ACKNOWLEDGMENTS xi
FOREWORD xiii
WELCOMING REMARKS xvii
INTRODUCTORY REMARKS xxi

Part I
CASE HISTORIES

**Density Dependence, Density Independence, and Recruitment in
the American Shad *(Alosa sapidissima)* Population of the
Connecticut River** 3
William C. Leggett
Introduction 3
Density-Dependent Factors 5
Density-Independent Factors 10
Discussion 13
Literature Cited 16

**Some Factors Regulating the Striped Bass Population in the
Sacramento–San Joaquin Estuary, California** 18
Harold K. Chadwick, Donald E. Stevens, and Lee W. Miller
Introduction 18
Environmental Setting 19
Striped Bass Survival 22
Factors Influencing Survival of Young 28
Discussion 33
Literature Cited 34

Man's Impact on the Columbia River Stocks of Salmon 36
Ernest O. Salo and Quentin J. Stober
Introduction 36
History of the Columbia River Salmon Runs 37
Limitations of Impact Assessment and Mitigative Efforts 40
Literature Cited 45

**Population Dynamics of Young-of-the-Year Fish in a
Reservoir Receiving Heated Effluent** 46
Richard Ruelle, William Lorenzen, and James Oliver
Introduction 46

Description of Study Area 47
Materials and Methods 49
Results 51
Discussion 61
Summary 65
Literature Cited 66

Part II
ESTIMATING ABUNDANCE, PRODUCTION, AND MORTALITY RATES OF YOUNG FISH

**Estimating the Size of Juvenile Fish Populations in Southeastern
   Coastal-Plain Estuaries** 71
Martin A. Kjelson
Introduction 71
Southeastern Estuarine Nursery Areas and
    Dominant Fish Species 73
Estimation of Absolute Fish Abundance 74
Quantitative Sampling Apparatus 76
Recommendations for Improving Abundance Estimates
    by Trawls and Seines 79
Literature Cited 86

**Striped Bass (*Morone saxatilis*) Monitoring Techniques in the
   Sacramento–San Joaquin Estuary** 91
Donald E. Stevens
Introduction 91
Study Area 92
Monitoring Techniques 92
Summary 107
Literature Cited 108

**Striped Bass Ichthyoplankton Abundance, Mortality, and
   Production Estimation for the Potomac River Population** 110
Tibor T. Polgar
Introduction 110
Ichthyoplankton Sampling 111
Estimation of Abundances 113
A Model with Uniform Age Distribution within Each Stage 115
A Model with Exponential Age Distribution within Each Stage 117
Discussion of Model Results 121
Literature Cited 125

**Methods for Calculating Natural Mortality Rate, Biomass
Production, and Proportion Entrained of Lacusterine
Ichthyoplankton** (Abstract) 127
Peter A. Hackney

**Confidence Intervals on Mortality Rates Based on the
Leslie Matrix** 128
Douglas S. Vaughan
Introduction 128
Methodology 129
Empirical Results 142
Normal Approximation 147
Summary 149
Literature Cited 149

Part III
COMPENSATION AND STOCK-RECRUITMENT RELATIONSHIPS

**An Argument Supporting the Reality of Compensation in Fish
Populations and a Plea to Let Them Exercise It** 153
James T. McFadden
Introduction 153
Historical Development of General Concept 154
A General Case Argument 158
Compensation in Fish Populations 161
Applicability of Fishery Exploitation Principles
to Power Plant Impacts 176
Literature Cited 178

**Impacts of Recent Power Plants on the Hudson River Striped
Bass (*Morone saxatilis*) Population** (Abstract) 184
K. Perry Campbell, Irvin R. Savidge, William P. Dey, and James B. McLaren

Modeling of Compensatory Response to Power Plant Impact
(Abstract) 185
John P. Lawler, Thomas L. Englert, Robert A. Norris, and C. Braxton Dew

**Assessing the Impact of Power Plant Mortality on the
Compensatory Reserve of Fish Populations** 186
C. Phillip Goodyear
Introduction 186
Compensation 187

Compensation Ratio 190
Compensation Ratio and Power Plant Mortality 193
Literature Cited 194

**Development of a Stock-Progeny Model for Assessing Power-Plant
Effects on Fish Populations** 196
Sigurd W. Christensen, Donald L. DeAngelis, and Andrew G. Clark
Introduction 196
Development of the Mathematical Model 199
Application of the Model 201
Results 206
Discussion 222
Literature Cited 224

Part IV
MONITORING PROGRAMS AND DATA ANALYSIS

**The Quality of Inferences Concerning the Effects of Nuclear
Power Plants on the Environment** 229
Donald A. McCaughran
Introduction 229
Minimum Detectable Differences 230
Application to Impact Studies 238
Literature Cited 242

**Factors to Consider in Monitoring Programs Suggested by
Statistical Analysis of Available Data** 243
John M. Thomas
Introduction 243
Some Changes in Current Practice 244
Designing and Conducting the Field Program 246
Some Aspects of the Statistical Analysis of Data 247
Sample Sizes 251
Literature Cited 254

**Estimation of Age Structure of Fish Populations from
Length-Frequency Data** 256
K. Deva Kumar and S. Marshall Adams
Introduction 256
Models for Fish Length 258
Estimation of the Parameters of the Model 262
Empirical Evaluation of the Estimation Technique 262

Simulation Results 267
Applications of the Model 268
Conclusion 279
Literature Cited 280

**Prediction of Fish Biomass, Harvest, and Prey-Predator Relations
in Reservoirs** 282
Robert M. Jenkins
Introduction 282
Prediction of Fish Standing Crop 283
Prey-Predator Relationships 290
Sport Fish Harvest and Angler Use Predictions 291
Conclusion 292
Literature Cited 293

Part V
USE OF POPULATION MODELS

**Effects of Power Station Mortality on Fish Population Stability in
Relationship to Life History Strategy** 297
Thomas J. Horst
Introduction 297
Impact Assessment and Mathematical Modeling 298
Life History Strategy 299
Population Model 299
Population Projection Matrices 300
Effects of Power Plant Impact 303
Population Simulation 306
Literature Cited 309

**Sensitivity Analysis Applied to a Matrix Model of the Hudson
River Striped Bass Population** 311
Saul B. Saila and Ernesto Lorda
Introduction 311
Materials and Methods 312
Sensitivity Analysis 312
Results and Discussion 327
Literature Cited 331

**Comparison of Simulation Models Used in Assessing the Effects of
Power-Plant-Induced Mortality on Fish Populations** 333
Gordon Swartzman, Rick Deriso, and Chris Cowan
Introduction 334

Models Reviewed 334
Approach Used in This Review 336
Systems Modeled 338
Young-of-the-Year Models 338
General Life Cycle Model Comparison 347
Conclusions 357
Literature Cited 359

Part VI
CONCLUSIONS AND RECOMMENDATIONS

**Conclusions and Recommendations for Assessing the
Population-Level Effects of Power Plant Exploitation:
The Optimist, the Pessimist, and the Realist** 365

Webster Van Winkle

Introduction 365
The Optimist 366
The Pessimist 368
The Realist 371
Literature Cited 372

SUBJECT INDEX 373

LIST OF CONTRIBUTORS

Adams, S. Marshall, Environmental Sciences Division, Oak Ridge National Laboratory, Oak Ridge, Tennessee 37830

Chadwick, Harold K., Bay-Delta Fishery Project, California Department of Fish and Game, Stockton, California 95205

Christensen, S. W., Environmental Sciences Division, Oak Ridge National Laboratory, Oak Ridge, Tennessee 37830

Clark, A. G., Department of Biology, Stanford University, Stanford, California 94305

Cowan, Chris, Center for Quantitative Science, University of Washington, Seattle, Washington 98195

DeAngelis, D. L., Environmental Sciences Division, Oak Ridge National Laboratory, Oak Ridge, Tennessee 37830

Deriso, Rick, Center for Quantitative Science, University of Washington, Seattle, Washington 98195

Dew, C. Braxton, Lawler, Matusky and Skelly Engineers, Tappan, New York 10983

Englert, Thomas L., Lawler, Matusky and Skelly Engineers, Tappan, New York 10983

Goodyear, C. Phillip, U.S. Fish and Wildlife Service, Office of Biological Services, National Power Plant Team, Ann Arbor, Michigan 48105

Hackney, P. A., Tennessee Valley Authority, Norris, Tennessee 37828

Horst, Thomas J., Consultant, Environmental Division, Stone & Webster Engineering Corporation, Boston, Massachusetts 02107

Jenkins, Robert M., National Reservoir Research Program, U.S. Fish and Wildlife Service, Fayetteville, Arkansas 72701

Kjelson, Martin A., National Marine Fisheries Service, Beaufort Laboratory, Beaufort, North Carolina 28516

Kumar, K. Deva, Environmental Sciences Division, Oak Ridge National Laboratory, Oak Ridge, Tennessee 37830

Lawler, John P., Lawler, Matusky and Skelly Engineers, Tappan, New York 10983

Leggett, William C., Department of Biology, McGill University, Montreal, Canada H3C − 3G1

Lorda, Ernesto, Graduate School of Oceanography, University of Rhode Island, Kingston, Rhode Island 02881

Lorenzen, William, U.S. Fish and Wildlife Service, Southeast Reservoir Investigations, Clemson, South Carolina 29631

McCaughran, Donald A., Center for Quantitative Science in Forestry, Fisheries and Wildlife, University of Washington, Seattle, Washington 98195

McFadden, James T., School of Natural Resources, University of Michigan, Ann Arbor, Michigan 48104

Miller, Lee W., Bay-Delta Fishery Project, California Department of Fish and Game, Stockton, California 95205

Norris, Robert A., Lawler, Matusky and Skelly Engineers, Tappan, New York 10983

Oliver, James, U.S. Fish and Wildlife Service, Southeast Reservoir Investigations, Clemson, South Carolina 29631

Polgar, Tibor T., Martin Marietta Corporation, Environmental Technology Center, Baltimore, Maryland 21227

Ruelle, Richard, U.S. Fish and Wildlife Service, Southeast Reservoir Investigations, Clemson, South Carolina 29631

Saila, S. B., Graduate School of Oceanography, University of Rhode Island, Kingston, Rhode Island 02881

Salo, Ernest O., Fisheries Research Institute, University of Washington, Seattle, Washington 98195

Stevens, Donald E., Bay-Delta Fishery Project, California Department of Fish and Game, Stockton, California 95205

Stober, Quentin J., Fisheries Research Institute, University of Washington, Seattle, Washington 98195

Swartzman, Gordon, Center for Quantitative Science, University of Washington, Seattle, Washington 98195

Thomas, John M., Battelle, Pacific Northwest Laboratories, Richland, Washington 99352

Vaughan, Douglas S., Marine Experiment Station, Graduate School of Oceanography, University of Rhode Island, Kingston, Rhode Island 02881

ACKNOWLEDGMENTS

Funds for this conference were provided by the U.S. Energy Research and Development Administration, Division of Biomedical and Environmental Research (ERDA/DBER), and the Electric Power Research Institute (EPRI). The Program Committee consisted of the following: W. Van Winkle (*chairman*) and R. W. Brocksen of Oak Ridge National Laboratory (ORNL), H. K. Chadwick of the California Department of Fish and Game, R. M. Jenkins of the U.S. Fish and Wildlife Service, H. A. Regier of the University of Toronto, and S. B. Saila of the University of Rhode Island. D. H. Hamilton (ERDA/DBER), R. A. Goldstein (EPRI), and R. L. Ballard and J. J. Davis of the U.S. Nuclear Regulatory Commission also contributed to the planning of this conference. Arrangements for the conference were handled by N. F. Callaham, W. H. Martin, and C. E. Normand of ORNL. Detailed editing of these proceedings was handled by L. F. Truett of ORNL. Secretarial assistance was provided by J. M. Davis, C. G. Gregory, and B. S. Taylor of ORNL.

In addition to the review and editing of manuscripts provided by W. Van Winkle, many helpful comments on a number of the manuscripts were provided by L. W. Barnthouse, S. W. Christensen, D. L. DeAngelis, W. R. Emanuel, and B. L. Kirk of the Environmental Sciences Division, ORNL. The authors of the various chapters are also to be thanked for their cooperation in working with the editor to revise manuscripts and still meet a very tight time schedule.

FOREWORD

Current events continue to emphasize that the issue of ecological effects of thermal power plant cooling systems on fisheries has clearly not been resolved. Continuing litigation concerning the question of cooling towers at the Brunswick Plant in North Carolina and the Indian Point Plant in New York are but two examples. Recently, the construction of a power plant at Seabrook, New Hampshire, has been delayed indefinitely – in part, as a result of questions concerning the ecological impact of its cooling system.

Concern in the United States regarding cooling systems historically centered on the potential impact of their heated discharge. However, more recently a major topic of concern has been the impact on fish populations resulting from mortality of fish eggs and larvae carried by the cooling waters through the plant (entrainment) and from mortality of juvenile fish colliding with the intake screens (impingement). How can the ultimate impact on adult fish populations as a result of mortality of this nature be assessed? The U.S. Energy Research and Development Administration and the Electric Power Research Institute supported this conference as a forum for discussion of this question. The published *Proceedings* provides a resource for examination and evaluation of the existing state of knowledge.

The Contents and Introductory Remarks indicate the thrust and scope of the conference. However, we think it helpful to call attention to several papers. Leggett's opening paper on the American shad presents a praiseworthy attempt to evaluate a population's compensatory potential from one of the longer records of population data available. An impressive review of the historical development of the theory of compensation, including an extensive bibliographic compilation of data supporting the existence and quantifying the magnitude of compensatory processes in fish populations, is given by McFadden. Leggett's cautious evaluation of the capacity of fish populations to compensate for heightened egg and larval mortality contrasts strongly with the arguments of McFadden. Jenkins identifies a large body of data, on reservoirs of the Eastern United States, which has tremendous potential for application to impact assessment. Christensen, DeAngelis, and Clark present a thought-provoking account of modeling and modeling inferences for single-species populations.

A variety of approaches for assessment of population effects is discussed. The approaches can be divided into two general, but not mutually exclusive, categories: application of empirical data bases and application of models. These two approaches have been applied to two distinct problems: quantitative prediction of effect where the operation of a power plant is proposed and quantitative measurement of effect where a power plant is operating. For the first problem, the state of the art, in general, is far from the point of development where acceptable predictions can be made; that is, for an arbitrary aquatic ecosystem and fish species there is no well-defined methodology for predicting effects resulting from cropping of early life stages. The state of the art (methodology) for measurement of effects when cropping is already taking place is clearly more advanced than the methodology for prediction. However, it is not clear from the papers presented how adequate the measurement methodology is and what the cost is, in terms of time and monies, to estimate effects at various levels of confidence.

Indicative of shortcomings in the state of the art is that few of the authors confronted the assessment of population effects in the abstract. Most of the papers addressed themselves to a specific ecosystem or a specific species, or dealt with only one element of the overall problem — for example, modeling or estimating abundance. The modeling papers made little attempt to define the effort needed to quantify the models. Hopefully, researchers will build upon the work reported in this volume to generalize assessment approaches and to broaden the scope of their research to include the entire assessment problem. In addition, most of the models currently being applied do not include explicit treatment of abiotic factors or interspecific interactions. To initially exclude some complicating factors is a valid approach within a research context of developing skills and methodologies for impact assessment. However, when the time comes to make the actual assessment, consideration of these complicating factors will very likely be unavoidable. Considerable evidence exists to demonstrate the importance of abiotic conditions for recruitment success. Single species models do not encompass many of the dynamic responses that can be observed in multi-species models. Observed stability of a single population may well be more a property of community dynamics than of the dynamics of the species in isolation.

It is important to improve the ability to assess population effects associated with cropping of early life stages, if we are to develop technically sound methods of ecosystem management to meet the diverse societal demands that are being placed on aquatic ecosystems. This improvement needs to be based on continuing research. In the meantime, how can we handle assessments that are needed today? A large body of ecological monitoring data for operating power plants has been collected that could be used to identify siting and design characteristics for which significant detrimental impact to populations is improbable. This information should be used in planning new power plants. In addition, considerable information has been developed over the last several years regarding siting and design of power plant intakes to minimize entrainment and impingement mortality, for example, development of new types of fish diversions. This information should be considered when designing new power plants, and further research in this area should be encouraged.

We hope that the results of this conference will contribute to increased understanding of the effects of mortality of early life stages of fish on adult population size. The papers were generally relevant, clearly presented, and current. The housekeeping and logistical chores that contribute so much to success or failure were particularly well done. All of us who attended owe the staff of the Environmental Sciences Division, Oak Ridge National Laboratory, and Webb Van Winkle, the conference chairman, a sincere vote of thanks for the atmosphere, the entertainment, and the efficient conduct of the meeting. Discussions following presented papers and within small groups assembled on the last day of the conference were highly enlightening and stimulating. We regret that we could not have made transcripts of these discussions available in the published *Proceedings*.

Robert A. Goldstein
Electric Power Research Institute
Palo Alto, California

Heyward Hamilton
Division of Biomedical and Environmental Research
U.S. Energy Research and Development Administration
Washington, D.C.

WELCOMING REMARKS

Mr. Chairman, Gentlemen, and Ladies:

In welcoming you, I would like to throw out a few thoughts for your consideration. Recently a new term has been added to the well-furbished lexicon of ecology. This term is "paradigm." Simply stated and as defined by Kuhn (1970), a paradigm is a viewpoint or body of ideas sufficiently coherent or unprecedented to attract widespread acceptance, as in a branch of science; the paradigm is typical and exemplary of a discipline. Explicitly stated, paradigms in science are needed to provide clearly the conceptual and contextual status and the assumptions upon which that branch of science depends. Two scholarly articles have just been published this year dealing with the philosophical question of paradigms in ecology. One, entitled "An Ecosystem Paradigm for Ecology" (P. L. Johnson 1977), is the result of a workshop organized by Oak Ridge Associated Universities and The Institute of Ecology. Note its title: "An Ecosystem Paradigm." The second article is an eloquent essay by George Woodwell (1976) entitled "A Comparison of Paradigms." Woodwell makes the following strong statement: "If we look for emergent theory in ecology that is useful in management, immediately applicable in government, simple and clear, the most powerful advances have been made in recent years around the concept of the ecosystem. The concept has been battered and reconstructed but persists as the most promising emergent model in ecology." These are powerful words, and we believe that the Environmental Sciences Division at ORNL has helped to contribute to the development of this paradigm.

Nevertheless, rather than view this as dogma, one should reflect on it as a challenge, especially those of us involved in the arena of ecological impact and environmental assessment. Why do I consider it a challenge?

When NEPA was enacted into law, a number of us felt that it represented the potential opportunity of the century for ecologists — because if strongly implemented it would require the development and strengthening of ecology at a scale unprecedented in our experience. Consequently, after the famous Calvert Cliffs decision by Judge Skelly Wright, when Oak Ridge National Laboratory was mandated to assist in the preparation of environmental impact statements, I readily agreed to turn our ecologist from their experimental research tasks to the analyses of impacts and the preparation of impact statements.

My reasons were simple. I believe that by involving ourselves in such an endeavor we would as ecologists be forced increasingly to come to grips with the complex problems of ecology in a context that demanded dedicated, persistent, and highly critical analysis. I and my management colleagues also felt that ecological analyses for impact assessment would uncover rich and poorly mined areas of research.

Because we are an ecosystem research unit, our early assessment philosophy was oriented in that direction. Our doubts began with our involvement in the Indian Point Nuclear Power Station controversy − the controversy which probably is the principal progenitor of this conference. Rather quickly our focus shifted to more classical problems of population ecology and its challenges. The nature of the problem, the adversary processes involved in nuclear power plant licensing, and the legalistic interpretations and regulations derived from NEPA have all served to substantiate the need for a focus on critical populations with a concomitant need to strengthen many facets of population ecology.

I am not one of that school which holds to the belief that the loss of a particular species represents a form of lasting damage to an ecosystem. On the other hand, I suffer, as I am sure many of you do, with a feeling of inadequacy when faced with the need for interpreting such a loss in a broad ecosystem context.

Many years ago, in their classic treatise, "The Distribution and Abundance of Animals," Andrewartha and Birch (1954) argued strongly that the study of populations should be the primary emphasis of ecology because all natural communities are composed of interacting populations. Ecosystem concepts were little known and even less understood 25 years ago.

But now we do know a great deal about such systems. Our researchers are beginning to demonstrate that ecosystems not only process nutrients, transfer energy, and cycle carbon but do this through the medium of interacting producer, consumer, and decomposer populations. We do not see the ecosystem as a form of super organism; rather all of its processes are mediated or regulated through the actions of populations acting as multiple interactive units, not the single species populations of classical ecological population ecology. These interactions involving competition, predation, herbivory, etc., are beginning to receive the increasing attention of ecosystem theorists. As yet, however, there is much more of a hiatus than a connection between population ecology and ecosystem ecology. One of our challenges is to contribute toward building a bridge between these two areas of ecology, so that ultimately it may be possible to assess ecological impact at the ecosystem level.

This conference exemplifies another kind of challenge. For many years, resource management ecology, including fisheries biology, was considered by academic ecologists to be among the more advanced aspects of our science. I am sure that many of you, like my colleagues and myself, were at the beginning of intense NEPA activities provoked by individuals whose questions all had a common theme − namely, "You ecologists are supposed to know all about these resources; why can't you provide good, solid, interpretative answers to our problems." There was no easy way to explain and obtain their acceptance and understanding of the weakness of ecological science at the beginning of this decade. The Indian Point controversy has served us well in this stead. It has brought into sharp focus the difficulties ecologists face as well as the weak data bases underlying our theoretical concepts, even in resource management fields. But the result has been an unprecedented support and intensification of research which will add immeasurably to ecology. Moreover, the population questions imposed by the impact controversy are being attacked in a healthy manner. Instead of one or two institutions being exclusively involved in these problems, there are many individuals, as this audience attests, interested in and working on aspects of a particular problem. It is this multiple approach, one that is stimulative, interactive, and yet also healthily competitive, which produces the real advances in science and which prior to the 1970s was largely lacking in the field of ecology. This is why you have another opportunity and challenge — not only to contribute to the resolution of complex environmental problems but also to demonstrate to many of our more academically oriented colleagues that real progress in ecological science can be achieved by individuals who have been directed by choice or circumstance to work on problems related to society's needs.

I urge you to consider this conference as a starting point for the development of personal communication links that will continue to enhance your endeavors.

On behalf of the sponsors, the Electric Power Research Institute, the Energy Research and Development Administration, and the Oak Ridge National Laboratory, it is my pleasure to welcome you officially to this conference in the heart of the Great Smoky Mountains National Park — one of the major pieces of undisturbed landscape left in the eastern United States. I hope you will not let your backgrounds as aquatic ecologists preclude your visiting parts of the magnificent terrestrial ecosystems that make up this Park.

Welcoming you is also a matter of considerable satisfaction and privilege for me. When the mandate to involve our ORNL ecological research staff in NEPA activities was issued, its enforcement became my managerial responsibility. Over the period of the last several years, it involved a considerable amount of tough decision-making, often with much trauma for my staff and myself as we endeavored to maintain our personal and scientific integrity in the belief that, in spite of weaknesses in knowledge, there is a right and proper way to go about the assessment of ecological impact. This conference and your attendance here more than amply recompense me for some of the difficulties of the past several years, and for this reason I not only personally welcome you but thank each of you for coming. I know you will have a good meeting.

S. I. Auerbach
Environmental Sciences Division
Oak Ridge National Laboratory
Oak Ridge, Tennessee

LITERATURE CITED

Andrewartha, H. G., and L. C. Birch. 1954. The distribution and abundance of animals. University of Chicago Press, Chicago.

Johnson, P. L. [ed.] 1977. An ecosystem paradigm for ecology. Oak Ridge Associated Universities, Oak Ridge, Tennessee.

Kuhn, T. H. 1970. The structure of scientific revolutions. (2nd ed.) University of Chicago Press, Chicago.

Woodwell, G. M. 1976. A confusion of paradigms (musings of a president-elect). Bull. Ecol. Soc. Am. **57**(4):8–10.

INTRODUCTORY REMARKS

The idea for this conference was conceived during a lunchtime conversation with H. A. Regier of the University of Toronto while we were attending the Workshop on the Biological Significance of Environmental Impacts in June 1975 in Ann Arbor, Michigan (Sharma et al. 1976). As indicated by its title, the theme of that workshop was rather broad, although some of the papers dealt with assessing the effects of environmental impacts on fish populations (e.g., Christensen et al. 1976 and McFadden 1976). Subsequent discussions during the fall of 1975 with colleagues at Oak Ridge National Laboratory (ORNL) and with S. B. Saila of the University of Rhode Island convinced me that this was an appropriate time to hold a conference that focused on the problem of assessing the population-level effects of exploitation of young fish. Saila had just organized a symposium entitled "Fisheries and Energy Production" (Saila 1975), and some of the papers had dealt with assessing the population-level effects of the exploitation of young fish (e.g., Hess et al. 1975 and Horst 1975); however, the majority of the papers dealt with impacts at the individual level.

A Program Committee was selected consisting of W. Van Winkle (*chairman*) and R. W. Brocksen of Oak Ridge National Laboratory, H. K. Chadwick of the California Department of Fish and Game, R. M. Jenkins of the U.S. Fish and Wildlife Service, H. A. Regier of the University of Toronto, and S. B. Saila of the University of Rhode Island. This committee met in February 1976 and recommended a tentative title, statement of purpose, and list of potential participants.

The statement of purpose (in its final form) was as follows:

> The fundamental question being asked at this conference is "How do mortality rates imposed by power plants on young fish affect adult population size?" The following areas of activity, each of which is germane to addressing this question, will be considered during the conference: (*a*) case histories; (*b*) estimating population sizes and natural mortality rates, especially for young-of-the-year fish; (*c*) evidence for and magnitude of compensation; (*d*) design of monitoring programs and statistical analysis of data; and (*e*) assessing power plant impacts with simulation models.

The participants were chosen because of their research contributions in one or more of these five areas. An attempt was made to select individuals from various parts of the country and from consulting firms, universities, and government research groups. One of our stated goals was to bring together scientists involved in assessing the effects of power plant exploitation of fish populations and scientists involved in assessing the effects of fishery exploitation. Both groups are involved in studies in the above five areas, and both groups are concerned with optimizing the long-term use of fishery resources.

xxi

In addition to the conference and session titles, the authors were given the following guidelines, which more clearly define the scope of this conference.

> Important invertebrates and forage fish, as well as sport and commercial fish species, may be considered. Participants should assume that estimates (with a measure of uncertainty) of power plant mortality by life stage and/or age class are available, and they should not deal with the technical details of estimating entrainment or impingement mortalities per se. In addition, except in passing, multispecies interactions and ecosystem effects, stocking as a mitigation measure, and power plant impacts other than mortality should not be considered.

The guidelines to authors excluded entrainment and impingement mortality because we considered these sources of mortality in and of themselves to be an impact that power plants may have at the individual level, whereas we were interested at this conference in assessing the subsequent population-level effects. The Third National Workshop in Entrainment and Impingement, February 2–4, 1976, in New York City further emphasized the need for this guideline. The focus of that workshop was very definitely the entrainment and impingement phenomena and not the subsequent population-level effects.

Multispecies interactions and ecosystem effects were excluded, not because the program committee judged these areas of research and assessment to be unimportant, but rather because the committee felt that assessment at the population level was presently the most effective approach. For most assessments a sequence of broader and broader "so what" questions can be asked, starting at the level of the individual organism (e.g., LD_{50}) and extending to the ecosystem level. Although questions at the individual level can frequently be answered with relatively high precision and accuracy, commonly they are not the critical questions and answers that can solely and effectively serve as a basis for decision making. On the other hand, the questions that can be asked at the ecosystem level certainly are critical, but these questions can rarely be answered in a sufficiently quantitative manner and with sufficient certainty to be of appreciable value for decision making (Christensen et al. 1976 and McFadden 1976).

A final guideline for this conference merits discussion. In presenting a synthetic or operational definition of biological significance in his summary paper for the Ann Arbor Workshop, Buffington (1976) indicated the need to distinguish between the concepts of "significance" and "acceptability." He viewed significance as relating to ecological factors and acceptability as relating to value systems and social factors. Van Winkle et al. (1976), in the same vein and motivated by the Ann Arbor Workshop, discussed the following two roles of an ecologist in defining and determining the acceptability of environmental impacts: (1) to supply scientifically sound and objective predictions of the potential impact of a project on an ecosystem (or component thereof) and (2) to reach conclusions, based on his own value system, concerning the acceptability of the predicted impact. Although several participants in the present conference are active in legal proceedings that involve both of these roles, the purpose of this conference is not to reach conclusions concerning acceptability.

The papers presented at this conference represent the state of the art in our ability to assess the population-level effects of power plant exploitation on fish populations. Extensions of existing methodologies and development of new methodologies have emerged in the past five years in the areas of data collection, data analysis, stock-recruitment relationships, and computer simulation models. Although the need to assess the effects of power plant mortality has provided the stimulus for most of the papers, the problems, methodologies, and answers are relevant to assessing the effects of any source of man-induced mortality on fish populations.

W. Van Winkle
Environmental Sciences Division
Oak Ridge National Laboratory
May 1977

LITERATURE CITED

Buffington, J. D. 1976. A synthetic definition of biological significance, p. 319–327. *In* R. K. Sharma, J. D. Buffington, and J. T. McFadden [eds.] Proc. Workshop Biol. Significance Environ. Impacts, NR-CONF-002. U.S. Nuclear Regulatory Commission, Washington, D.C.

Christensen, S. W., W. Van Winkle, and J. S. Mattice. 1976. Defining and determining the significance of impacts: concepts and methods, p. 191–219. *In* R. K. Sharma, J. D. Buffington, and J. T. McFadden [eds.] Proc. Workshop Biol. Significance Environ. Impacts, NR-CONF-002. U.S. Nuclear Regulatory Commission, Washington, D.C.

Hess, K. W., M. P. Sissenwine, and S. B. Baila. 1975. Simulating the impact of entrainment of winter flounder larvae, p. 1–30. *In* S. B. Saila [ed.] Fisheries and energy production: a symposium. D. C. Heath and Company, Lexington, Massachusetts.

Horst, T. J. 1975. The assessment of impact due to entrainment of ichthyoplankton, p. 107–118. *In* S. B. Saila [ed.] Fisheries and energy production: a symposium. D. C. Heath and Company, Lexington, Massachusetts.

McFadden, J. T. 1976. Environmental impact assessment for fish populations, p. 89–137. *In* R. K. Sharma, J. D. Buffington, and J. T. McFadden [eds.] Proc. Workshop Biol. Significance Environ. Impacts, NR-CONF-002. U.S. Nuclear Regulatory Commission, Washington, D.C.

Saila, S. B. [ed.] 1975. Fisheries and energy production: a symposium. D. C. Heath and Co., Lexington, Massachusetts. 300 p.

Sharma, R. K., J. D. Buffington, and J. T. McFadden [eds.] 1976. Proceedings of the workshop on the biological significance of environmental impacts. NR-CONF-002. U.S. Nuclear Regulatory Commission, Washington, D.C. 327 p.

Van Winkle, W., S. W. Christensen, and J. S. Mattice. 1976. Two roles of ecologists in defining and determining the acceptability of environmental impacts. Int. J. Environ. Stud. 9:247–254.

Part I

Case Histories

Density Dependence, Density Independence, and Recruitment in the American Shad (*Alosa sapidissima*) Population of the Conneticut River

William C. Leggett

Department of Biology
McGill University
Montreal, Canada

ABSTRACT

The role of density-dependent and density-independent factors in the regulation of the stock-recruitment relationship of the American shad (*Alosa sapidissima*) population of the Connecticut River was investigated. Significant reductions in egg-to-adult survival and juvenile growth rates occurred in the Holyoke–Turners Falls region in response to increases in the intensity of spawning in this area. For the Connecticut River population as a whole, egg-to-adult survival was estimated to be 0.00056% at replacement levels, and 0.00083% at the point of maximum population growth. Density-independent factors result in significant annual deviations from recruitment levels predicted by the density-dependent model. Temperature and flow regimes during spawning and early larval development are involved, but they explain only a small portion (<16%) of the total variation. In spite of an extensive data base, the accuracy of predictions concerning the potential effects of additional mortality to pre-recruit stages is low. The implications of these findings for environmental impact assessment are discussed.

Key words: American shad, Connecticut River, density dependence, density independence, egg-to-adult survival, flow regime, juvenile growth rates, recruitment, stock-recruitment relationship, temperature regime

INTRODUCTION

Knowledge of the stock-recruitment relationships of fish populations and of the density-dependent and density-independent factors influencing these relationships is becoming increasingly important in assessments of the potential population effects of mortality imposed on eggs, larvae, and juveniles by power plants and other industrial water users. This paper examines the role of density-dependent and density-independent factors in the regulation of the stock-recruitment relationship of the American shad (*Alosa sapidissima*) population of the Connecticut River.

3

The American shad, largest member of the family Clupeidae, is native to the Atlantic coast of North America (Hildebrand 1963). Anadromous in habit, shad spawn in fresh water in the spring and descend to the sea in the fall of their first year where they remain until reaching maturity three to five years later. Mature shad return to their natal rivers to spawn (Carscadden and Leggett 1975, Hollis 1948). The timing of the spawning migration varies with the latitude of the home river, being earliest in the south. This timing is highly correlated with water temperature, which ensures that the majority of the adults arrive on the spawning grounds when temperature is optimum for egg and larval survival (Leggett and Whitney 1972). In the Connecticut River the peak of spawning activity occurs in late May and June, and the young begin their active seaward migration in October (Marcy 1976a).

DENSITY-DEPENDENT FACTORS

The influence of density-dependent factors on recruitment to the Connecticut River population is evident in the decline in egg-to-adult survival and in the decline of juvenile growth rates associated with the increasing number of shad spawning in the river between the Holyoke and Turners Falls dams (Fig. 1) since 1955. Historically, spawning shad ascended the Connecticut River as far as Bellows Falls. However, a series of dams constructed at Vernon, Turners Falls, and Holyoke during the 18th and 19th centuries blocked the upriver migration, and no spawning occurred above Holyoke between 1849 and 1955. In 1955 a successful fish lift was installed at the Holyoke Dam, and the first shad in over a century spawned in the river between Holyoke and Turners Falls. The number of shad spawning in this area has increased steadily since that time (Fig. 2b).

Annual counts of the number of shad passed over the Holyoke Dam since 1955 were used to estimate egg-to-adult survival rates corresponding to different levels of egg deposition above Holyoke. Annual egg deposition was determined by multiplying the number of females lifted into the area by the average number of eggs spawned per female. Estimates of average fecundity of females spawning above Holyoke were derived from data provided by Foote (1976), who determined the fecundity, egg retention, sex ratio, and age-class composition of shad passed over the Holyoke Dam in 1975. Analysis of these data indicated the average egg deposition per female to be approximately 185,600 eggs.

During the interval 1969–1975 females have accounted for approximately 25% of the shad passed over the Holyoke Dam (Watson 1970, Shearer 1974, Foote 1976, R. Reed, *personal communication*). Thus, the number of eggs spawned in year x, E_x, can be estimated as

$$E_x = N_x \times 0.25 \times 185,600 , \qquad (1)$$

where N_x = number of shad passed over Holyoke Dam in year x.

In the Connecticut River the majority of males mature at age 4 and the majority of females mature at age 5 (Leggett 1969, Glebe and Leggett 1976). Thus, recruitment of sexually mature males and females from eggs spawned above Holyoke in year x, R_x, during 1955 to 1970 was estimated as

$$R_x = 0.75(N_{x+4}) + 0.25(N_{x+5}) , \qquad (2)$$

where x = year of spawning, N_{x+4} = number of shad lifted at Holyoke in year $x + 4$, and N_{x+5} = number of shad lifted at Holyoke in year $x + 5$.

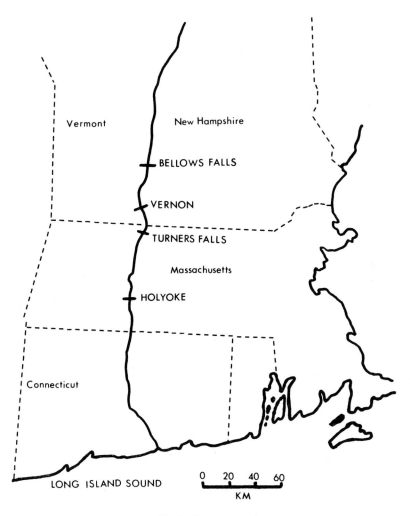

Fig. 1. Location map.

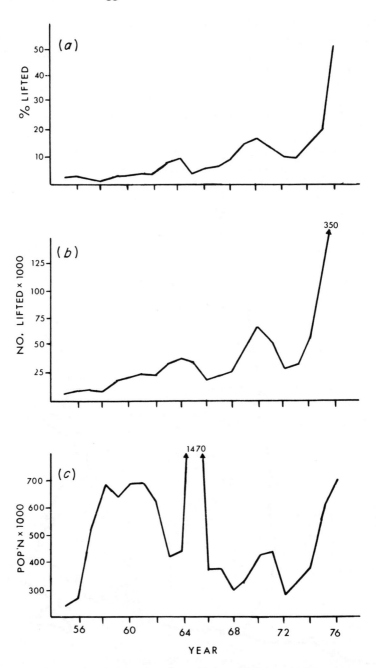

Fig. 2. (*a*) Percent of the total population lifted at Holyoke, 1955–1976, (*b*) number of shad lifted over the Holyoke Dam, and (*c*) size of total Connecticut River adult (i.e., sexually mature) shad population.

This method of estimation assumes 100% passage of recruits returning to the Holyoke Dam. However, this assumption was not valid for the period in question. Minor modifications to the lift structure were made on several occasions between 1955 and 1970. No quantitative estimate of the effect of these modifications on lift efficiency is available. However, in 1976 major modifications designed to facilitate the passage of shad previously attracted to, and trapped at, the base of the dam were completed. These changes, plus an approximate doubling of the lift capacity, resulted in a tripling of the number of shad lifted over Holyoke in 1976 relative to 1975 (Fig. 2b), in spite of the fact that the total number of shad entering the river to spawn increased only slightly (Fig. 2c). It is clear that a major improvement in lift efficiency was achieved in 1976 (Fig. 2a). It would appear that the lift efficiency prior to 1976 probably did not exceed 50%. For this reason all recruitment estimates, R_x, used in the analysis were doubled to provide a more realistic, although still minimum, estimate of survival rate.

The method of estimation of R_x (Eq. 2) also assumes that the increase in the number of shad spawning in the Holyoke—Turners Falls region since 1955 is due to the return of fish spawned in that area. There is no direct evidence that this is the case, although several factors support the hypothesis. First, Carscadden and Leggett (1975) have presented evidence which indicates that shad home to tributaries within a given river system with the same precision that they return to major rivers. Second, the number of shad passed over Holyoke (Fig. 2b) remained relatively stable during the first four years of lift operation, increased sharply in the fifth when the first recruitment from the 1955 year class (the first to be spawned above the dam) occurred, and continued to show some evidence of a four- to five-year periodicity, as would be expected if homing to the approximate natal spawning area were occurring. Finally, the relatively low proportion of repeat spawners moving over Holyoke, as compared to the proportion of repeat spawners below Holyoke, would not be expected if significant numbers of fish spawned in the lower river were entering the Holyoke pool area. This expectation is based on the fact that adult shad spawning above Holyoke experience heavy mortality resulting from downstream passage at Holyoke and from the high energy costs of the spawning migration to the Holyoke area and back to the sea (Glebe and Leggett 1976). Shad spawning below Holyoke experience greater survival, and therefore, they exhibit a greater frequency of repeat reproduction. If the choice of spawning locations were random, one would not observe the considerable differential in repeat spawning that exists above and below Holyoke.

Survival from egg to recruitment for year x, S_x, for the Holyoke pool was calculated as

$$S_x = R_x/E_x .$$ (3)

The negative relationship (r^2 = 0.59) between the number of eggs spawned above Holyoke, E, and the resulting egg-to-adult survival, S, is illustrated in Fig. 3. The position and curvature of this regression line are influenced directly by the accuracy of the values for lift efficiency. If efficiency prior to 1976 was greater than 50%, the resulting survival estimates are too high, whereas if the efficiency was less than 50%, the estimates are too low. As indicated previously, some improvement in lift efficiency occurred between 1955 and 1975. This improvement results in an underestimate of survival in the early years (low egg deposition) relative to the later years (high egg deposition). Correction for this bias, if possible, would increase the curvature of the regression line in Fig. 3. Thus, until further information is available, these values should be regarded as estimates only.

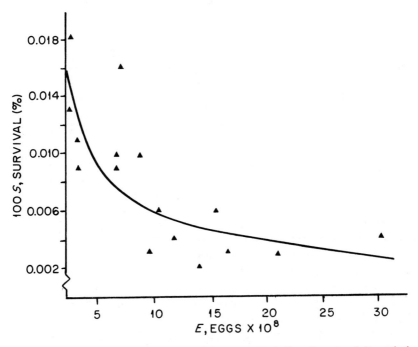

Fig. 3. Relationship between number of eggs spawned, E, and egg-to-adult survival, S; Holyoke–Turners Falls, 1955–1970. $S = 0.030E^{-0.70}$ (r^2 = 0.59; n = 16).

The growth of juvenile shad in the Holyoke–Turners Falls area has been monitored by personnel of the Massachusetts Cooperative Fishery Unit, Amherst, Massachusetts, each year since 1966 (Watson 1968, 1970, Shearer 1974, Foote 1976, R. Reed, *personal communication*). In these studies daily growth was estimated by determining the length of juveniles emigrating from the Holyoke pool on a given day in the fall, and then dividing this value by the number of days between emigration and the date when 90% of the adult shad spawning above Holyoke in that year had been lifted over the dam (day 1). The mean daily growth rate, G, of juvenile shad utilizing the Holyoke–Turners Falls region of the Connecticut River as a nursery area exhibits a strong negative relationship ($r^2 = 0.96$) with the number of adults, R, spawning in this area during the period 1966–1976 (Fig. 4).

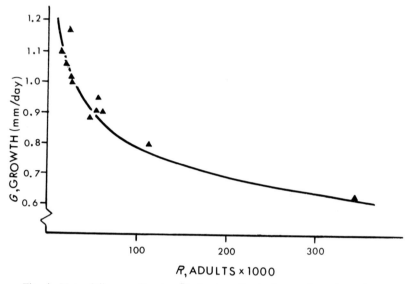

Fig. 4. Mean daily growth rates, G, for juvenile shad at various sizes of spawning stock, R; Holyoke–Turners Falls, 1966–1976. $G = 6.18R^{-0.18}$ ($r^2 = 0.96; n = 11$).

This finding suggests there has been a significant increase in intraspecific competition for resources since 1955. A negative linear relationship was also found between the mean daily growth rate of juvenile shad in the Connecticut River below Holyoke (Marcy 1976a, Table 92) and the number of adults spawning below Holyoke during the period 1966–1972 ($r = 0.77; n = 7; p < 0.05$).

Additional evidence of the role of density-dependent factors in the regulation of survival to the adult stage has been reported by Leggett (1976), who examined the relationship between the size of the spawning stock and resulting recruitment in the total Connecticut River shad population during the period 1935–1967. This relationship is described by a standard Ricker function (Ricker 1958) in the form

$$F = P \exp[0.7118(1 - P/86.59)] , \qquad (4)$$

where F is recruitment measured in number of eggs potentially available for spawning from the recruits and P is measured in number of eggs spawned by the parent stock.

Analysis of this relationship indicates that maximum survival and recruitment occur at approximately 40×10^9 eggs, at which point egg-to-adult survival approximates 0.00083%. At higher P values, survival declines steadily, reaching replacement (i.e., $F = P$) at approximately 85×10^9 eggs. Egg-to-adult survival at replacement population levels is estimated to be 0.00056%. Thus, a very small change in egg-to-adult survival has a major influence on recruitment, the difference in survival rates between the point of maximum population growth and replacement being only 0.00027%.

DENSITY-INDEPENDENT FACTORS

The major influence of very small changes in egg-to-adult survival rates on recruitment suggests that density-independent mortality factors may play a major role in determining individual year-class strengths. To examine this hypothesis, the observed recruitment of each year class from 1935–1970 (Leggett 1976, Table 140; R. A. Jones, Connecticut Department of Environmental Protection, *personal communication*), denoted R_x, was divided by the corresponding recruitment predicted by the Ricker function (Eq. 4 converted from number of eggs to number of recruits), denoted $\hat{R}_x$, thereby providing an index of survival from density-independent factors, denoted Z_x. An index value greater than 1.0 indicates higher-than-expected survival and is assumed to result from favorable environmental factors. An index value of less than 1.0 has the opposite meaning. As indicated in Fig. 5, deviations from the recruitment predicted solely on the basis of density-dependent factors are

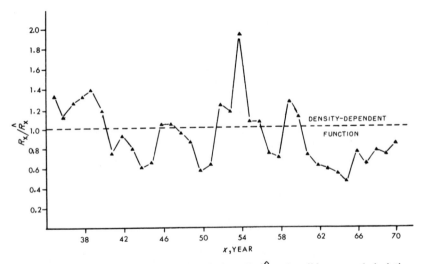

Fig. 5. Density-independent survival index, $R_x/\hat{R}_x$, describing annual deviations from density-dependent recruitment, $\hat{R}_x$; Connecticut River, 1935–1970.

common and frequently large. The mean value of the index for 1935–1970 is 0.93 (SD = 0.31). An important characteristic of these deviations is the fact that they are nonrandom; there is a considerable tendency for several years of below-average survival to occur in succession, followed by several years of above-average survival.

Marcy (1976a) demonstrated that parental stock size, water temperature, and river discharge levels during June (the month of maximum shad spawning activity in the Connecticut River) were important regulators of juvenile year-class strength below Holyoke during the years 1966–1973. The relationship between these variables is expressed by the equation

$$Y = 233.11 + 0.00021P + 7.4T + 0.057D \quad (r^2 = 0.86; p < 0.05), \quad (5)$$

where Y = index of juvenile year-class strength (number of juveniles per seine haul), P = parental stock size, T = mean June water temperature (°C), and D = mean June discharge (m³/sec).

Marcy's equation, based on data for 1966–1973, was used to develop estimates of juvenile year-class strength, Y, for the years 1958–1964 (Table 1). Spawning escapements were derived from Leggett (1976, Table 140).

Table 1. Environmental and population parameters utilized in estimating the relationship between year-class strength and recruitment

Year	Parent stock × 1000, P^a	Mean June temperature, T (°C)	Mean June discharge, D (m³/sec)	Index of year-class strength, Y^b	Recruitment × 1000, R^c
1958	480	18.8	293	23.5	325
1959	473	22.5	234	55.4	308
1960	493	22.9	397	62.5	566
1961	483	20.4	413	42.8	486
1962	435	23.8	165	43.8	302
1963	290	22.3	165	2.2	201
1964	324	21.5	138	1.9	199
1965	1297	24.4	166		
1966	287	22.4	275	8.7	251
1967	292	20.5	518	9.1	202
1968	222	20.1	711	3.6	201
1969	264	22.3	408	7.9	221
1970	348	21.6	306	11.6	303

[a]Values for 1958–1965 from Leggett (1976), Table 140; values for 1966–1970 from Marcy (1976a), Table 97.

[b]Values for 1958–1964 from the regression equation $Y = 233.11 + 0.00021P + 7.4T + 0.057D$; $r^2 = 0.86$; $F(3, 4) = 7.9$; $p < 0.05$ (Marcy 1976a). Values for 1965–1970 from Marcy (1976a), Table 97.

[c]Values for 1958–1967 from Leggett (1976), Table 140; values for 1968–1970 from data provided by R. Jones (Connecticut Department of Environmental Protection).

Mean June temperature for these years was calculated from intake water temperature records provided by the Hartford Electric Light Company's Middletown station. June discharge values were supplied by the U.S. Geological Survey, Hartford, Connecticut. No estimate of year-class strength was made for 1965 because spawning escapement in that year was 3.7 times greater than the largest value used by Marcy in the derivation of his equation (Table 1). These predicted values of Y, plus observed values of Y provided by Marcy (1976a, Table 97), were used to examine, in turn, the relationship between juvenile year-class strength, Y, and recruitment, R, during the years 1958–1970 (Fig. 6). The statistically significant correlation coefficient indicates that temperature and flow, acting through their influence on egg and larval survival, may significantly affect recruitment.

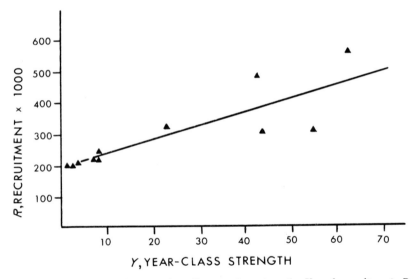

Fig. 6. Relationship between juvenile year-class strength, Y, and recruitment, R; Connecticut River, 1958–1970. $R = 197.5 + 4.38Y$ ($r = 0.82$; $n = 12$; $p < 0.01$). Y values are derived as indicated in Table 1.

Temperature records necessary to fully test this hypothesis are unavailable for the years prior to 1958. However, adequate discharge measurements (U.S. Geological Survey, Hartford, Connecticut) are available for the full time interval for which recruitment data exist (1935–1970). To test the influence of June flow on recruitment over this period, a yearly discharge index, $D_x{}^*$, was developed by dividing the mean June discharge for each year, D_x, by the 36-year mean June discharge $\bar{D}$. This index of relative discharge provides a measure of the deviation of the mean June discharge for each year from the long-term average. Then a regression analysis was performed using the model $Z = a + bD^*$, where $Z_x = R_x/\hat{R}_x$. This analysis was first conducted on the complete data set (1935–1970) and was then repeated using only years when discharge values, D_x, deviated 2, 3, and 4 standard errors (SE), respectively, from the long-term average June discharge, $\overline{D}$.

Data set	Number of years	r^2
Total	36	0.007
2 SE	24	0.01
3 SE	19	0.02
4 SE	14	0.03

As the values for the coefficient of determination, r^2, indicate, there was an increase in the proportion of the total density-independent variation in egg-to-adult survival, Z, accounted for by flow as the deviations of the yearly June flows from the long-term mean increased. However, even at very large deviations from the mean (>4SE), only 3% of the variation in survival was accounted for. A stepwise multiple regression analysis of the influence of temperature and flow on survival during the shorter time interval of 1958–1970 indicated that these two factors accounted for approximately 16% of the variation in survival attributed to density-independent factors. Yearly changes in the intensity of interspecific competition and of predation may influence the variability in shad survival rates. Marcy (1976a) noted that shad utilizing nursery areas below Enfield must compete with very large numbers of other plankton-feeding clupeids. During the time interval 1965–1969, the ratio of blueback herring and alewife larvae to shad larvae in biweekly ichthyoplankton collections conducted from March to September was 2095:1 (Marcy 1976b). The corresponding ratio for juveniles of these three species in 1008 bag-seine collections made between 1965 and 1971 was 43.6:1 (Marcy 1976a). Marcy (1976a) attributed the significant reduction in daily growth of juvenile shad collected below Enfield Dam (which serves as a barrier to migrating bluebacks and alewives), relative to juvenile shad collected above the dam, to the increased competition from juvenile blueback herring and alewife. Annual differences in the level of competition from these species and annual differences in the intensity of predation by species such as white perch, yellow perch, white and brown bullheads, and northern pike during the freshwater residence, and by bluefish and striped bass during the seaward migration may also contribute significantly to variation in egg-to-adult survival. The data required to test this hypothesis are not presently available.

DISCUSSION

The available evidence indicates that in the Connecticut River American shad population density-dependent compensation plays a major role in determining the relationship between the size of the reproductive stock and the resulting recruitment. This compensation is reflected both in the egg-to-adult survival rates and in the growth rates of juvenile shad.

In the Holyoke–Turners Falls area, where the size of the spawning stock has increased dramatically since 1955, both egg-to-adult survival rates and juvenile growth rates have shown a strong negative correlation with stock size. In the river below Holyoke, juvenile growth rates are also inversely proportional to juvenile year-class strength. It is not presently known whether juvenile growth and survival are directly linked in shad. However, Foerster (1954) and Burgner (1962) found strong positive correlations between smolt size and recruitment in sockeye and pink salmon, possibly because of a greater ability of larger smolts to escape predation, and Hiyama et al. (1972) found that for chum salmon large fry experienced significantly less predation than small fry during their seaward migration. It is possible that annual changes in growth rates and in size at migration in shad, resulting from differences in year-class strength, may be important factors in the observed density-dependent compensation in the Connecticut River shad population.

One possible conclusion, based on these findings, is that mortalities imposed on populations of American shad by industrial water uses would have no detectable influence on recruitment because the probability of survival of the remaining individuals would be correspondingly increased through compensation. However, the validity of this conclusion is, in part, dependent on the developmental stages at which density-dependent compensation is maximum. If compensation is maximum at the egg and larval stages and is low at the juvenile stage, then additional mortality at the juvenile stage could result in an impact on recruitment that would be quite direct. There is, at present, no direct evidence of the mechanisms or timing of density-dependent compensation for shad.

Density-independent mortality has also been shown to be an important regulator of juvenile year-class strength and of recruitment in the Connecticut River shad population. The annual deviations from predicted recruitment are large, and the causes of these deviations are poorly understood. Temperature and flow regimes during spawning and early development are clearly involved, but the means by which they operate are not yet known. Nonetheless, significant changes in temperature and/or discharge regimes resulting from industrial water uses, flood control programs, or diversions carry with them the potential for altering the levels of recruitment to the adult stock.

The very low egg-to-adult survival rates experienced by shad and the equally small difference in the survival rates yielding replacement vs maximum population growth suggest that the greatest threat to the continued viability of the Connecticut River shad population stems from changes in the general characteristics of the spawning and nursery environment that influence egg-to-adult survival rates. Small, long-term reductions in survival resulting from such changes would have a major influence on the ability of the population to recover from several consecutive years of below-average recruitment.

The structure of the adult population is also important in determining its ability to withstand additional mortality in the pre-recruit stages. For that portion of the Connecticut River population spawning above Holyoke, the level of egg-to-adult survival required for replacement is approximately 0.0011%, whereas for the population as a whole, replacement is achieved at a survival of approximately 0.00056%. This ability of the population spawning below Holyoke to withstand a higher level of mortality stems from a higher average lifetime fecundity, which is due to a relatively high (35%, Leggett 1969) frequency of repeat spawning. Repeat spawning also dampens year-to-year fluctuations in population size by increasing the number of year classes contributing to the reproductive stock. Relatively small changes in overall survival rates induced by environmental alteration could, therefore, assume considerable importance in the event of significant changes in adult population structure resulting from exploitation, environmental degradation, or natural causes.

The above considerations suggest that, in spite of the very strong data base available for the Connecticut River American shad population, the potential for error in numerical predictions of the effect of proposed levels of increased mortality on pre-recruit stages is large, while the biologically acceptable range of error is small. This problem is compounded severalfold in populations for which the relevant life-cycle parameters are less well defined, which includes the majority of commercially important fish stocks and virtually all noncommercial stocks. Until a better understanding is achieved of the interacting roles of density-dependent and density-independent factors in regulating population size and stability, and until a much better data base is available for the majority of the stocks subject to mortality from industrial activity, precise numerical predictions of the impact of this incremental mortality on adult stocks should be interpreted with great caution.

Unfortunately, with few exceptions, the existing approach to environmental impact assessment involves intensive, short-term surveys of virtually everything in the aquatic environment, frequently with inadequate time devoted to rigorous analysis and interpretation of the data. This approach is contributing little to the data base required for predictive purposes or to a general understanding of the functioning of aquatic ecosystems. This approach serves, rather, to create the illusion of progress. This problem has recently been addressed by Schindler (1976). If we are truly interested in protecting our resources for the future, this cookbook approach to impact assessment must be abandoned. Instead, we need a more rational program that (1) matches the nature of the studies undertaken more precisely with the nature of the problem presented; (2) demands, and provides, an adequate time frame for rigorous analysis of the data obtained; and (3) allocates a greater proportion of the available resources to studies designed to achieve a better understanding of important processes and mechanisms of natural systems that are critical to the development of accurate predictive capabilities.

ACKNOWLEDGMENTS

This paper was prepared while on sabbatical leave at the Department of Biology, Laval University, Quebec, to which I am indebted for space and technical assistance. Financial support was provided by the National Research Council of Canada (Grant No. A6513).

LITERATURE CITED

Bradford, A. D., J. G. Miller, and K. Buss. 1968. Bio-assays of eggs and larval stages of American shad, *Alosa sapidissima. In* Suitability of the Susquehanna River for restoration of shad. U.S. Department of the Interior, Washington, D.C. 60 p.

Burgner, R. L. 1962. Studies of red salmon smolts from the wood lakes, Alaska, p. 248–314. *In* T. S. Moo [ed.] Studies of Alaska red salmon. Univ. Wash. Publ. Fish. New Ser. 1, Seattle.

Carscadden, J. E., and W. C. Leggett. 1975. Meristic differences in spawning populations of American shad, *Alosa sapidissima:* evidence for homing to tributaries in the St. John River, New Brunswick. J. Fish. Res. Board Can. **32**:653–660.

Foote, P. 1976. Blood lactic acid levels and age structure of American shad, *Alosa sapidissima* (Wilson), utilizing the Holyoke fish lift, Massachusetts. M. S. Thesis, Univ. Massachusetts, Amherst.

Foerster, R. E. 1954. On the relation of adult sockeye salmon (*Oncoryhnchus nerka*) returns to known smolt seaward migrations. J. Fish. Res. Board Can. **11**:339–350.

Glebe, B. D., and W. C. Leggett. 1976. Weight loss and energy expenditure of American shad during the freshwater migration. Final Rep., Proj. AFC-8 (Connecticut), U.S. Natl. Mar. Fish. Serv. 110 p.

Hildebrand, S. F. 1963. Family clupeidae. Fishes of the Western North Atlantic. Mem. Sears Found. Mar. Res. **1**(3):257–454.

Hiyama, Y., Y. Nose, M. Shimizu, T. Ishiata, H. Abe, R. Sato, T. Maiwa, and T. Kajihara. 1972. Predation of chum salmon fry during the course of its seaward migration. II. Otsuchi River investigations 1964 and 1965. Bull. Jap. Soc. Sci. Fish. **38**(3):223.

Hollis, E. H. 1948. The homing tendency of shad. Science **108**:332–333.

Leggett, W. C. 1969. Studies of the reproductive biology of the American shad (*Alosa sapidissima*, Wilson). A comparison of populations from four rivers of the Atlantic seaboard. Ph.D. Thesis, McGill Univ., Montreal, Canada. 125 p.

———. 1976. The American shad (*Alosa sapidissima*), with special reference to its migration and population dynamics in the Connecticut River, p. 169–225. *In* D. Merriman and L. M. Thorpe [eds.] The

Connecticut River ecological study: the impact of a nuclear power plant. Am. Fish. Soc. Monogr. I. 252 p.

Leggett, W. C., and R. R. Whitney. 1972. Water temperature and the migrations of American shad. U.S. Fish Wild. Serv., Fish. Bull. **70**(3):659−670.

Marcy, B. C., Jr. 1976a. Early life history studies of American shad in the lower Connecticut River and the effects of the Connecticut Yankee plant, p. 141−168. *In* D. Merriman and L. M. Thorpe [eds.] The Connecticut River ecological study: the impact of a nuclear power plant. Am. Fish. Soc. Monogr. I. 252 p.

————. 1976b. Planktonic fish eggs and larvae of the lower Connecticut River and the effects of the Connecticut Yankee plant including entrainment, p. 115−139. *In* D. Merriman and L. M. Thorpe [eds.] The Connecticut River ecological study: the impact of a nuclear power plant. Am. Fish. Soc. Monogr. I. 252 p.

Ricker, W. E. 1958. Handbook of computations for biological statistics of fish populations. Fish. Res. Board Can. Bull. **119**:1−300.

Shearer, M. D. 1974. Analysis of factors affecting passage of American shad (*Alosa sapidissima*, Wilson) at Holyoke Dam, Massachusetts, and assessment of juvenile growth and distribution above the dam. Ph.D. Thesis, Univ. Massachusetts, Amherst. 224 p.

Schindler, D. W. 1976. The impact statement boondoggle. Science **192**(4239). Editorial.

Watson, J. F. 1968. The early life history of the American shad, *Alosa sapidissima* (Wilson), in the Connecticut River above Holyoke, Massachusetts. M.S. Thesis, Univ. Massachusetts, Amherst. 55 p.

————. 1970. Distribution and population dynamics of American shad, *Alosa sapidissima* (Wilson), in the Connecticut River above Holyoke Dam, Massachusetts. Ph.D. Thesis, Univ. Massachusetts, Amherst. 105 p.

Some Factors Regulating the Striped Bass Population in the Sacramento–San Joaquin Estuary, California*

Harold K. Chadwick, Donald E. Stevens and Lee W. Miller

Bay-Delta Fishery Project
California Department of Fish and Game
Stockton, California

ABSTRACT

The abundance of young and adult striped bass (*Morone saxatilis*) in the Sacramento–San Joaquin Estuary is related to the magnitudes of water diversions and of water flows in the estuary. Principal variations in survival occur during the first two months of life. Density-independent mortality caused by the loss of young bass in water diversions is a major factor regulating population size. Population size also is directly related to flow rates, which serve to control the transport of young bass to suitable nursery areas, which in turn influence survival through factors such as food availability.

Key words: estuary, *Morone saxatilis*, population dynamics, striped bass, survival

INTRODUCTION

Striped bass, *Morone saxatilis*, were introduced into the Sacramento–San Joaquin Estuary in 1879. The population expanded rapidly and soon supported a sport and a commercial fishery. The commercial fishery was closed in 1935 due to competition with the sport fishery, which has continued as the dominant sport fishery in the estuary.

Recent information indicates that the survival of young fish and the recruitment of adults vary in response to annual variations in environmental conditions (Turner and Chadwick 1972, Stevens 1977a). The purpose of this

*Funds for this investigation were provided by Dingell-Johnson project California F-9-R, "A Study of Sturgeon and Striped Bass," supported by Federal Aid to Fish Restoration funds, by the U.S. Bureau of Reclamation and by the California Department of Water Resources.

paper is to examine the sources of mortality responsible for the annual variations in survival, with particular reference to the relative importance of density-dependent (i.e., compensatory) and density-independent mechanisms in regulating population abundance.

ENVIRONMENTAL SETTING

The Sacramento—San Joaquin Estuary is formed by the Sacramento and San Joaquin rivers joining and flowing through a series of embayments to the Pacific Ocean (Fig. 1). The largest and most downstream embayment is San Francisco Bay. Some 1130 km of channels interlace the triangular area formed by the junction of the two rivers, and this region is known as the Sacramento—San Joaquin Delta. Tidal action occurs to the upstream limits of the delta or slightly farther, but ocean salts intrude no further than the western delta. Kelley (1966) describes the estuary in more detail.

California is a semiarid state. To accommodate the growth in human population and agriculture, vast water-development projects have been built, resulting in major changes in stream flows. One result is that the freshwater flow to the ocean from the Sacramento—San Joaquin system has been approximately halved since the 1800s.

The San Joaquin system has been almost completely developed for upstream use. Much of the development of the Sacramento system has been designed to transport water through the delta for use in the more arid, southern portions of the state. As a result, from 1970—1975 an average of 88% of the inflow to the delta came from the Sacramento River.

Water is exported to the south via two large pumping plants in the southwestern delta (Fig. 1). One is a 130-m^3/s pumping plant built by the U.S. Bureau of Reclamation in 1951, and the other is a 170-m^3/s plant completed by the California Department of Water Resources in 1968.

Typical export rates substantially exceed the flow of the San Joaquin River; therefore, most water must come from the Sacramento River. Approximately the first 100 m^3/s of flow from the Sacramento River crosses the delta through channels upstream from the mouth of the San Joaquin River. These channels are too small to carry greater flows, so at higher export rates water is drawn up the San Joaquin River from its junction with the Sacramento River (Fig. 2). Such net upstream flows in the San Joaquin River are typical in the spring, except in wet years, and in the summer and fall of all years.

The changes in magnitude and pattern of flow affect striped bass survival and are the principal factors considered in this paper.

Striped bass spawn during April, May, and June. About one-half to two-thirds spawn in the Sacramento River upstream from the delta, while most others spawn in the lower San Joaquin River from Antioch upstream for 16 to 24 km (Turner 1976). Most eggs spawned in the Sacramento River reach

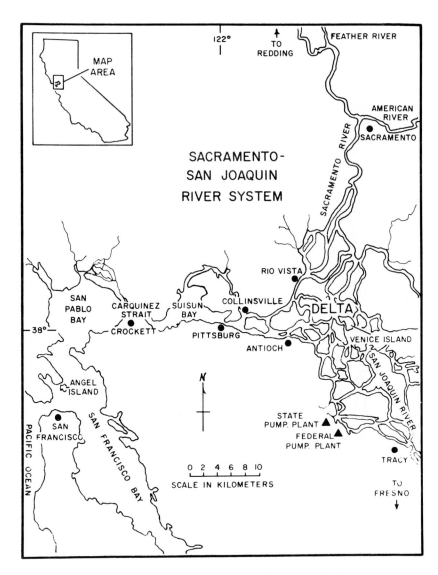

Fig. 1. The Sacramento–San Joaquin Estuary, California.

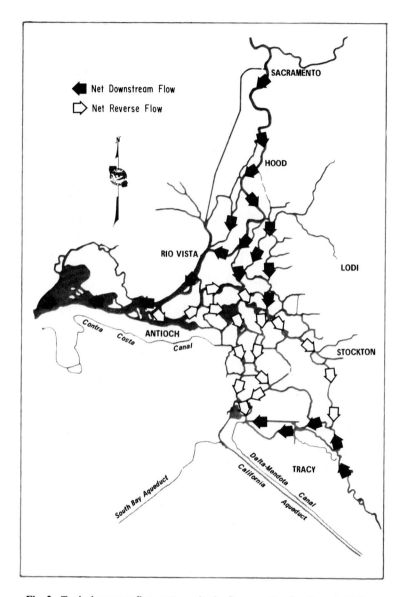

Fig. 2. Typical summer flow patterns in the Sacramento–San Joaquin Delta.

the delta as eggs or as larvae less than 6 mm long. Most young spend their first summer either in the delta or immediately downstream from there in Suisun Bay.

The flow patterns created by the Federal and State water exports transport many young bass to the diversion intakes. From 10 to 42 million bass have been salvaged annually since 1968 at fish screens in the plant intakes. These bass are returned to the estuary near Antioch. Skinner (1974) estimated that 69% of the bass approaching the State facility in 1970 were salvaged. That percentage is probably an overestimate, due to an underestimate of the abundance of young bass.

Young bass are also lost through local diversions (Allen 1975). About 110 m^3/s of water is diverted through hundreds of siphons and pumps, mostly 15 to 30 cm in diameter, for local use during the summer, and lesser amounts are diverted during the rest of the year. Furthermore, two fossil fuel power plants at Antioch and Pittsburg divert a total of about 90 m^3/s of water for once-through cooling.

STRIPED BASS SURVIVAL

Early Survival of Young

The abundance of young striped bass has been measured annually since 1959, except for 1966, in the estuary from Carquinez Strait upstream through the delta. An index of the number surviving when the mean length of the population is 38 mm has been developed from these measurements. The techniques used are described by Chadwick (1964), Turner and Chadwick (1972), and Stevens (1977b).

Turner and Chadwick (1972) reported correlation coefficients of +0.89 and −0.90 for relationships between this index and the mean daily June—July delta outflow* (Fig. 3) and the mean percent of June—July delta inflow diverted for export and local consumption respectively. These correlations describe observations from 1959 through 1970. During that period, diversions were rather constant, so the percentage diverted had an almost perfect inverse correlation to the outflow. Hence, both correlations reflect the same relationship, although they suggest different cause and effect mechanisms. More specifically, the percent-diverted correlation suggests that the loss of young in diversions influences survival, while the outflow correlation suggests that control is by some environmental factor dependent on hydraulic conditions.

---

*Delta outflow is the estimated net Sacramento River flow near Pittsburg (Fig. 1) and is the measure generally used to describe freshwater flow to the ocean.

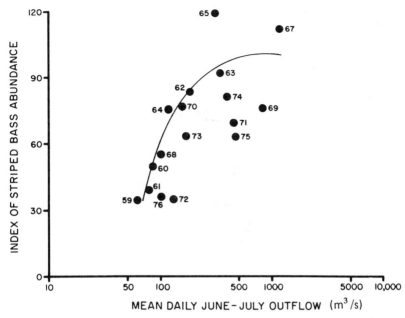

Fig. 3. Relationship between abundance of young striped bass from Carquinez Strait upstream through the delta, Y, and mean daily delta outflow during June and July. Numbers adjacent to points indicate year over the period of 1959–1976. The regression equation for the years 1959–1970 is $Y = -488.3 + 429.2$(log mean daily June–July outflow) $- 77.9$(log mean daily June–July outflow)^2; $R = 0.892$.

Turner and Chadwick (1972) discussed potential cause and effect mechanisms but had insufficient information to determine which mechanisms were most important.

It is now apparent that survival since 1970 has consistently been poorer than would be expected from the 1959–1970 relationship (Fig. 3). Nonetheless, the relationship with outflow has not changed since 1970 for that portion of the young bass population in Suisun Bay (Fig. 4). Thus, the decrease in survival has occurred in the delta portion of the population, apparently as a result of greater water exports during May, June, and July. Mean 1971–1976 exports for these months were 83%, 60%, and 52% greater, respectively, than the mean 1959–1970 exports. Variations in the index of abundance in the delta are described best by a regression relationship with the magnitude of diversions and outflow during May and June ($R = 0.83$). Indices predicted from this relationship and the actual indices of abundance of young bass in the delta are compared in Fig. 5. Variations in the magnitude of diversions are determined chiefly by export rates, because local diversion rates change very little annually.

This relationship for the delta index indicates that diversions are a major factor controlling the survival of young bass in the delta. Both the absolute

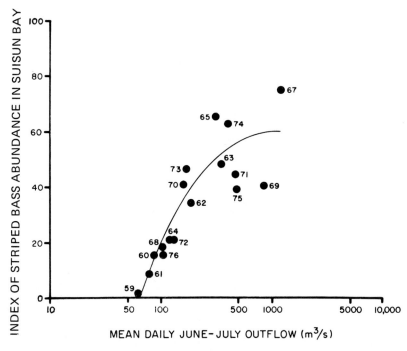

Fig. 4. Relationship between abundance of young striped bass downstream from Collinsville (Suisun Bay index), Y, and mean daily delta outflow during June and July. Numbers adjacent to points indicate year over the period 1959–1976. The regression equation for the years 1959–1970 is $Y = -294.3 + 234.8(\log$ mean daily June–July outflow$) - 39.0(\log$ mean daily June–July outflow$)^2$;$R = 0.876$.

magnitude of diversions and their relationship to the magnitude of delta inflow appear to be important, which is reasonable since the size of the area having a net flow toward the export pumps is primarily a function of export rates rather than of percent diversions.

Taken collectively, the indices of young striped bass abundance when the population mean length is 38 mm indicate that annual abundance has varied by a factor of about 3.4 between 1959 and 1976. Most of this variation is associated with variations in exports and delta outflow in May, June, and July.

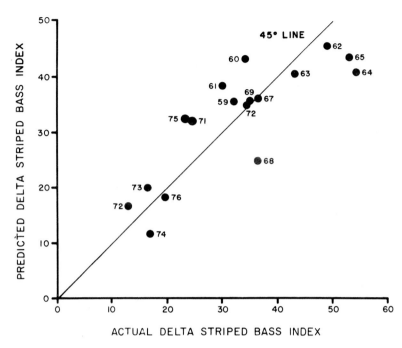

Fig. 5. Relationship between actual abundance of young striped bass in the delta (X-axis variable) and abundance predicted from May–June diversions and delta outflow (Y-axis variable). Numbers adjacent to points indicate year over the period 1959–1976. The regression equation used to obtain the Y-axis coordinates of the points plotted in this figure is $Y = -202.7 - 0.25$(mean daily May–June diversions) $+ 225.9$(log mean daily May–June outflow) $- 43.36$(log mean daily May–June outflow)^2; $R = 0.831$. Outflows are in cubic meters per second.

Survival from Young to Adult

Two indices of recruitment to that part of the population available to the sport fishery have proven reasonably satisfactory (Stevens 1977a). An index of abundance of two-year-old bass in the sport fishery in Suisun Bay has been developed for the period 1938–1954 based on catch rates. However, starting in 1956, recruitment to the fishery was dominated by three-year-olds, due to an increase in the minimum legal length. Consequently, an index of abundance of three-year-old bass has been developed for the 1956–1971 year classes based on harvest and survival estimates from tagging studies and on catches in a portion of the fishery.

Excluding 1939 and 1941, the simple correlation coefficient is 0.91 for the relationship between the abundance index for two-year-olds and the mean June–July outflow two years earlier. Considering all years, about 70% of the variation in the abundance index is described by a multiple correlation between the index, June–July outflow two years earlier, outflow during the fishing period, and the ocean water temperatures. These predicted values for catch per day are compared with the actual values in Fig. 6.

Similarly, excluding 1967, about 70% of the variation in the 1956–1971 year-class index is described by a correlation between the index and June–July delta outflows three years earlier (Fig. 7). Water was exported from the delta throughout the 1956–1971 period. The correlation between the index and fraction of delta inflow which is diverted in June and July is somewhat higher ($r = 0.87$) than the correlation with delta outflow ($r = 0.83$). It remains to be seen, as data for post-1971 year classes become available, whether the relationship between this year-class index and flows and diversions changes in the same manner as did the analogous relationship for the 38-mm index.

The index of three-year olds is not significantly correlated with the young-of-the-year index, even though both are significantly correlated with flow. In part, this lack of correlation is due to the curvilinear nature of the relationship between the young-of-the-year index and outflow (Figs. 3 and 4).

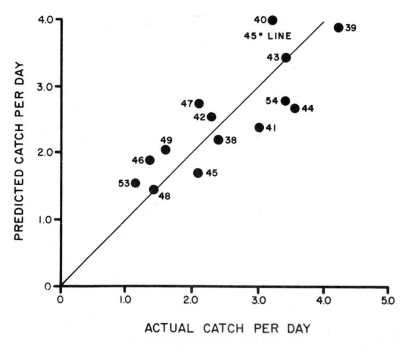

Fig. 6. Predicted versus actual fishing success on party boats in Suisun Bay (Stevens 1977a). Numbers adjacent to points indicate fishing year over the period 1938–1954. The regression equation used to obtain the Y-axis coordinates of the points plotted in this figure is $Y = -13.11 + 0.91$(log mean daily June–July outflow two years earlier) $- 0.0069$(mean daily delta outflow during the August–November fishing period for the given year) $+ 0.88$(mean annual sea temperature at La Jolla, California); $R = 0.843$. Outflows are in cubic meters per second and temperatures are in degrees Celsius.

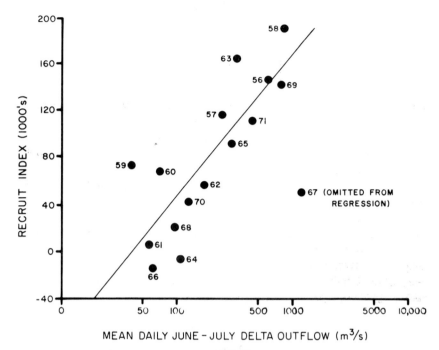

Fig. 7. Relationship between index of abundance of three-year-old bass, Y, and mean daily delta outflow during June and July three years earlier (adapted from Stevens 1977a). Numbers adjacent to points indicate year classes over the period 1956–1971. The regression equation is $Y = -196,000 + 122,000(\log \text{mean daily June–July outflow})$; $r = 0.829$.

The shape of this curve probably results from significant numbers of young bass being transported to portions of the estuary downstream from the survey area during high-flow years. If one assumes a straight-line relationship between the young index and outflow and then estimates values for the young index for high-flow years from this regression, a significant although relatively low correlation ($r = 0.61$) exists between the indices for three-year-olds and young.

In summary, the abundance indices for young and adults indicate that most of the variation in the survival of striped bass in the Sacramento–San Joaquin Estuary occurs during the first two months or so of life. The relationship between the index of two-year-olds and outflow probably indicates a cause and effect between flow and survival, as the relationship existed even prior to significant exports from the delta. The relationship does not prove this point, however, as local diversions existed at about their present level. The most recent information on the abundance of young indicates that diversions now play a major role in determining survival.

FACTORS INFLUENCING SURVIVAL OF YOUNG

Timing

The correlations described in the previous section indicate that most of the variability in survival occurs before the population mean length reaches 38 mm. In an attempt to measure and describe mortality during this period, striped bass eggs and larvae have been sampled every other day at 32 or more locations throughout the estuary for most years since 1967 (Stevens 1977b). Rates of change in abundance have been difficult to interpret, as migration and mortality cannot be distinguished from each other. The abundance changes do not show any consistent pattern which would indicate when the greatest variation in mortality occurs.

Mortality rates also have been estimated from the survey data used in computing indices of abundance at a population mean length of 38 mm. Estimated mean daily loss rates during the time that the population mean length is increasing from 25.4 to 40.6 mm range from 2.9 to 7.5%. Preliminary analysis indicates no significant correlations between these rates and either population density or environmental conditions.

Diversions

Numbers of fish salvaged at the fish screens at the intakes to the export pumps are estimated from subsamples. These estimates and approximate screen efficiencies for various sizes of striped bass have been used to estimate the number of young striped bass passing through the screens and down the export canals (California Department of Fish and Game et al. 1976). This number is correlated ($r = 0.68$) with the export rate during the 40-day period before the population mean length reaches 38 mm (Fig. 8). While this linear correlation is significant, the grouping of points below and above the regression line at export rates below and above 110 m³/s suggests a discontinuity in the relationship. Such a discontinuity is logical, since at typical early-summer flows and local diversion rates, the net flow in the lower San Joaquin River changes from the downstream to the upstream direction when export rates exceed about 110 m³/s. Such a flow reversal would tend to expose a much larger proportion of the population to the export pumps. In that case, the asymptotic upper portion of the curve would presumably indicate a small incremental impact from increasing exports beyond 140 m³/s.

The diversion of 40 to 80 million young bass in some years (Fig. 8) probably represents a significant fraction of the population. For example, the population of young bass in the estuary can be estimated crudely from sampling done to develop the young-of-the-year (38 mm) index. One month before the population mean reaches 38 mm, such estimates of population size

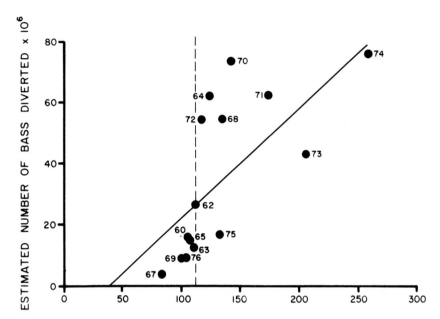

MEAN DAILY FEDERAL AND STATE EXPORT RATE (m^3/s) DURING 40-DAY PERIOD BEFORE BASS MEAN LENGTH REACHES 38 mm

Fig. 8. Relationship between estimated total number of bass diverted within 40 days prior to the date on which the 38-mm index of abundance is set, Y, and mean daily Federal and State exports during the same time. Numbers adjacent to points indicate year over the period 1960–1975. The regression equation is $Y = -15.91 + 0.38$(mean daily export during 40-day period); $r = 0.68$.

range from 70 million to 260 million bass. The true population is probably significantly higher, since the efficiency of the sampling net was likely over-estimated and since sampling does not include the relatively dense bass populations along the shoreline.

Angler catches in the export system, principally in the San Luis Reservoir complex, also document the fact that many bass are diverted from the estuary. Postcard questionnaires in 1971, 1972, 1973, and 1974 indicated that anglers caught from about 25 to 75% as many bass in the San Luis Reservoir complex as they did in the Sacramento–San Joaquin system. A large proportion of the bass caught in San Luis, however, are shorter than the minimal legal length in the estuary.

A final means of indicating the potential magnitude of losses in diversions is the magnitude of the diversions. The percentage of inflow being exported (Table 1) indicates the maximum percentage of eggs and young, spawned upstream from the delta, which might be diverted. Again, most upstream spawning is in the Sacramento River.

The percentage of inflow diverted, however, is a poor description of potential loss of eggs and young spawned in the lower San Joaquin River.

Table 1. Diversion of delta inflow for local use and export

Year	Mean monthly percent diverted								
	May			June			July		
	Local	Export	Total	Local	Export	Total	Local	Export	Total
1959	19	21	40	42	41	83	39	36[a]	75
1960	14	16	30	33	34	67	41	35	76
1961	18	20	38	33	35[a]	68	40	43[a]	83
1962	11	13	23	21	21	42	38	37[a]	75
1963	4	5	9	14	14	28	31	30[a]	60
1964	16	21	38	31	32	63	36	38[a]	74
1965	7	9	15	16	16	32	30	30[a]	60
1966	14	19	33	37	40[a]	77	36	38[a]	74
1967	3	2	6	6	3	9	14	9	23
1968	17	39[a]	56	31	39[a]	70	33	39[a]	72
1969	4	5	8	7	5	12	21	18	39
1970	14	23[a]	38	25	34[a]	59	29	35[a]	64
1959–1970 mean	12	16	28	25	26	51	32	33	65
1971	8	14[a]	21	12	19[a]	31	19	28[a]	48
1972	18	45[a]	63	26	36[a]	62	28	32[a]	60
1973	12	31[a]	43	20	40[a]	60	26	46[a]	72
1974	7	20[a]	27	13	31[a]	43	18	44[a]	62
1975	7	16[a]	22	12	14[a]	26	21	25[a]	46
1976	22	48[a]	70	33	38[a]	70	33	33[a]	66
1971–1976 mean	12	29	41	19	40	59	24	35	69

[a]Months when mean export rate exceeded 110 m^3/s.

Tidal dispersion undoubtedly exposes some of these eggs and larvae to export even when net downstream flows occur. When a net upstream flow occurs, all western delta bass are potentially subject to diversion, and thus, the percentage of inflow exported then underestimates the proportion of the population potentially subject to diversion. Since 1970, May through July export rates have always been above the 110 m^3/s rate which typically causes such net upstream flow to occur (Table 1, Fig. 2).

The actual proportion of the population reaching the export points is a complex function of several variables, including export rates, flow rates, and the degree to which flow rate determines bass movement. Striped bass less than 20 mm long often concentrate in shallow areas along the shore (Chadwick 1964), and net flow rates in the lower San Joaquin River typically are less than 6 cm/s. Simulation modeling of the striped bass population in the Hudson River indicated that the downstream transport of early life stages was significantly slower than the downstream flow rate (Eraslan et al. 1976). All of this information indicates that the proportion of bass reaching the export pumps is probably substantially less than the potential maximum.

Bass losses in local diversions cannot be estimated precisely. Eggs and larvae are locally diverted roughly in proportion to their density in the immediate vicinity (Allen 1975). A substantial percentage of local diversions occurs where bass densities are low, and the local diversions do not affect flow patterns substantially. Thus, given diversions of equal magnitude, the impact of local diversions is probably less than the impact of export diversions.

Power Plants

Pacific Gas and Electric Company has power plants located at Antioch and Pittsburg with a combined once-through-cooling water demand of 90 m^3/s. These power plants are obviously another source of loss of young bass. Kerr (1953) and Chadwick (1974) concluded that these plants had minimal impact on striped bass, as screens effectively prevented significant entrainment losses of striped bass longer than about 50 mm, and shorter bass could pass through the plants with minimal mortality. The conclusion that mortality was minimal was based on thermal tolerance tests showing that, for the short exposure to elevated temperatures existing at these plants, few bass were killed at temperatures below 33°C.

The recent evidence that losses in water-project diversions have a major impact on striped bass abundance raised the obvious question as to whether earlier laboratory studies were indicative of actual operating experience at the power plants. If mechanical stress or some other factor caused substantial mortality of bass passing through the power plants, this loss would obviously be a significant contribution to the overall diversion loss. Field tests in 1976 indicated that mechanical damage was small and that through-plant mortality was a temperature-related function similar to that measured in the laboratory experiments (Finlayson and Stevens MS). These results have not yet been used to estimate overall impacts of the power plants.

Power plant effects are not evident in the statistical relationships between striped bass abundance and environmental conditions (Figs. 3–5). Of the present 88 m^3/s cooling-water demand, 52% existed in 1959 when indexing of striped bass abundance started, 23% was added in 1961, and 25% was added in 1964.

Outflow

The magnitude of outflow determines the distribution of young bass in the nursery area, with the proportion in downstream areas increasing as flow increases (Chadwick 1964). Thus high outflow in May, June, and July makes the population less vulnerable to diversions. However, this is not likely to be the only way outflow affects abundance. The best evidence supporting this conclusion is the relationship between two-year-old recruits and flow (see caption for Fig. 6). This relationship existed before the years of significant exports, and it seems improbable that local diversions would have caused the relationship.

Turner and Chadwick (1972) considered the availability of food (reflecting some combination of the effects of flow on bass distribution, on detritus-nutrient input, and on spawning time) to be the most probable mechanism relating survival to flow. Except for 1976, however, evidence since 1971 has not given any indication of a positive relationship between flow and primary or secondary producers, which if present, would indicate an important effect of flow on detritus or nutrient input. Conversely, standing crops of phytoplankton in the principal striped bass nursery area generally decreased as flows increased (Ball MS), and average June–October standing crops of *Neomysis mercedis*, the principal food of striped bass, varied only by a factor of 1.6 and were not correlated with flow from 1968 to 1975 (unpublished data). Partially analyzed data indicate that populations of phytoplankton, zooplankton, and *Neomysis* were unusually low in 1976, a year when delta outflows were at controlled minimum levels continuously after January. This was the longest period of controlled minimum flow in history. Inorganic nutrients appear not to have been limiting in 1976 (*unpublished data*), so something other than insufficient nutrient input was probably responsible for the low productivity.

It now appears that the direct effect of flow on the spatial distribution of young bass in the estuary is a major mechanism controlling the flow-survival relationship (Stevens 1977a). As flows increase, young bass become distributed more widely in the estuary (Turner and Chadwick 1972), with some remaining in the delta, but the majority being transported to the larger downstream embayments, particularly Suisun Bay. At low flows their distribution is largely restricted to the smaller channels of the delta.

DISCUSSION

The measurements of young striped bass abundance and of adult recruitment indicate that the principal variations in survival occur within about the first two months of life. These variations are correlated with, and apparently largely controlled by, the magnitude of water diversions from the estuary and magnitude of delta outflow.

Diversions appear to act as a density-independent source of mortality, literally cropping a portion of the population during the first two months of life. Data are insufficient to estimate the mortality rate attributable to that cropping. Delta outflow probably has its effect on survival within the first two months of life, at least in part, by controlling the transport of young bass to suitable nursery areas which, in turn, influence survival through factors such as food availability.

The observed statistical relationships and abundance measurements suggest that some of the variability in survival prior to recruitment is not associated with diversions and outflow and that other sources of mortality must be of considerable magnitude. Those other sources of mortality are largely unknown, but one source — cannibalism — has been well documented (Stevens 1966, Thomas 1967), is substantial, and is compensatory.

Considering the large reproductive potential of striped bass and the known and suspected compensatory processes, our judgment until recently was that compensatory processes were dominant in controlling population size. It is now clear that density-independent processes, particularly mortality due to losses in water diversions from the delta, play a major role in controlling population size.

Such losses in diversions are, of course, partially analogous to losses at electrical generating plants. They differ, however, in that 100% of all fish diverted are unquestionably lost as far as the estuarine population is concerned. In contrast, at the two electrical generating power plants affecting striped bass in the Sacramento—San Joaquin Estuary, mortality of entrained bass is apparently 10% or less for many conditions and approaches 100% only at temperatures exceeding 30°C. Also, impingement of striped bass is insignificant at these two power plants. It seems reasonable to infer from the significance of losses in diversions in the Sacramento—San Joaquin Estuary, however, that density-independent losses due to power plants could be an important factor controlling striped bass abundance, if a major fraction of the young bass were lost due to entrainment and/or impingement.

LITERATURE CITED

Allen, D. H. 1975. Loss of striped bass (*Morone saxatilis*) eggs and young through small, agricultural diversions in the Sacramento—San Joaquin Delta. Calif. Fish Game, Anadromous Fish. Branch, Admin. Rep. No. 75-3. 12 p. (*Mimeo*)

Ball, M. D. MS. Chlorophyll levels in the Sacramento—San Joaquin Delta through San Pablo Bay. Interagency Ecological Study Program for the Sacramento—San Joaquin Estuary, Tech. Rep. No. 1.

California Department of Fish and Game, California Department of Water Resources, U.S. Fish and Wildlife Service, and U.S. Bureau of Reclamation. 1976. Interagency ecological study program for the Sacramento—San Joaquin Estuary. Annu. Rep. No. 5. 115 p.

Chadwick, H. K. 1964. Annual abundance of young striped bass, *Roccus saxatilis*, in the Sacramento—San Joaquin Delta, California. Calif. Fish Game **50**(2):69—99.

————. 1974. Entrainment and thermal effects on the mysid shrimp and striped bass in the Sacramento—San Joaquin Delta, p. 23—30. *In* L. D. Jensen [ed.] Proc. of the second entrainment and intake screening workshop. The Johns Hopkins University Cooling Water Research Project, Rep. No. 15.

Eraslan, A. H., W. Van Winkle, R. D. Sharp, S. W. Christensen, C. P. Goodyear, R. M. Rush, and W. Fulkerson. 1976. A computer simulation model for the striped bass young-of-the-year population in the Hudson River. ORNL/NUREG-8, ESD-766, Oak Ridge National Laboratory, Oak Ridge, Tennessee. 208 p.

Finlayson, B. J., and D. E. Stevens. MS. Mortality-temperature relationship for young striped bass (*Morone saxatilis*) entrained at two power plants in the Sacramento—San Joaquin Delta, California. Calif. Fish Game, Anadromous Fish. Branch, Admin. Rep.

Kelley, D. W. 1966. Description of the Sacramento—San Joaquin Estuary. Calif. Dep. Fish Game Fish, Bull. **133**:8—17.

Kerr, J. E. 1953. Studies on fish preservation at the Contra Costa Steam Plant of the Pacific Gas and Electric Company. Calif. Dep. Fish Game Fish, Bull. **92**:1—66.

Skinner, J. E. 1974. A functional evaluation of a large louver screen installation and fish facilities research on California water diversion project, p. 225—249. *In* L. D. Jensen [ed.] Proc. of the second entrainment and intake screening workshop. The Johns Hopkins University Cooling Water Research Project, Rep. No. 15.

Stevens, D. E. 1966. Food habits of striped bass, *Roccus saxatilis*, in the Sacramento—San Joaquin Delta. Calif. Dep. Fish Game Fish, Bull. **136**:68—96.

————. 1977a. Striped bass (*Morone saxatilis*) year class strength in relation to river flow in the Sacramento—San Joaquin Estuary, California. Trans. Am. Fish. Soc. **106**(1):34—42.

————. 1977b. Striped bass (*Morone saxatilis*) monitoring techniques in the Sacramento—San Joaquin Estuary. (*In this volume*)

Thomas, J. L. 1967. The diet of juvenile and adult striped bass, *Roccus saxatilis*, in the Sacramento—San Joaquin River system. Calif. Fish Game **53**(1):49—62.

Turner, J. L. 1976. Striped bass spawning in the Sacramento and San Joaquin rivers in Central California from 1963 to 1972. Calif. Fish Game **62**(2):106—118.

Turner, J. L., and H. K. Chadwick. 1972. Distribution and abundance of young-of-the-year striped bass, *Morone saxatilis*, in relation to river flow in the Sacramento—San Joaquin Estuary. Trans. Am. Fish. Soc. **101**(3):442—452.

Man's Impact on the Columbia River Stocks of Salmon*

Ernest O. Salo and Quentin J. Stober

Fisheries Research Institute
University of Washington
Seattle, Washington

ABSTRACT

The impact of man on the salmon and steelhead of the Columbia River includes 35 years of effort to prevent declines in the stocks. Between 1866 and 1940, the commercial fishery harvested an annual average of 29 million pounds of salmon and steelhead, but in the next 35 years the annual average harvest dropped to 8 million pounds. During these 35 years the habitat for anadromous fishes dropped from 163,200 to 72,800 square miles. Various approaches to mitigation were tried, but the cumulative effects of the impacts were not compensated for and the runs declined. The productivity in terms of returning adults per spawner has become less than 1 for the upper river stocks. The management flexibility of allowing additional losses at any stage in the life history of these stocks has been lost. Thus, a policy of zero impact for all new developments has been adopted.

Key words: allowable impact, Columbia River, hatchery releases, mitigation, returning adults per spawner, salmon, steelhead

INTRODUCTION

The impact of man on the salmon and steelhead populations of the Columbia River includes 35 years of efforts to prevent declines in the stocks. During this time, the life histories of the species have become well known. The mortality rates for the successive stages — egg-to-fry, fry-to-smolt, salt-water residence, and adult migration — have been estimated, and these estimates have been used in developing mitigative efforts. Nevertheless, the results of mitigation have not been adequate, because declines have not been prevented.

Significant impacts due to human activities have been identified for each stage of the salmon life history. However, on the Columbia River, multiple impacts resulting in cumulative effects have not been considered in each mitigative effort. The terms mitigation, compensation, and enhancement have

*This research was supported in part by a contract with the Nuclear Regulatory Commission.

been used interchangeably in the literature on the Columbia River salmon. The lack of differentiation of these terms reflects the case-by-case approach taken in attempts to maintain the stocks of salmon. These terms are referred to collectively in this paper as mitigative efforts.

HISTORY OF THE COLUMBIA RIVER SALMON RUNS

Because the numbers of salmon and steelhead returning to the Columbia River and its principal tributary, the Snake River, were never originally described, estimates are only conjecture. Anthropological studies have estimated the Indian population in the Columbia basin at the time of the Lewis and Clark Expedition at about 50,000. In the early 1800s, an estimated 18 million pounds of fish were taken annually by the Indians for food and trading (Beiningen 1976). Early explorers noted that there were many more fish than the Indians could use.

In the 75 years from 1866 to 1940, the commercial fisheries harvested an annual average of 29 million pounds (approximately 2.2 million salmon of all species and steelhead per year). Catches of chinook in some years exceeded 40 million pounds. However, in the following 35 years (1940 to 1974), the average commercial catch dropped to 8 million pounds per year of all species combined (Netboy 1974).

The period 1933 to the present was the time of extensive development of hydroelectric power, logging, agriculture involving irrigation, and other radical changes in land use. Rock Island, Bonneville, and Grand Coulee dams began operation in 1933, 1938, and 1941 respectively (Fig. 1), and these dams were followed rapidly by additional dams as well as by the extensive plutonium production and nuclear-power facilities at Hanford. Until the development of the lower-river dams, only the catch of the fisheries scattered from the Columbia River mouth to mid-Idaho gave any indication of the timing and relative abundance of the various stocks of salmon and steelhead. The fish counts at the different dams pointed out for the first time the variation in numbers, timing, size, and quality of the individuals comprising the different stocks of spring, summer, and fall chinook and of summer and winter steelhead.

The commercial fishery from the 1860s to 1889 concentrated on the prime quality chinook that appeared in late June and July. As the stocks of summer chinook declined, the fishery shifted to the lesser quality spring and fall chinook and then to the other species, adding sockeye and steelhead in 1889, coho in 1892, and chum in 1893. The shifting of effort to the other species maintained the total catch at a high level for many years (Fig. 2). The decline in total catch began in the early 1930s and ironically, each successive dam served as an excellent counting fence recording the demise of the upriver stocks.

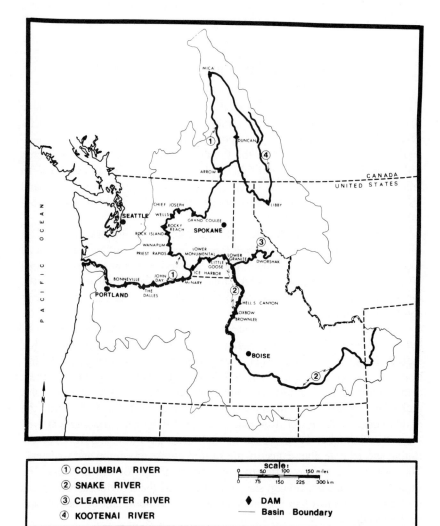

Fig. 1. Map of the Columbia basin, showing the major tributaries and the locations of the major dams. Source: Boyer 1974.

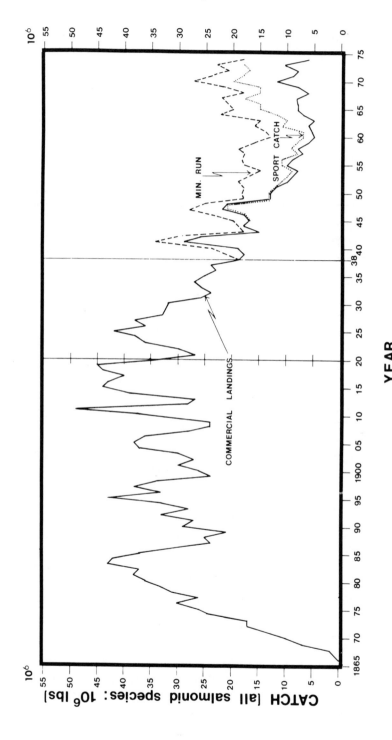

Fig. 2. Columbia River commercial landings since 1866, total minimum run since 1938, and sport catch since 1946. Source: modified from Beiningen 1976.

Rather rapidly, the 163,200 square miles of habitat for anadromous fishes was reduced to 72,800 square miles. Fortunately, salmon and steelhead are remarkably resilient and some of the stocks adapted to new timing and habitats of lesser quality. However, the cumulative effects of mounting environmental hazards are presently exceeding the levels to which salmon and steelhead can adapt.

LIMITATIONS OF IMPACT ASSESSMENT AND MITIGATIVE EFFORTS

Assuming an adequate knowledge of the life history of a species, an impact may be assessed initially by its effect upon the life history stages impacted. For example, if an impact affects salmon eggs incubating in the nests and if the number of fry expected to emerge under normal conditions is known, deviations from this expected number of fry can be mitigated. In fact, monetary values have been placed on salmon in different stages of development. For juveniles (i.e., smolts) there is a predicted loss of 15% of the downstream migrants at each dam. Obviously, only a few such impacts would be tolerable, so mitigative efforts include capturing the smolts at the upper dams and either trucking or flying them to the lower river. Survival of returning adults, once they reach the river, is more predictable at the lower dams than at the upper dams, and the mitigative efforts of fish ladders, hauling, hatcheries, sanctuaries, and restrictions on fishing have been moderately successful. However, the upriver stocks have not responded as hoped.

Monetary values based on life expectancy have been placed on the successive life stages. Thus, mitigative efforts, including financial compensation, have been applied to the impacts encountered. The following is an example of a mitigative effort. A typical chinook salmon has a fecundity of about 5000 eggs. Assuming a 33% survival from egg-to-fry, each female would produce 1650 fry. If the fry-to-smolt survival (three months) is 50%, 825 smolts would migrate. With a 15% mortality at each of four dams, 431 smolts would reach the sea. A marine survival of 1.5%, including the fishery, would result in six adult migrants returning to the river. Assuming an average survival of 80% for the upstream passage of adults at each dam, it is apparent that no more than four dams can be ascended to allow two spawners (a male and female) to reach the spawning grounds to maintain the run. Theoretically, knowing the limits of variation in survival from year to year, from stock to stock, and from life stage to life stage, it should be possible to specify an allowable impact, for any cause, at each life history stage and the balancing mitigative efforts required. During low-flow years, (1) special efforts are made to capture the downstream migrants at the dam immediately below the tributaries used for spawning, (2) adequate flows are maintained for mainstream spawners, (3) hatchery fish are substituted for smolts that are lost, and (4) the fishery is restricted when the numbers of returning adults are known to be low. The inadequacy of these methods is apparent but is not fully understood because each successive impact, although apparently similar, has a compounding effect.

The status of the Columbia River stocks is that the effect of the impacts has increased with the addition of each new dam, while the resiliency of these species has steadily declined. The changes in environment have reduced productivity of salmon until recruitment (in terms of spawners) has fallen below replacement. The upriver salmon is in a tenuous condition, and it is doubtful that all combined mitigative efforts can ensure the survival of these stocks. On the other hand, similar efforts to mitigate impacts on lower-river stocks appear likely to succeed.

The intolerance of the upriver stocks is reflected in the spawner-recruit curves, which have become increasingly flattened with the completion of each new dam. In recent years, the Ricker-type, spawner-recruit curve has dropped below the replacement line, allowing for no fishery (Salo 1974a). Figures 3 and 4 show the decline in number of adult salmon returning four years after the spawning escapement. The spawning escapement is the number of salmon (male and female) that ascend the river to spawn. Production rate, P.R., has been calculated as the number of adult salmon returning to the Columbia River in year t divided by the spawning escapement in year $t - 4$; that is, P.R. is the production of returning adults per spawner. A P.R. of 1.0 is the line of equal replacement (Figs. 3 and 4). When the P.R. of spring chinook is greater than 2.5, a substantial fishery (up to 200,000 fish) may be allowed. When the P.R. drops to between 1.5 and 2.5, the fishery must be reduced (50,000–100,000 fish), and as the P.R. drops to near or below 1.0, no fishery is allowed. The drop in production has been more dramatic for the summer chinook (Fig. 4), and no fishery has been allowed in recent years.

The plans of relocation, hatcheries, fish ladders, turbine bypasses, spawning channels, trucking of downstream migrants, and flip-lips for nitrogen control, have not functioned in concert as hoped. All such efforts have contributed to the maintenance of the remnants of the wild stocks, but 50% of the total salmonid population is now produced by hatcheries.

In many cases, where mitigation either failed or was only partially successful, the protective measures were reduced or abandoned. The responsibility of the developer was temporary, usually only for the initial impact, and then the costs of mitigation were transferred to the rate payer or taxpayer, or were absorbed in the overall budget of the management agency.

Until recently, there was no continuing responsibility, legal or otherwise, recognized by the power producer toward anadromous fish in the upper watershed. In effect, the $400 million expended to date amounts to a one-time-cost to the water users. Historically, once a portion of the resource was lost, it was no longer taken into consideration in subsequent mitigative efforts for additional losses caused by further developments of the watershed.

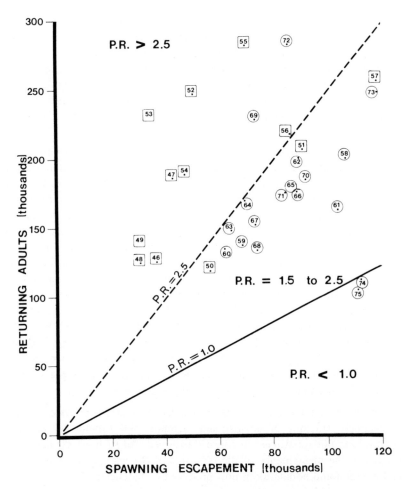

Fig. 3. Number of adult spring chinook salmon returning to the Columbia River in the indicated year (1946 to 1975) as a function of the spawning escapement four years earlier. Runs after 1957 are circled and the earlier runs are in squares (from Beiningen 1976). P.R. = production rate or production of returning adults per spawner — that is, the ratio of the Y-axis variable to the X-axis variable.

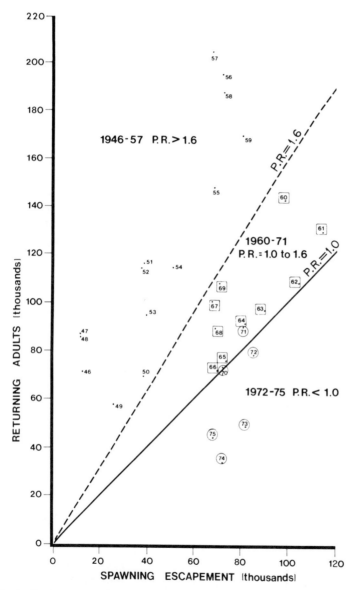

Fig. 4. Number of adult summer chinook salmon returning to the Columbia River in the indicated year (1946 to 1975) as a function of the spawning escapement four years earlier. Runs during the 1960s are squared and runs during the 1970s are circled. P.R. = production rate or production of returning adults per spawner — that is, the ratio of the Y-axis variable to the X-axis variable. Source: Beiningen 1976.

Fisheries reparation has been received only for stocks completely blocked by dams or for spawning areas inundated; no reparation has ever been received for extensive passage mortalities caused by mainstem dams.

Historically, there was an overconfidence in the early hatcheries, followed by a fairly long period (1930s to 1950s) of general lack of confidence in hatchery production as a mitigation tool. This was followed by an attitude that even partial mitigation for environmental damage was a substantial accomplishment. The production by modern hatcheries is significant (Fig. 5), and the severe problems of fishing mixed stocks of hatchery and wild fish are realized (Salo 1974b). The maintenance of wild stocks in many streams has become problematical, as the returns-per-spawner are much greater for hatchery stocks. The hatchery fish are not, however, replacing the wild fish "in kind and equal value." The valuable upper river stocks have been replaced by mass production of salmon with common life histories, morphology, and genetic characteristics. Now that the production of hatchery and wild fish is approximately 50/50, the remaining wild stocks must be conserved.

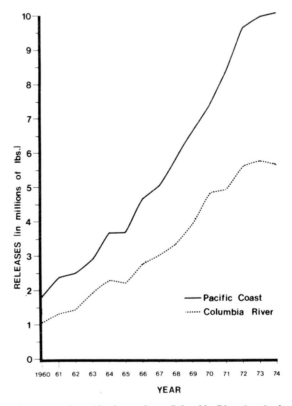

Fig. 5. **Anadromous salmonid releases from Columbia River hatcheries and from all Pacific Coast facilities from 1960 to 1974.** Source: modified from Beiningen 1976.

At the present time, there are two nuclear plants (Hanford and Trojan) in operation on the Columbia River, and three others are in the planning stages. Hearings for construction permits have been underway for two others (Skagit and Satsop) on two additional river systems of Washington State. Other sites have been studied, some quite extensively [e.g., Kiket Island (Stober and Salo 1973)], and these sites either have been abandoned or a construction decision has been delayed. In all of these instances, the policy essentially has been to allow no mortality of juvenile salmonids beyond that presently existing. Mitigation by hatchery production is no longer a substitute, although a utility may apply and under certain circumstances (Skagit) may be allowed to enhance production.

In summary, the policy in the Pacific Northwest is that "wherever possible," losses of juvenile salmon are not acceptable. When losses are unavoidable, mitigation is based on the original potential of the salmon rather than on the present production. If, based on post-operational data, the design concept used does not protect the salmon, the management agencies reserve the right to require redesign and backfit. Perhaps there are other fishery resources in the United States that are in a condition analogous to that of the salmon.

ACKNOWLEDGMENTS

We thank J. A. Lasater of the Washington State Department of Fisheries for his views on the policies of the State of Washington; also, W. Van Winkle's editing was greatly appreciated.

LITERATURE CITED

Beiningen, K. T. 1976. Fish runs. *In* Investigative reports of Columbia River fisheries project. Prepared for the Pacific Northwest Regional Commission. p. E-1 to E-65.

Boyer, P. B. 1974. Lower Columbia and Lower Snake Rivers. Nitrogen (gas) supersaturation and related data analysis and interpretation. North Pacific Division, Corps of Engineers, Portland, Oregon.

Netboy, A. 1974. The salmon: their fight for survival. Houghton Mifflin, Boston. 613 p.

Salo, E. O. 1974a. Special report to the U.S. Army Corps of Engineers on two reports concerning proposed compensation for losses of fish caused by Ice Harbor, Lower Monumental, Little Goose, and Lower Granite locks and dam projects, Washington and Idaho. Fisheries Research Institute, College of Fisheries, University of Washington, Seattle. 47 p.

————. 1974b. Anadromous fishes. Salmonid management, supplement to Trout 15(1):12–21.

Stober, Q. J., and E. O. Salo. 1973. Ecological studies of the proposed Kiket Island nuclear power site, final report, September 1969 to February 1973. FRI-UW-7304. Fisheries Research Institute, College of Fisheries, University of Washington, Seattle. 537 p.

Population Dynamics of Young-of-the-Year Fish in a Reservoir Receiving Heated Effluent

Richard Ruelle, William Lorenzen, and James Oliver

U.S. Fish and Wildlife Service
Southeast Reservoir Investigations
Clemson, South Carolina

ABSTRACT

Limnetic young-of-the-year fish populations were sampled by trawl in 1973, before significant heated discharges began from a South Carolina nuclear power plant, and in 1974–1976, after the plant was in operation. Three endemic forms were yellow perch (*Perca flavescens*), black crappie (*Pomoxis nigromaculatus*), and sunfishes (mostly bluegill, *Lepomis macrochirus*). Threadfin shad (*Dorosoma petenense*) were introduced in early 1974 and were included in the study. Increases in water temperatures were in direct proportion to the amount of power generated by the plant and were greatest in the deeper water. Catches of yellow perch, black crappies, and sunfishes declined between 1973 and 1975. Catches of yellow perch and black crappies were highest at the control station and lowest at the heated water discharge. Sunfish catches were low at all sampling stations after power generation began, but it is unclear whether the decrease in numbers resulted from aging of the reservoir and other natural changes or from the man-induced increase in water temperatures. Threadfin shad catches did not change noticeably during the study.

Key words: black crappie, heated water discharge, multiple regression analysis, population dynamics, reservoir, sunfishes, threadfin shad, yellow perch, young-of-the-year

INTRODUCTION

Demands for electrical power are increasing throughout the world, and an increasing number of electrical generating plants are being constructed. Of the various difficulties caused by power plants, one of the most important problems arises from the discharge of large volumes of heated water into the environment. The effects of these heated water discharges are not fully known.

Although many studies have been conducted on the effect of heated effluent from nuclear power plants on adult fish, only a few studies of changes in young-of-the-year fish populations have been completed. The results of most of these studies have been inconclusive for several reasons: (1) the studies were started after heated water was being discharged into the receiving water; (2) the studies were conducted for one year or less; (3) standardized sampling methods were not used; (4) net mesh sizes generally were too large to sample newly hatched fish; and (5) samples were taken only from a small area near a power plant.

We found few published reports on the vertical distribution of young-of-the-year fish in a deep reservoir receiving heated effluent. We believe that the discharge of heated water into a reservoir alters the distribution and abundance of young-of-the-year fish. Some species of larval fish have a lower thermal tolerance than juvenile and adult fish have (Banner and Van Arman 1973), and larval fish are unable to move out of the plume because they lack the mobility of adult fish. Consequently they may suffer unusually high mortality, which in turn may result in a damaging reduction in recruitment to the adult population.

The objective of this paper is to describe changes in relative abundance of limnetic populations of young-of-the-year (here termed "young" unless otherwise specified) fish in Keowee Reservoir before and after operation of the Oconee Nuclear Station near Clemson, South Carolina. Fish were sampled by trawl from March through September 1973–1976. The order in which the species spawned and in which the young became available to the trawl as the season advanced were as follows: yellow perch (*Perca flavescens*), black crappie (*Pomoxis nigromaculatus*), threadfin shad (*Dorosoma petenense*), and sunfishes (mostly bluegill, *Lepomis macrochirus*).

Because changes in the fish populations resulting from heated effluents from the Oconee Nuclear Station are still occurring in Keowee Reservoir, the total impact of the plant's operation on young fish stocks cannot yet be assessed.

DESCRIPTION OF THE STUDY AREA

Keowee Reservoir, which is in northwestern South Carolina and was filled between 1968 and 1971, provides cooling water for Duke Power Company's Oconee Nuclear Station. The reservoir covers 7435 ha and has a mean depth of 15.8 m at full pool (243.7 m, mean sea level). The shape of the reservoir resembles an hour glass, with upper and lower sections of nearly equal area (Fig. 1). The 2658 MW Oconee Nuclear Station is located between the two sections; cooling water is drawn from the lower (south) section and is discharged into the upper (north) section. Cooling water passes under a skimmer wall 19.8–27.4 m below full-pool surface level, flows over a submerged weir, and then travels 1.5 km down a canal to the Oconee Station.

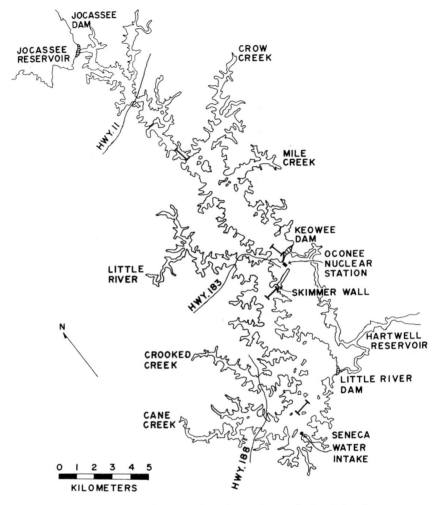

Fig. 1. Trawling stations (⊢———⊣) on Keowee Reservoir, South Carolina.

Heated water from the plant is discharged into the upper section of Keowee Reservoir, 9 to 12 m below the full-pool surface level. Maximum cooling water flow is 134 m³/s, and mean flow is 99.1 m³/s at 80% generating capacity.

The first of three nuclear reactors began operating in May 1973, the second in December 1973, and the third in September 1974. During the March—September fish sampling period, power production averaged 8.4% of capacity in 1973, 31.1% in 1974, 72.9% in 1975, and 56.8% in 1976. The decline in power production in 1976 resulted from the shutdown of two reactors at alternate times for refueling during the summer.

Jocassee Dam, which impounds Jocassee Reservoir and houses the Jocassee pumped-storage hydroelectric generators, is 15.7 km upstream from the Oconee Nuclear Station. This reservoir serves as the upper pool and Keowee Reservoir as tne lower pool for tne pumped-storage operation. During periods of high electrical need, the Jocassee hydroelectric plant generates electricity, and during periods of low demand, the reversible turbines pump water from Keowee back into Jocassee.

MATERIAL AND METHODS

Field Collections

Limnetic larval and juvenile fish were sampled by frame trawl in 1973–1976 in Keowee Reservoir. The trawl consisted of a 1.3-m square aluminum frame with a 6.1-m-long, 0.8-mm-mesh net. The four trawling stations (and their distance from the heated water discharge) were Crow Creek, north section of the reservoir immediately upstream from the confluence of the Crow Creek arm and the main reservoir (8 km); heated water discharge (0 km); skimmer wall, in front of the skimmer wall (3 km); and south area, south section of the reservoir near the Seneca water intake (11 km) (Fig. 1).

Samples were collected during daylight once each week from March through September. Two 10-min hauls at a constant engine speed were made at 1 and 5.5 m at each station. Inasmuch as the young of endemic species in Keowee Reservoir were captured in largest numbers near the surface in other studies (Faber 1967, Noble 1970, and Amundrud et al. 1974), the largest numbers of trawl hauls were taken at the shallower depths. The possibility existed that some fish might be in the cooler water below 5.5 m after heated water was added to the reservoir; therefore, one 10-min haul was taken at mid-depths of 10.5 and 15.5 m at each station every two weeks from March to September of 1973 and 1976. During the same period in 1976, one 5-min haul was also taken at a depth of 20.5 m immediately in front of the skimmer wall. Samples of young fish were preserved in 10% formalin.

Temperatures were recorded weekly at depths of 1 and 6 m at each station from 1973 through 1976, every two weeks at 11 and 16 m at each station in 1973 and 1976, and at 21 m every two weeks in front of the skimmer wall in 1976. Additional temperature profiles were taken each month from the surface to a depth of 35 m. Three-point moving averages were used to smooth the weekly temperature curves for 1973 and 1975 (Figs. 2 and 3).

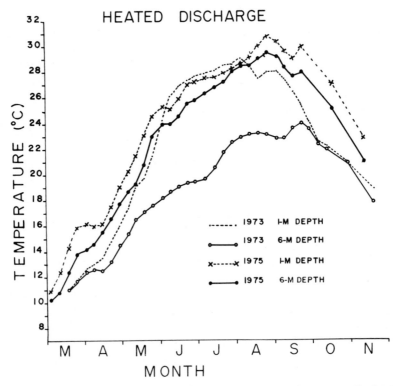

Fig. 2. Three-point moving average of weekly water temperatures at the heated discharge station on Keowee Reservoir, March–September, at 1 and 6 m for the baseline year (1973) and the warmest year (1975).

Laboratory Analyses

Centrarchids, other than black crappies, that were shorter than 10 mm (total length) could not be positively identified and were classified as sunfishes. Most of these were probably bluegills, since adult bluegills were the most abundant sunfish in the reservoir. Redbreast sunfish (*Lepomis auritus*), green sunfish (*Lepomis cyanellus*), warmouth (*Lepomis gulosus*), spotted sunfish (*Lepomis punctatus*), and redear sunfish (*Lepomis microlophus*) also were present in the reservoir. The young fish were identified, counted, and measured (total length in mm) in the laboratory.

Statistical Analysis

Multiple regression techniques (Neter and Wasserman 1974) were used to compare the "mean responses" of weekly temperature regressed on week and year for each depth at each station and the mean response of weekly tempera-

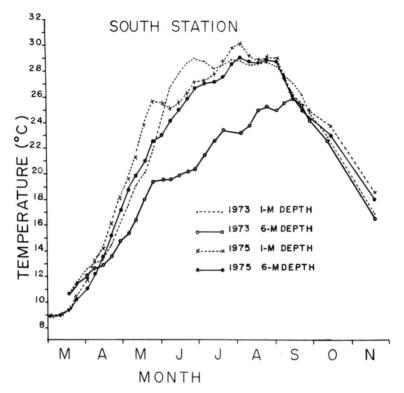

Fig. 3. Three-point moving average of weekly water temperatures at the control (south) trawling station on Keowee Reservoir, March–September, at 1 and 6 m for the baseline year (1973) and the warmest year (1975).

ture regressed on week and station for each depth for each year. Multiple regression techniques were also used for each of four species to compare mean responses of weekly catch of young fish at 1 and 6 m combined regressed on weekly temperature and year at each station and also regressed on weekly temperature and station for each year. Differences between mean responses for years or stations were tested with interaction terms. Results of statistical analyses were considered significant at $\alpha = 0.05$ and highly significant at $\alpha = 0.01$.

RESULTS

Water Temperature

Keowee Reservoir is a warm monomictic reservoir. Before the operation of the Oconee Nuclear Station, the reservoir began warming in March and stratification was completed by June. The thermocline was between a depth of 4 and 8 m in June and July, began moving downward in August (5–8 m) and September (8–9 m), and was below 15 m by October.

The discharge of heated water from the power plant affected the reservoir in two major ways: the reservoir warmed earlier in the year and cooled later, and water temperature increases were greater at 6 m than at 1m.

Temperatures at the heated water discharge station at both 1 and 6 m were increased 1.5 to 4.2°C in March, April, and May 1975 as compared with the same period in 1973 (Fig. 2). The difference in water temperatures between 1973 and 1975 was only 0.2 to 0.3°C at 1 m from the second week of June through the first week of August; however, temperatures were 2 to 4°C higher in 1975 than in 1973 from mid-August to mid-November.

Generally, the spring and fall water temperatures were highest at all stations and depths in 1975, and lowest at all stations and depths in 1973; midsummer temperatures were closely similar in the two years. The highest water temperature at 1 m at the heated water discharge station was 29.0°C in 1973 and 31.6°C in 1975. The greatest temperature difference between the two years at this station was at 6 m, where the highest temperature recorded was 24.3°C in 1973 and 30.1°C in 1975.

In 1975, when all three units were operating, a volume of water equal to that of the hypolimnion was circulated through the plant during the course of the summer, and water temperatures were nearly homothermous from the surface to 20 m throughout the reservoir by September. Temperatures of water discharged from the plant averaged 4 to 5°C higher than intake water temperatures in 1973 and 1975. Water discharged from the plant was increased a maximum of 9.5°C above the temperature of the intake water in 1973 and 7.6°C in 1975. However, the temperature of the incoming cooling water differed between the two years: the maximum temperature of intake water was about 10.0°C higher (and maximum temperature of water discharged from the plant was about 8.0°C higher) in 1975 than in 1973.

The south station was the farthest from the heated water discharge. Water temperatures at this station were only slightly higher in 1975 than in 1973 at 1 and 6 m from March through April and from mid-September through mid-November (Fig. 3). From early June through mid-July, the temperatures at 1 m were 0.4 to 0.7°C lower in 1975 than in 1973. The greatest difference in temperature between 1973 and 1975 was at 6 m during the critical period from mid-May through mid-September. Water temperatures during this period at this depth were 2.1 to 6.2°C (average 4.7°C) higher in 1975 than in 1973. The maximum water temperature at 6 m was 26.0°C in 1973 and 29.1°C in 1975.

Hot water discharged from the Oconee Station flowed into the Crow Creek sampling area. Maximum water temperature at 1 m was 28.0°C in 1973 and 31.0°C in 1975. The largest difference in maximum temperatures (6.5°C), observed at 6 m between 1973 and 1975, was at Crow Creek: the maximum temperature was 23.5°C in 1973 and 30.0°C in 1975.

Water temperature ranges (°C) for all sampling stations during the March—September trawling period at 1 and 6 m were: 1973, 8.7—30.0; 1974, 11.5—31.0; 1975, 8.9—31.6; and 1976, 11.9—29.7.

Water temperatures were higher at 1 and 6 m in 1975 than in any other year and lower in 1973, with one exception (Table 1). There were no statistically significant differences between water temperatures at 1 m in 1975 and 1976 at any of the stations, or at 6 m between 1974, 1975, and 1976 at the south station. However, water temperatures in 1974, 1975, and 1976 differed significantly at the skimmer wall, heated discharge, and Crow Creek stations.

Water temperatures were highest at the south station and lowest at Crow Creek in 1973 (Table 2). There were no statistically significant differences in water temperature among the stations in 1974. Design of the plant, water intake facilities, and plant operation were probably responsible for the homo-thermous conditions between 1 and 6 m in 1974, even with the plant operating at only 31.4% of capacity. Hypolimnetic water used for cooling the plant passed through the cooling condensers of the plant and was warmed to near-surface water temperature before being discharged into the reservoir. Water temperatures at the heated water discharge station at 1 and 6 m were significantly higher than temperatures at all other stations in 1975 and 1976. Water temperature was lowest at the south station in 1975 and 1976. The addition of heated effluent can change this pattern of thermocline movement. In Keowee Reservoir, for example, a well-defined thermocline was present between 1 and 6 m at all four stations in 1973, with the maximum tempera-

Table 1. Comparison, among years, of the mean responses of weekly temperature at 1- and 6-m depths for each of the four stations on Keowee Reservoir, March—September 1973—1976

Station	Depth (m)	Years[a]			
South	1	1973	1976	1974	1975
	6	1973	1976	1974	1975
Skimmer wall	1	1974	1973[b]	1976	1975
	6	1973	1974	1976	1975
Heated discharge	1	1973	1974	1976	1975[b]
	6	1973	1974	1976	1975
Crow Creek	1	1973	1974	1976	1975
	6	1973	1974	1976	1975

[a]Years are ordered within station and depth with increasing temperature from left to right. The line under two or more adjacent years means that the mean responses of weekly temperature for these years did not differ significantly from each other at the 0.05 probability level.

[b]Significantly different from 1976 at the 0.10 probability level.

Table 2. Comparison, among stations, of the mean responses of weekly temperature
at 1- and 6-m depths for each of four years on Keowee Reservoir,
March–September 1973–1976

Year	Depth (m)	Stations[a]			
1973	1	Crow	Heated[b]	Skimmer wall	South
	6	Crow	Heated	Skimmer wall	South
1974	1	Skimmer wall	South	Heated	Crow
	6	Crow	Heated	Skimmer wall	South
1975	1	South	Skimmer wall	Crow	Heated
	6	South	Skimmer wall	Crow	Heated
1976	1	South	Skimmer wall	Crow	Heated
	6	South	Crow	Skimmer wall	Heated

[a]Stations are ordered within year and depth with increasing temperature from left to right. The line under two or more adjacent stations means that the mean responses of weekly temperature for these stations did not differ significantly from each other at the 0.05 probability level.

[b]Significantly different from the south station at the 0.10 probability level.

ture difference between 1 and 6 m being 9.0°C. In contrast, in 1975 when Oconee was operating most of the time and large volumes of heated water were being added to the reservoir, the depth of the thermocline rapidly decreased to below 6 m, was less well defined throughout the summer, and was absent at all four sampling stations by September.

Abundance of Young-of-the-Year Fish

Young limnetic fish were captured by trawl in the same sequence each year. Yellow perch were captured in March; black crappies in May 1973, in April 1974 and 1975, and March 1976; threadfin shad in May 1974 and 1975 and April 1976; and sunfishes in May of each year.

Yellow Perch

A total of 8891 young yellow perch were captured at all stations in 1973; catches declined steadily to a low of 233 in 1976. The decline was greatest at the heated water discharge station (Table 3) and lowest at the south station between 1973 and 1975. Yellow perch were more abundant at 1 m than at 5.5 m at all sampling sites in all years. They were captured in larger numbers near the surface even in years when water temperatures at the surface were higher than at 6 m.

Table 3. Mean numbers of young-of-the-year yellow perch caught per haul of a frame trawl at 1 m, at 5.5 m, and at both depths combined, from four areas of Keowee Reservoir, 1973–1976

Depth[a] (m)	Station				
	South	Skimmer wall	Heated discharge	Crow Creek	All stations
1973					
1	22.2	14.3	36.3	32.2	26.2
5.5	10.6	6.4	9.8	21.6	12.1
Combined	16.4	10.3	23.0	26.9	19.2
1974					
1	12.3	14.2	1.1	19.4	11.8
5.5	5.7	3.7	1.1	6.7	4.3
Combined	9.0	9.0	1.1	13.0	8.0
1975					
1	2.9	0.8	0.1	0.7	1.1
5.5	1.5	0.3	0.1	0.1	0.5
Combined	2.2	0.5	0.1	0.4	0.8
1976					
1	1.5	0.8	0.5	0.5	0.8
5.5	0.4	0.2	0.2	0.1	0.2
Combined	1.0	0.5	0.3	0.3	0.5

[a]Number of hauls per depth per year from each station ranged from 53 to 58.

A regression of the logarithm of total annual catch on the mean of the two lowest consecutive monthly winter water temperatures at 2 m indicated that catches of young yellow perch were inversely related to winter water temperatures ($r = 0.81, p < 0.01$). In 1976, when mean monthly winter water temperatures were above 11.0°C, total yellow perch catches at all sampling stations from March–September were only about one-fortieth as large as in 1973, when the mean monthly winter water temperature was 8.7°C.

Mean responses of weekly catch of young fish at 1 and 5.5 m combined for each of the four species for 1974, 1975, and 1976 were compared with the mean response of the baseline year, 1973 (Table 4). Yellow perch mean responses were significantly different $\alpha = 0.0001$) between 1973 and 1975 and between 1973 and 1976 at the Crow Creek, heated discharge, and skimmer wall stations. There were no significant differences in the mean responses at the south station between 1973 and 1974 or between 1973 and 1975 or at Crow Creek between 1973 and 1974. Once the yellow perch population reached a low level in 1975, it remained low in 1976;

Table 4. Probability levels for the observed difference between years
in the mean response of weekly catch of young fish at 1 and 5.5 m
combined for each of three species at each of four stations in
Keowee Reservoir, 1973–1976

Years compared	Yellow perch	Black crappie	Sunfishes
Crow Creek			
1973–1974	NS[a]	0.05	NS
1973–1975	0.0001	0.05	0.0001
1973–1976	0.0001	NS	0.0001
1975–1976	NS	NS	NS
R^2	0.68	0.34	0.72
Heated discharge			
1973–1974	0.05	NS	NS
1973–1975	0.0001	0.10	0.001
1973–1976	0.0001	0.05	0.0001
1975–1976	NS	NS	0.10
R^2	0.86	0.38	0.59
Skimmer wall			
1973–1974	0.001	NS	NS
1973–1975	0.0001	0.10	0.05
1973–1976	0.0001	0.05	0.0001
1975–1976	0.05	NS	NS
R^2	0.70	0.42	0.39
South			
1973–1974	NS	NS	NS
1973–1975	NS	0.10	0.001
1973–1976	0.001	NS	0.0001
1975–1976	NS	NS	NS
R^2	0.53	0.20	0.48

[a]NS = no statistically significant difference.

consequently there were no significant differences at the Crow Creek, heated water, or south stations between 1975 and 1976.

Mean responses of weekly catch of young fish at 1 and 5.5 m combined for each of the four species for the Crow Creek, heated and skimmer wall stations were compared with those of the control (south) station (Table 5). There were no significant differences in yellow perch mean responses between the south station and the other three stations in 1974 and 1976 (Table 5). There was a significant difference ($\alpha = 0.0001$) between the mean responses of the two areas closest to the power plant (heated water discharge and skimmer wall) and the south area in 1975, but not between Crow Creek and

Table 5. Probability levels for the observed difference between the south (control) station and the other three stations in the mean response of weekly catch of young fish at 1 and 5.5 m combined for each of four species in Keowee Reservoir, 1973–1976

Stations compared	Yellow perch	Black crappie	Sunfishes	Threadfin shad
		1973		
South–Crow Creek	0.05	NS[a]	0.05	
South–heated	NS	0.05	0.10	
South–skimmer wall	NS	NS	NS	
R^2	0.62	0.34	0.19	
		1974		
South–Crow Creek	NS	NS	NS	NS
South–heated	NS	NS	NS	NS
South–skimmer wall	NS	NS	NS	NS
R^2	0.30	0.11	0.04	0.18
		1975		
South–Crow Creek	NS	NS	NS	NS
South–heated	0.001	0.10	NS	0.10
South–skimmer wall	0.001	NS	NS	NS
R^2	0.38	0.26	0.04	0.06
		1976		
South–Crow Creek	NS	NS[b]	NS	NS
South–heated	NS	0.10	NS	NS
South–skimmer wall	NS	NS	0.05	NS
R^2	0.25	0.38	0.20	0.03

[a] NS = no statistically significant difference.

[b] Linear interaction at the 0.05 significance level.

the south area. In 1973, yellow perch catches were significantly higher at the south station than at Crow Creek, but there was no significant difference between the mean responses for the south station and the heated water discharge, or between the south station and the skimmer wall.

Black Crappie

Time of spawning for black crappies changed more during the study period than for any other species. The length at first capture was the same in all years, but the young were captured 44 days earlier in 1976 than in 1973.

Numbers of limnetic young black crappies captured at 5.5 m were about two to three times higher than at 1 m in 1973, when a well-defined thermocline was present in the reservoir (Table 6). When the thermocline was absent or not well defined, as in 1974, 1975, and 1976, catches were more variable but generally greater at 1 m than at 5.5 m.

Numbers of black crappies were inversely related to power plant generation. The total catch declined from 1480 in 1973 to 756 in 1976. The decline was also evident in catch values at all depths and stations between 1973 and 1975 (Table 6). Generally, black crappie catches were lowest at the heated water discharge station and highest at the south stations for all years and depths. Black crappie mean responses were statistically different between the south and heated water areas at the 0.05 significance level in 1973 and at the 0.10 significance level in 1975 and 1976 (Table 5).

There were significant differences in black crappie mean responses between the years 1973 and 1975 at all sampling stations (Table 4). Mean responses were also significantly different between 1973 and 1976 at the heated water discharge and skimmer wall stations but not at Crow Creek or the south station.

Table 6. Mean numbers of young-of-the-year black crappie caught per haul of a frame trawl at 1 m, at 5.5 m, and at both depths combined, from four areas of Keowee Reservoir, 1973–1976

Depth[a] (m)	Station				
	South	Skimmer wall	Heated discharge	Crow Creek	All stations
1973					
1	3.3	2.3	1.3	1.6	2.1
5.5	9.6	8.6	2.3	5.1	6.4
Combined	6.4	5.5	1.8	3.3	4.3
1974					
1	1.9	1.1	0.9	2.6	1.6
5.5	3.1	1.0	0.7	1.5	1.6
Combined	2.5	1.0	0.8	2.0	1.6
1975					
1	1.5	0.4	0.1	0.3	0.6
5.5	0.9	0.3	0.3	0.2	0.4
Combined	1.2	0.4	0.2	0.2	0.5
1976					
1	3.8	2.2	1.2	1.2	2.1
5.5	2.1	1.6	0.4	1.0	1.3
Combined	3.0	1.9	0.8	1.1	1.7

[a]Number of hauls per depth per year from each station ranged from 53 to 58.

Threadfin Shad

Threadfin shad were introduced into the reservoir in January 1974 by the South Carolina Wildlife and Marine Resources Department. They reproduced the first year and were common in the reservoir in 1976.

Threadfin shad 4–5 mm long were first caught about 14 days earlier in 1976 than in 1974. The total number of threadfin shad captured in 1974–1976 remained about the same (a range of 735–842). Catch by depth varied; larger numbers of fish were captured at 1 m more often than at 5.5 m (Table 7). Highest average catches at both depths were at Crow Creek when 4.5 fish were captured in 1974 and 2.4 fish in 1975. Catches at the heated water discharge at both 1 and 5.5 m were the lowest of all sampling stations (less than 1 fish per haul) in 1974 and 1975 but were the highest in 1976 (Table 7).

Sunfishes

Populations of young sunfish declined more than did those of young perch, crappies, or threadfin shad between 1973 and 1976. A total of 9963 sunfishes were captured in 1973, but only 89 were captured in 1976. Sunfish

Table 7. **Mean numbers of young-of-the-year threadfin shad caught per haul of a frame trawl at 1 m, at 5.5 m, and at both depths combined, from four areas of Keowee Reservoir, 1974–1976**

Depth[a] (m)	Station				
	South	Skimmer wall	Heated discharge	Crow Creek	All stations
1974					
1	0.4	1.0	0.9	3.2	1.4
5.5	0.9	0.9	0.6	5.8	1.9
Combined	0.6	0.9	0.8	4.5	1.6
1975					
1	1.3	3.0	1.3	3.5	2.2
5.5	3.0	1.6	0.3	1.4	1.6
Combined	2.1	2.3	0.8	2.4	1.9
1976					
1	0.4	3.1	5.5	3.4	3.1
5.5	0.6	0.7	0.5	1.1	0.7
Combined	0.5	1.9	3.0	2.2	1.9

[a]Number of hauls per depth per year from each station ranged from 53 to 58.

catches declined sharply between 1973 and 1974 and again between 1974 and 1975. They continued to decline and reached their lowest level in 1976, even though the power plant operated at a lower level in 1976 than in 1975 (Table 8). The highest catch (48.1) was recorded at the 1-m depth at Crow Creek in 1973 and the lowest catch (0) came from the same depth at the heated discharge in 1976.

Although trawl catches declined, there were no significant differences in sunfish mean responses at any of the stations between 1973 and 1974 (Table 4). Mean responses were significantly different at all stations between 1973 and 1975 and between 1973 and 1976. Mean responses between 1975 and 1976 were not significantly different at Crow Creek, skimmer wall, or south stations but were significantly different ($\alpha = 0.10$) at the heated water discharge station.

There was only one significant difference in mean responses between stations from 1974 through 1976 — that between the south station and skimmer wall in 1976 (Table 5). This uniformity indicates that sunfish catches

Table 8. **Mean numbers of young-of-the-year sunfishes caught per haul of a frame trawl at 1 m, at 5.5 m, and at both depths combined, from four areas of Keowee Reservoir, 1973–1976**

Depth[a] (m)	Stations				
	South	Skimmer wall	Heated discharge	Crow Creek	All stations
1973					
1	30.3	26.9	29.7	48.1	33.7
5.5	7.7	8.6	12.1	8.4	9.2
Combined	19.0	17.7	20.9	28.3	21.5
1974					
1	8.1	11.4	14.3	6.3	10.0
5.5	8.8	9.6	4.3	2.3	6.2
Combined	8.4	10.5	9.3	4.3	8.1
1975					
1	0.9	1.2	2.5	0.7	1.4
5.5	0.6	0.3	0.2	0.2	0.3
Combined	0.8	0.8	1.3	0.5	0.8
1976					
1	0.1	0.4	0.3	0.3	0.3
5.5	0.3	0.1	0.0	0.1	0.1
Combined	0.2	0.3	0.1	0.2	0.2

[a]Number of hauls per depth per year from each station ranged from 53 to 58.

were equally low at all stations; although after 1974, catches at all stations within any one year may have been too low to permit detection of statistically significant differences.

Semimonthly Deep Water Samples

Generally, catches of all species at depths of 10.5 and 15.5 m were lower in 1976 than in 1973. Catches of yellow perch were lower than those of crappies, sunfishes, or shad at 10.5, 15.5, 20.5 m. Average catches of yellow perch were 0.7 at 10.5 m and 0.2 at 15.5 m in 1973, and 0.2 at 10.5 m, 0.3 at 15.5 m, and none at 20.5 m in 1976.

Black crappies were captured in larger numbers at the greater depths than were the other forms in both 1973 and 1976. In 1973, black crappie catches were 8.4 at 10.5 m and 3.4 at 15.5 m. Catches were lower in 1976 — 3.6 at 10.5 m and 1.2 at 15.5 m; three black crappies were captured at 20.5 m in 1976.

Threadfin shad were slightly more abundant at 20.5 m than at either 10.5 or 15.5 m in 1976. Threadfin shad catches were 1.1 in the 5-min hauls at 20.5 m, compared with 0.5 at 10.5 m and 0.6 at 15.5 m in 10-min hauls.

Sunfish numbers declined at the deeper depths, as they did at the 1- and 5.5-m depths, between 1973 and 1976. The sunfish catches at 10.5 m were 6.3 in 1973 and only 0.2 in 1976. Catches were lowest at 15.5 m in both years (3.0 in 1973 and 0.2 in 1976). No sunfish were captured at 20.5 m in 1976.

Thus fewer fish of all four forms were taken at 10.5–20.5 m in 1976 than in 1973. No yellow perch or sunfish and only a few black crappies and threadfin shad were captured at 20.5 m, suggesting that few if any of these fish entered the plant's water intake.

DISCUSSION

The decline in trawl catch of the three endemic species between 1973 and 1976 appear to be related to increased water temperatures. Although the water never reached the lethal temperature for the species studied (Reutter and Herdendorf 1975, Banner and Van Arman 1973), the changes in temperature may have been sufficient to alter reproductive or other behavioral patterns and to reduce the numbers of larval fish in the open-water areas of the reservoir. Declining numbers of young also may have resulted from trends repeatedly demonstrated to be associated with the aging of a reservoir (Jenkins 1970). Other shifts in abundance may be inherent in a particular species.

Yellow perch are typically associated with relatively cold northern waters. Keowee Reservoir is near the southern limit for the species, although a population exists in Par Pond, 158 km further to the south, near Aiken, South Carolina (Clugston 1973). Spawning success of yellow perch is increased if water temperatures fall to 4.0 to 6.0°C (Jones et al.*). Low populations of young yellow perch in 1974–1976 were possibly related to higher midwinter water temperatures that reduced the number of ova brought to maturity, or were possibly related to the absence of mature perch from the main body of the reservoir during the spawning season.

Barons (1972) reported that, in laboratory studies, temperatures selected by adult and by young yellow perch from Lake Erie were significantly different. During all seasons young yellow perch selected temperatures 4.0°C or more above the 21.0–27.0°C summer or 13.0–29.0°C fall temperature preference of adults, and adults avoided temperatures above 29.0°C in summer. Water temperatures at a depth of 1 m in the discharge area of Keowee Reservoir were above 29.0°C for five consecutive weeks and above 29.0°C at all other stations for at least two weeks during summer 1975. Part of the decline of young yellow perch in the discharge area after the plant began operation may have resulted from adult fish avoiding the high temperatures of the discharge in the summer and fall and not returning to the area to spawn the following spring.

In samples from coves of Keowee Reservoir treated with rotenone, young yellow perch shorter than 63 mm were taken at rates of 475 per hectare in 1975 and 235 per hectare in 1976 (Southeast Reservoir Investigations, *unpublished data*). This relatively large number suggests that most young yellow perch may have been in coves and not available to the open-water trawl.

Water temperatures did not exceed spawning temperatures for yellow perch (8.9–15.8°C) in any part of the reservoir in March of any year. The temperatures in the heated discharge in late summer of 1975 were just below the levels (32.0°C) that have been reported to cause deformations or mortality among young yellow perch (McCormick 1976). Even though temperatures in the heated discharge area of Keowee Reservoir were below the reported lethal temperatures, they exceeded the temperatures that yellow perch reportedly avoid (Barons 1972). The water temperatures at all other sampling stations in all years were below the levels at which temperature-related mortality could be expected. We do not believe that the decline in young yellow perch abundance in the main body of Keowee Reservoir was directly related to mortalities caused by high water temperatures.

Water temperatures in all areas of Keowee Reservoir in all years were well below the lethal level for black crappies. Ruetter and Herdendorf (1975)

---

*B. R. Jones, K. E. F. Hokanson, and J. H. McCormick. *In preparation*. Temperature requirements of maturation and spawning of yellow perch, *Perca flavescens* (Mitchell) [*sic*]. Environmental Research Laboratory, Duluth, Minnesota.

simulated fish swimming into a plume by raising the water temperature rapidly and were able to determine the critical thermal maximum (C.T.M.). In adult black crappies the C.T.M. ranged from 19.5 to 33.5°C in December, January, and February (depending on acclimation temperature) and from 33.2 to 35.8°C in June and July. Hokanson* found an upper incipient lethal temperature of 32.5°C for juvenile black crappies, an optimum temperature of 22.0–25.0°C for growth, and an upper temperature limit of 30.0°C for zero growth.

In studies relating black crappie movement to abnormally high water temperatures, Miller and DuMont (1972) showed that adults avoided and did not spawn in heated water discharge areas in Lake Norman, North Carolina. Large numbers of larval black crappies were captured in control areas with townets, but none were found in the heated water discharge.

The highest temperature at which young black crappies were captured in the present study was 28.0°C. When these fish were captured, this was the lowest water temperature between 1 and 6 m at any trawling station. The thermal tolerance or preferred temperature for adult black crappies may be lower than that of the other three species under study in Keowee Reservoir; Siler (1975) reported that the maximum temperature at capture was 31.0°C and that most catches occurred at 27.0–29.0°C in another South Carolina cooling reservoir. The seasonal temperature preference range for black crappies may be narrow; Reutter and Herdendorf (1974) showed that the adults preferred temperatures between 20.5 and 22.2°C during all seasons of the year.

Large numbers of adult black crappies may have moved away from the heated water discharge and into coves and arms of incoming streams after the plant began operation. Some of these fish may have remained in the coves until the spring spawning season, resulting in reduced numbers of young black crappies at the heated water discharge station and larger numbers at the south station each year.

Crappie populations are known to fluctuate widely, but the reasons for these fluctuations are not completely understood. Strong year classes form every three to five years (Swingle and Swingle 1967), but the age at which strong year-class formation is detectable appears to be unknown. Nelson et al. (1967) speculated that year-class strength is established during the first year of life. Year-class strength may be related to the strength of preceding year classes. When dominant year classes of age I and II crappie are present in a reservoir, few young fish survive, or the adults may not reproduce (Swingle and Swingle 1967). Inasmuch as black crappies generally made up only 0.1 to

---

*K. E. F. Hokanson. 1976. *Personal communication*. U.S. Environmental Protection Agency, Monticello Ecological Research Station, Monticello, Minnesota.

6.6% of the adult fish during the present study, it appears unlikely that they were abundant enough to reduce later year classes.

Threadfin shad normally are found at relatively high water temperatures. The population appeared to expand or remain the same as heated water was added to the reservoir. Threadfin shad survive best when winter water temperatures are between 10.0 and 15.5°C, and suffer high mortality at 7.0°C (Parsons and Kimsey 1954). However, little is known about their upper temperature limit. Adults have been captured consistently from areas where the temperature exceeds 30.0°C (McNeely and Pearson 1974). Maximum temperatures at which eggs develop are between 34.2 and 34.4°C (Hubbs and Bryan 1974). The highest water temperatures observed in Keowee Reservoir were within the known survival limits of threadfin shad eggs and adults. Large numbers of young threadfin shad were captured in the present study when the water temperature was 27.4°C. This temperature could be near their preferred summer temperature, since warmer and cooler waters were available in other sections of the reservoir at the time of collection.

Young sunfishes were found in large numbers in the open water areas of Keowee Reservoir in 1973. Trawl catches yielded peak catches of young sunfishes in August 1973, but there were no August peaks in 1974, 1975, and 1976. Small bluegills commonly inhabit the open water areas of lakes and reservoirs. Bluegills (10–25 mm long) were captured from the limnetic areas of an Indiana lake over a six- to seven-week period and started moving into the littoral zones at lengths of 21 to 25 mm (Werner 1967). These fish were captured in larger numbers at 1 m than at 5.5 m.

Numbers of sunfishes of all ages declined rapidly and steadily in all areas of Keowee Reservoir from 1973 through 1976. By 1976, almost all sunfish in rotenone samples were less than 100 mm long; only a few were 150–200 mm long (Southeast Reservoir Investigations, *unpublished data*). The decline in abundance of young sunfishes in Keowee Reservoir between 1973 and 1976 may have been caused by the decline in the adult population or by poor spawning conditions. Reservoir age has been strongly associated with population declines. Jenkins (1970) reported that sunfish populations were negatively correlated with reservoir age. The young sunfish in Keowee Reservoir grew slowly, which may have made them more susceptible to predation for a sustained period of time (Southeast Reservoir Investigations, *unpublished data*). They fed entirely on zooplankton and insects during their first growing season, and may have competed poorly with other plankton feeders such as threadfin shad. Young shad have a size advantage over the young sunfishes, because the species spawns earlier in the year. Clugston (1973) found bluegill nests and fry at temperatures ranging between 17.5 and 36.5°C in a cooling reservoir that had been receiving heated effluent for over 15 years. He reported many instances where young bluegills 10–25 mm long were active at temperatures above 30.0°C.

The largest decline in the abundance of all species of young fish between 1973 and 1976 occurred near the heated discharge. Avoidance by adults was suggested as a reason for the reduced number of young at this site. Simple flushing of the area by a large volume of relatively fish-free water may be another reason. The skimmer wall apparently serves as a barrier, and few young fish pass under it. Duke Power Company biologists,* who sampled the flowing intake water weekly from March through September 1976, collected only one black crappie and two threadfin shad. Further, our semimonthly samples at 20.5 m near the skimmer wall indicated that few young fish were near the intake.

Temperature changes may have affected the vertical or horizontal distribution of young fish during the study period. Although oblique hauls were not taken, additional samples at 10.5, 15.5, and 20.5 m did not reveal large concentrations of young fish. Six streams flow into Keowee Reservoir through long irregularly shaped coves. Possibly the upper sections of these coves served as major spawning areas and sanctuaries for young fish during the later years of the study. The best test of the validity of the reduced catch of young fish as a measure of reduction in year-class strength will be made by later sampling of the adult fish population. If all the trends suggested by the young fish sampling are detected later in the adults, this study will have shown a strong relation between the abundance of young-of-the-year fish and year-class strength.

SUMMARY

This paper describes the limnetic young-of-the-year fish populations in Keowee Reservoir, South Carolina, before and after the discharge of heated water from a nuclear power plant. After the nuclear plant went into operation the reservoir reached higher temperatures, warmed earlier in the spring, and remained warm later in the fall; also the thermocline sank at an accelerated rate during the summer.

The relative abundance of the young of three endemic forms declined during this study. Reduced numbers were measured in all areas of the reservoir, but were greatest in areas closest to the heated water discharge. Numbers of threadfin shad, an introduced species, changed little during the three years during which heated water was discharged.

It appears that the decline in abundance of young fish was not caused by lethal water temperatures or by entrainment of young fish in the nuclear plant cooling water. Since the smallest numbers of fish were caught near the heated discharge, temperature avoidance by adults could contribute to reduced spawning and young in that area. Since temperatures above preferred

---

*M. Killough and J. Sevic, Station Biologists, Duke Power Company, Oconee Nuclear Station, Seneca, South Carolina.

temperatures for some species were measured elsewhere in the reservoir, temperature-dependent behavorial changes may have been sufficient to alter reproductive success or distribution of young in the reservoir. Determining the causes of reduced numbers of young is confounded by possible reduction normally associated with the aging of a reservoir. Adult fish numbers of the three endemic species in future years should reflect the reduction described for the young during this study period.

LITERATURE CITED

Amundrud, J. R., D. J. Faber, and A. Keast. 1974. Seasonal succession of free-swimming perciform larvae in Lake Opinicon, Ontario. J. Fish. Res. Board Can. **31**(10):1661–1665.

Banner, A., and J. A. Van Arman. 1973. Thermal effects on eggs, larvae and juveniles of bluegill sunfish. U.S. Environmental Protection Agency, Ecological Research Series EPA-R3-73-041. Office of Research Monitoring, Washington, D.C. 111 p.

Barans, C. A. 1972. Seasonal temperature selections of white bass, yellow perch, emerald shiners, and smallmouth bass from western Lake Erie. Ph.D. Thesis, Ohio State Univ., Columbus. 88 p.

Clugston, J. P. 1973. The effects of heated effluent from a nuclear reactor on species diversity, abundance, reproduction, and movement of fish. Ph.D. Thesis, Univ. Georgia, Athens. 104 p.

Faber, D. J. 1967. Limnetic larval fish in northern Wisconsin lakes. J. Fish. Res. Board Can. **24**(5):927–937.

Hubbs, C., and C. Bryan. 1974. Maximum incubation temperature of the threadfin shad, *Dorosoma petenense*. Trans. Am. Fish. Soc. **103**(2):369–371.

Jenkins, R. M. 1970. The influence of engineering design and operation and other environmental factors on reservoir fishery resources. Water Resour. Bull. J. Am. Water Resour. Assoc. **6**(1):110–119.

McCormick, J. H. 1976. Temperature effects on young yellow perch, *Perca flavescens* (Mitchell) [*sic*]. U.S. Environmental Protection Agency, Ecological Research Series EPA-600/3-76-057. Duluth, Minnesota. 18 p.

McNeely, D. L., and W. D. Pearson. 1974. Distribution and condition of fishes in a small reservoir receiving heated waters. Trans. Am. Fish. Soc. **103**(3):518–530.

Miller, R. W., and D. J. DeMont. 1972. Effects of thermal pollution upon Lake Norman fishes. North Carolina Federal Aid Project F-19-4. Final Completion Report. North Carolina Wildlife Resources Commission, Raleigh. 32 p. (*Mimeo*)

Neter, J., and W. Wasserman. 1974. Applied linear statistical models, regression, analysis of variance, and experimental designs. Richard D. Irwin, Inc., Homewood, Illinois. 842 p.

Nelson, W. R., R. E. Siefert, and D. V. Swedberg. 1967. Studies of the early life history of reservoir fishes, p. 374–385. *In* Reservoir Fishery Resources Symposium, Southern Division, Am. Fish. Soc. Univ. Georgia Press, Athens.

Noble, R. L. 1970. Evaluation of the Miller high-speed sampler for sampling yellow perch and walleye fry. J. Fish. Res. Board Can. 27(6):1033–1044.

Parsons, J. W., and J. B. Kimsey. 1954. A report on the Mississippi threadfin shad. Prog. Fish-Cult. 16(4):179–181.

Reutter, J. M., and C. E. Herdendorf. 1974. Laboratory estimates of the seasonal final temperature preferenda of some Lake Erie fish. Proc. Conf. Great Lakes Res. (Int. Assoc. Great Lakes Res.) 17:59–67.

Reutter, J. M., and C. E. Herdendorf. 1975. Laboratory estimates of fish response to the heated discharge from the Davis-Besse reactor, Lake Erie, Ohio. Federal Aid Project F-41-R-6. Annual Performance Report, Center for Lake Erie Area Research, Ohio State Univ., Columbus. 54 p. (*Mimeo*)

Siler, J. R. 1975. The distribution of fishes in two cooling reservoirs with different heat loads. M.S. Thesis, Univ. Georgia, Athens. 94 p.

Swingle, H. S., and W. E. Swingle. 1967. Problems in dynamics of fish populations in reservoirs, p. 229–243. *In* Reservoir Fishery Resources Symposium, Southern Division, Am. Fish. Soc. Univ. Georgia Press, Athens.

Werner, R. G. 1967. Intralacustrine movement of bluegill fry in Crane Lake, Indiana. Trans. Am. Fish. Soc. 96(4):416–420.

Part II

Estimating Abundance, Production, and Mortality Rates of Young Fish

Estimating the Size of Juvenile Fish Populations in Southeastern Coastal-Plain Estuaries

Martin A. Kjelson

National Marine Fisheries Service
Southeast Fisheries Center
Beaufort Laboratory
Beaufort, North Carolina

ABSTRACT

Understanding the ecological significance of man's activities upon fishery resources requires information on the size of affected fish stocks. The objective of this paper is to provide information to evaluate and plan sampling programs designed to obtain accurate and precise estimates of fish abundance. Nursery habitats, as marsh—tidal creeks and submerged grass beds, offer the optimal conditions for estimating natural mortality rates for young-of-the-year fish in Atlantic and Gulf of Mexico coast estuaries. The area-density method of abundance estimation using quantitative gears is more feasible than either mark-recapture or direct-count techniques. The blockage method provides the most accurate estimates, while encircling devices enable highly mobile species found in open water to be captured. Drop nets and lift nets allow samples to be taken in obstructed sites, but trawls and seines are the most economical gears. Replicate samples are necessary to improve the precision of density estimates, while evaluation and use of gear-catch efficiencies is feasible and required to improve the accuracy of density estimates. Coefficients of variation for replicate trawl samples range from 50 to 150%, while catch efficiencies for both trawls and seines for many juvenile fishes range from approximately 30 to 70%.

Key words: catch efficiencies, estuary, fishes, juveniles, marine, number of samples, nursery grounds, population size, sampling gear, sample variability

INTRODUCTION

Estimation of the population size of young-of-the-year fish is an important requirement for the assessment of their natural mortality rates, since survival statistics normally are based upon estimates of population abundance at two successive intervals of time. The ecological significance of mortalities directly due to man's activities is usually difficult to interpret unless this mortality can be evaluated against some measure of the natural mortality and the size of the affected fish stock. Hence, estimates of natural mortality of

71

young-of-the-year fish are particularly relevant to the question: "How do mortality rates imposed by power plants on young fish affect adult population size?"

Based on these considerations, estimates of fish abundance must be both accurate and precise. The accuracy of fish abundance estimates is influenced by biases introduced through errors in measurement and the inaccessibility of certain portions of the population being assessed. The ability of fish to avoid the gear, the selective nature of fishing gear relative to the size and species collected, and the accessibility of the fish to the gear all may affect the accuracy of the abundance estimate. The precision of the estimate is influenced by the number of samples taken, the manner in which the samples are drawn, and the way the data are analyzed. The highly nonrandom distribution of fishes in their environment results in estimates of low precision unless sample effort is large. More lengthy reviews concerning the practical problems associated with the sampling of fish populations are provided by Allen, DeLacy, and Gotshall (1960), Macketts (1973), and Watt (1968).

The above sampling problems emphasize that to achieve a usable estimate of population size an investigator must consider the physical geography of the study area, life history of the species considered, the varied factors influencing fish distribution, and the characteristics of the available fish sampling gear. The objective of this paper is to indicate how such information can be used to plan sampling programs designed to achieve accurate and precise estimates of population size for juvenile fishes in southeastern (Atlantic and Gulf) coastal-plain estuaries and to encourage the improvement of sampling methodologies.

Refined estimates of population abundance in southeastern estuaries often are particularly difficult to achieve because of temporal-spatial movements. Both marine- and estuarine-spawned fishes utilize these systems as nursery areas, with spawning and nursery grounds often separated by considerable distances (Gunter 1967, McHugh 1967). Hence, adult, juvenile, and larval populations often exhibit large-scale seasonal migrations between ocean and estuary and within estuaries (Copeland and Bechtel 1974). Several individual size classes of a given year class, of Atlantic menhaden, *Brevoortia tyrannus*, for example, may be found in a variety of different estuarine or nearshore oceanic habitats during their migration from offshore spawning sites to low-saline, estuarine nursery grounds, and during their return to coastal waters as one-year-old fish (Reintjes and Pacheco 1966). These movements greatly increase the costs of assessing population size within these coastal areas.

Such conditions force the development of particular strategies to ensure that population abundance estimates are well characterized relative to the population being studied. In other words, the relationship between the time frame and spatial distribution of the sample population and the true, total biological population that may inhabit the geographical region under consideration must be clearly defined.

These considerations are particularly relevant to our basic concern of determining the natural mortality of juvenile fish in estuarine systems, since the inherent need is the estimation of population size at two successive intervals of time. For the two estimates to be comparable they must be based on the same sample population. Secondly, the estimates must be characterized by sufficiently high precision to enable changes in abundance (possibly small), due to mortality, to be observed. Cole (1973) provides a useful review of the problems involved in assessing natural mortality in young-of-the-year estuarine fishes. The migratory behavior of many estuarine fishes greatly complicates matters and restricts the time and place where usable mortality estimates may be realistically achieved. The population must be stable in its movements for a sufficient period of time so that mortality can be assessed, and the area inhabited by the population must be small enough to be practically or economically sampled.

It appears that the most suitable life stage to assess juvenile mortality in southeastern estuaries is when the fishes are inhabiting estuarine nursery areas, since at that time their movements are relatively localized compared to older (late juveniles and yearlings) and younger (larval and postlarval) stages which exhibit migratory behavior. Many species in the southeast appear to inhabit relatively small estuarine areas for periods of three to four months, although as growth occurs the larger forms are observed to move out of these areas (Hansen 1969, Herke 1971). In addition, southeastern estuarine nursery grounds are made up of relatively uniform physical habitats and are often distinct geographical areas, for example, tidal creeks or submerged grass beds, that are easy to locate and identify (Lippson 1973). These conditions enable the researcher to restrict his study area to a single habitat type (thus eliminating the need for further stratification), and to restrict the study area to a feasible size.

The remaining text will be restricted to brief discussions concerning (1) characterization of southeastern estuarine nursery areas and their dominant fishes, (2) basic methods used to estimate absolute fish abundance, (3) quantitative sampling apparatus, and (4) recommendations for the improvement of fish abundance estimates.

SOUTHEASTERN ESTUARINE NURSERY AREAS AND DOMINANT FISH SPECIES

Joseph (1973) stressed that estuarine fish nursery grounds must provide an abundant, suitable food supply, a degree of protection from predation, and a physiologically suitable area in terms of chemical and physical features. These requirements tend to restrict most southeastern nursery areas to regions of low salinity, shallow depths, and low water-current velocities. In addition, many nursery grounds are characterized by high turbidities and soft bottoms of small sediment particle size. Specific habitat types may vary greatly in terms of floral composition. The major fish nursery habitats found in southeastern estuarine systems, classified by dominant adjacent vegetation, are the following: hardwood swamp, tidal creek; *Juncus* (needlerush) marsh, tidal creek; *Juncus* marsh, open-water embayment; mangrove swamp, tidal creek; mangrove swamp, open-water embayment; open water, *Ruppia* (widgeongrass) bed; open water, *Thalassia* (turtlegrass) bed; open water, *Zostera* (eelgrass) bed; *Spartina* (smooth cordgrass) marsh, tidal creek; *Spartina* marsh, open-water embayment. As nursery habitat varies, we can expect the composition of the respective fish communities to change. Some of the common fishes generally present in southeastern estuarine nursery areas are shown in Table 1. Proper choice of sampling methods depends on both habitat type and the species present.

ESTIMATION OF ABSOLUTE FISH ABUNDANCE

Several basic methods for obtaining absolute abundance of fish populations were described by McFadden (1975). These include the area-density, mark-recapture, and direct-count methods.

The area-density method consists of randomly sampling a plot of known area or volume and capturing some or all of the fish present. The average number of fish per unit area or volume is then calculated and multiplied by the total area (volume) to obtain an estimate of the size of the entire population. The area-density method is a refinement over the general catch-per-unit-effort approach in that the area from which the sample is obtained is known, thus providing at least a minimal measure of absolute abundance, as opposed to relative abundance. By correcting for underestimation of abundance, due typically to mesh selection or net avoidance, an estimate of absolute abundance is obtained.

The mark-recapture method consists of collecting a known number of fish from a population, marking them, and then releasing them to mix with the total population. Next, a sample is withdrawn and an estimate of the fraction of marked fish in the sample is computed. The number of marked fish released divided by the fraction of marked fish in the sample provides an estimate of the total size of the fish stock. Using a refinement of this technique, A. L. Pacheco (NMFS, Sandy Hook Laboratory, Highlands, New Jersey, *personal communication*) estimated the juvenile Atlantic menhaden population in White Creek, Delaware (4.5 km long by 300 m wide) to be 410,844 fish with a 95% confidence interval of 351,010 to 495,270. The field study utilized a crew of two, working full time for four weeks. Several other investigators used mark-recapture techniques to estimate the size of fish populations in estuaries (Saila 1961, Turner 1973, Hoss 1974) and met with varied degrees of success. Unfortunately, too often the assumptions inherent in this method are difficult to meet and precision is low, making resulting estimates

Table 1. Some numerically important fishes found as juveniles in southeast Atlantic and Gulf coast estuarine nursery grounds, Classified by their typical vertical distribution in the water column

Distribution	Family	Representative species
Pelagic	Atherinidae	Atlantic, rough, and tidewater silverside
	Clupeidae	Atlantic and Gulf menhaden, American shad, river herring
	Engraulidae	Bay and striped anchovy
	Mugilidae	Striped and white mullet
Semidemersal	Ariidae	Gaftopsail and sea catfish
	Cyprinodontidae	Mummichog, sheephead minnow
	Gerreidae	Silver jenny, spotfin mojarra
	Percichthyidae	Striped bass, white perch
	Pomadasyidae	Pigfish, white grunt
	Sciaenidae	Atlantic croaker, silver perch, spot, star, and red drum
	Sparidae	Pigfish
	Stromateidae	Harvestfish, butterfish
Demersal	Bothidae	Fringed, southern, and summer flounder
	Soleidae	Hogchoker
	Cynoglossidae	Blackcheek tonguefish

biased and unusable. Secondly, the cost of this method often is higher than that of the area-density technique. More complete information regarding the use of mark-recapture techniques is provided by Seber (1973) and Jones (1976).

The third approach, use of direct counts of juveniles, is usually impractical due to the high turbidity encountered in most southeastern estuarine nursery areas. However, aerial surveys have been used to assess the abundance of pelagic juvenile menhaden with some success (Turner 1973).

The relatively new acoustical techniques also appear impractical at present for estimating juvenile abundance in nursery areas because of the shallow depths, the background noise caused by detritus and sediment particles, and the lack of information on target identification. However, acoustical gear, once technically refined, offers the most economical method for the assessment of population size, and research on this method should be encouraged.

Considering the limitations of the mark-recapture, direct-count, and acoustical methods, it appears that the most usable estimates of juvenile fish abundance will be achieved using the area-density method.

QUANTITATIVE SAMPLING APPARATUS

An important decision in the use of the area-density method is the choice of an appropriate sampling gear relative to the species of fish considered, the nursery habitat type involved, and the budgetary limitations. Table 2 provides a list of major quantitative sampling gears currently in use. Each gear category may contain many variations of the general form, with each designed for a specific sampling need.

Blockage methods trap the fish and prevent their escape, and thus, these methods provide some of the most accurate estimates of abundance. Following enclosure, the fish are collected by any one of a variety of techniques, all of which are usually designed to sacrifice the entire population in the sample area. Fish may be collected by repeated seining of the area until the area is swept clean or until enough samples are taken to estimate population density using the DeLury regression method (Lagler 1959, Zippin 1958). Alternatively, fish may be collected through the use of a poison (e.g., rotenone) combined with varied recovery techniques, or they may be collected with a tidal net in those situations where the water flows completely out of the enclosed sample site at low tide. This latter method is particularly suited for small tidal-marsh creeks, ponds, or embayments where the study area is small enough in size and depth to be covered by a single sweep of a seine or where trawling is not practical. A common problem with this technique, however, is the lack of total recovery of all enclosed fish, particularly where soft muddy bottoms, stumps, or submerged aquatic vegetation are present. The time necessary to obtain a single sample using the blocking method is usually great

Table 2. Quantitative sampling apparatus used for the assessment of
juvenile fish abundance in southeast Atlantic and Gulf coast estuaries

General method	Collection gear	Numbered references[a]
Blockage	Poison	49, 59
	Seine	33, 42
	Tidal net	4, 59
Encircling	Longhaul seine	34
	Purse seine	26, 27, 48
Enclosure	Drop-net quadrat	20, 23
	Lift net	37
	Portable drop net	1, 30, 33, 47
	Stationary drop net	35
Trawl	Surface	21, 43, 56
	Otter	6, 18, 24, 40, 51, 53, 58
	Beam	14, 36
Seine	Standard haul	10, 44, 58

[a]See numerical listing in literature cited.

(often measured in hours or days), unless the study area is very small. Replicate samples usually are not taken since the total population is often sampled; however, under certain conditions, as in estuarine marsh ditches (Marshall 1976), multiple areas may be blocked, thus yielding simultaneous multiple estimates of the population density.

Encircling methods, such as purse seines and long-haul seines, are designed to enclose very rapidly a known area of water from surface to bottom, and they are not dependent on other techniques to collect the fish as in the blocking method. Purse seines and long-haul seines of sufficient size to cover a large circular area are particularly suited for sampling fast-swimming, schooling, pelagic fishes in open-water habitats, where net avoidance is a major problem. The objective is to enclose a body of water before the fish within it realize they are enclosed. The area covered by these large seines varies with each sample but is most easily calculated when the net encloses a circle. Here, as with most blocking techniques, operation of the gear takes considerable effort in terms of both manpower and time, and thus, encircling methods prevent economical sample replication and the attainment of high precision. In addition, initial net costs are usually high. Information concerning the accuracy of estimates from long-haul and purse-seine samples is limited. Our studies for the long-haul seine on catch efficiency, defined as the percentage of fish in the sample area that are captured, indicate that the seine averaged 41% efficiency for pelagic fishes and 47% efficiency for semidemersal forms (Kjelson and Johnson 1975b).

Enclosure methods, e.g., drop nets and lift nets, also take instantaneous samples, but they usually sample a relatively small area (1 to 30 m^2). Hellier's (1959) drop-net quadrant, however, covered approximately 900 m^2, although this area had to be seined to collect the entrapped fish. It is expensive to construct a large mechanical device that drops or lifts rapidly, even if from an engineering point of view it undoubtedly could be done. Drop nets and lift nets appear to be highly useful in nursery habitats that have considerable submerged aquatic vegetation, soft sediment, or where other physical barriers make trawling or seining impractical. Many drop-net designs include a heavy dead line and pursing device that enables recovery of a high percentage of the entrapped fish even when submerged vegetation is present. Such designs also are useful in sampling demersal forms which may burrow into the bottom substrate, making them less vulnerable to conventional trawls and seines. For example, simultaneous samples taken with a 16 m^2 portable drop net and a 6.1 m otter trawl resulted in drop-net estimates of tongue fish (*Symphurus plagiusa*) density that were 15 times greater than predicted from the trawl data. Replicate drop-net and lift-net samples may be taken when using multiple gears set at various locations or through the use of portable forms, although such replication is costly in terms of time. The small area covered by such nets greatly limits their efficiency for highly mobile pelagic fishes (Kjelson and Johnson 1975a).

Trawls and short-length and medium-length haul seines (<30 m) probably are the most commonly used fish sampling devices. Their classification as "quantitative" sampling gear may be open to question. The area covered per sample can vary considerably, unless care is taken to standardize the operation and the conditions under which the gear is used, and unless the dimensions of the gear are evaluated during operation (i.e., trawl-mouth opening, width covered by the seine, etc.). Various studies (Edwards 1968, Hoese et al. 1968, Edwards 1973, Oviatt and Nixon 1973, Turner and Johnson 1973) have calculated the area sampled by trawls and seines. Certain subtle complications, such as the effect upon the size of the sampled area of trawl bridles, doors, and warps as herding devices, also must be considered (Alverson and Pereyra 1969). Nevertheless, a strong case easily is made for the economic, versatile nature of trawls and haul seines. With certain exceptions, most estuarine juvenile nursery habitats in the southeast can be sampled with these gear. Manpower needs for their operation seldom are more than two individuals, and vessel requirements for either trawls or seines are minimal, with a great many being designed for use with one or two small boats (<6 m in length) each powered by an outboard engine. Replicate samples are obtainable with a minimum of time and cost.

Haul seines are restricted to shallow water and substrate types that can be sampled, but many nursery grounds in the southeast are under 1 m in depth and have firm enough bottoms to be hauled. Trawls, conversely, may be designed to sample any portion of the water column. Modifications of otter trawls (and seines), such as adding tickler chains or heavy lead lines, disturb the bottom and thus increase collection of benthic forms such as flounders and soles. The efficiencies of trawls and seines for such fishes, however, probably are lower than the efficiencies for semidemersal and pelagic forms.

The above brief evaluation of quantitative sampling apparatus emphasizes that, for the present time, trawls and haul seines will continue to be the most economic and versatile means available to estimate juvenile fish abundance in southeastern estuarine nursery areas. Therefore, the next section will concern ways in which abundance estimates obtained by these two gears can be improved.

RECOMMENDATIONS FOR IMPROVING ABUNDANCE ESTIMATES BY TRAWLS AND SEINES

Two strategies are of high value for the improvement of fish abundance estimates: (1) obtaining a sufficient number of samples to achieve the desired level of precision, and (2) assessing gear catch efficiencies to improve accuracy. A practical approach for the implementation of these two strategies relative to sampling with trawls and haul seines follows.

Number of Samples

Once the sample area has been restricted to a uniform, nursery-ground habitat, replicate samples can be taken in a randomized pattern within the site to obtain estimates of the sampling error. This permits the establishment of a level of precision for the fish-abundance estimate for a given sampling effort (Cassie 1971).

Two major parameters are important in determining the precision of a given estimate of fish population abundance: (1) the observed sample variance, s^2, influenced primarily by the distribution pattern of the fish if standardized sampling techniques are used, and (2) the number of samples. The investigator has no control over the basic sample variability observed from a given population, but he can control the number of samples, and hence he can achieve the specified level of precision required to complete the study objectives. Depending upon the use of the population abundance estimate, a given level of acceptable error about the final mean may be chosen. A common measure of precision is the confidence limit about the mean, computed as

$$\bar{X} \pm t\sqrt{s^2/n} \text{ or } \bar{X} \pm L , \tag{1}$$

where $\bar{X}$ is the mean of the samples, n is the number of samples, and t is Student's t at the chosen level of probability. The half-width of the desired confidence interval, L, may be interpreted as the absolute amount of error around the mean that is acceptable relative to the intended use of the mean estimate of abundance. From Eq. (1) it is obvious that the precision of the estimate of a population mean is greater (i.e., the confidence limit is smaller) for large n. Using the equation

$$n = s^2 t^2 / L^2 ,$$ (2)

which may be derived from Eq. (1), the number of samples, n, required to achieve a given level of precision may be calculated. Although the value of t is dependent on n and therefore is unknown, it is usually set equal to 2.0 if n is expected to be greater than 30. The level of precision, P, may be specified by expressing the half-width of the confidence interval, L, as a fraction (or percentage) of the mean, that is, $P = 100L/\overline{X}$, and thus $L = P\overline{X}/100$. An estimate of the mean population abundance, $\overline{X}$, and of the population variance, s^2, may be obtained with relative ease through a pilot survey using replicate tows at the study site.

Table 3 presents mean population densities and their standard deviations for replicate trawl samples taken during a pilot survey of a North Carolina estuarine nursery ground. With the exception of the coefficient of variation for Atlantic menhaden, which may be abnormally low, the coefficients of variation range from approximately 50 to 150%. It is apparent from these data that achievement of high precision for trawl estimates will require a considerable number of samples. Using the values from the 6.1-m otter-trawl survey for the mean density and standard deviation of spot (Table 3) and using Eq. (2) with a value of 2.0 for t, a curivlinear relationship between the desired level of precision, P, and number of samples required, n, may be obtained (Fig. 1). As indicated, considerable effort is involved in decreasing the error about the mean from 40% to 30, 20, or 10%. In reality, it may be uneconomical to expend such high levels of sample effort (i.e., $n = 28, 64$, and 256, respectively) to achieve such a small improvement in precision.

Table 3. Observed variation in densities of juvenile fish of several species in nursery habitat in a North Carolina estuary as sampled with two types of gear

Gear	Species	N	$\overline{X}^a$	s	CV^b
6.1-m otter trawl	Bay anchovy	5	27.7	27.8	100
	Pinfish	10	1.1	0.6	55
	Spot	10	3.0	2.4	80
6.1-m surface trawl	Atlantic menhaden	6	23.0	3.5	15
	Bay anchovy	10	2.3	3.5	152
	Bluefish	6	1.0	1.0	100
	Rough silverside	5	101.9	141.3	139

[a] Number/100m^2.

[b] Coefficient of variation (%).

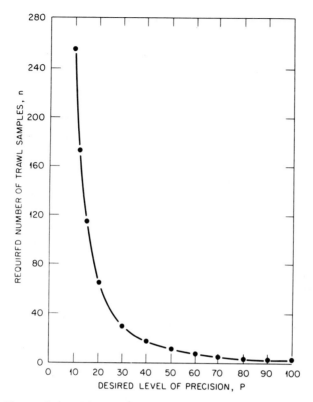

Fig. 1. The number of 6.1-m otter-trawl samples required to achieve a given level of precision for juvenile spot abundance.

Unfortunately, the assessment of natural mortality of juveniles in estuarine nursery areas through comparison of mean population abundances probably will require high levels of precision, since within the short time periods involved (a few months) only small changes in abundance may occur. Simply stated, considerable numbers of samples will be required to gain the required precision because of the high sample variability characterizing such fish data. In addition, care must be taken in designing the sample replication procedure so that the population is not harassed to the point of reducing their vulnerability to the sample gear or of forcing the fish to leave the site all together. Limiting sampling to a few short tows or hauls per day for several days may overcome this problem.

Thorough discussion of sampling designs, sample variability, and the use of statistics in the evaluation of ecological impacts is provided by Eberhardt (1975), Eberhardt and Gilbert (1975), and Zar (1975).

Assessment of Gear Catch Efficiency

Avoidance of sampling gear by fish causes abundance estimates to be biased. Various means may help overcome such bias — for example, increasing tow speed and the size of the area covered by the net, using a net of a color that is difficult for the fish to see, or simply restricting sampling to periods of darkness (Clutter and Anraku 1968). Sampling at night is one of the simplest and most successful ways of reducing gear avoidance. Unfortunately, problems of navigational safety and increased cost often make nighttime sampling impractical.

Both relative and absolute efficiencies are used to quantify bias in fish sampling. Relative catch efficiencies are obtained by comparing catches made with different gears, under different light conditions, at different tow speeds, etc. Such comparisons yield valuable information about sampling bias when absolute catch efficiencies cannot be practically assessed. The absolute catch efficiency, defined as the number of fish collected divided by the number actually present in the area or volume sampled, is required for estimates of absolute abundance (Grosslein 1971, Edwards 1968).

Some studies designed to obtain absolute, gear catch efficiencies for trawls and seines have been reported in the literature (Derickson and Price 1973, Edwards and Steele 1968, Edwards 1973, Jones 1974, and Kuipers 1975). A general review concerning the evaluation and use of gear catch efficiencies in the estimation of fish abundance is provided by Kjelson and Colby (1977).

Many factors may potentially influence the vulnerability of fish to capture. Therefore, practical evaluation of catch efficiencies for a given gear design must be restricted to a single species and size class. In addition, study conditions must be characterized as to habitat type, fish density, environmental conditions, etc., to enable resulting efficiency values to be used to correct estimates gained from natural sampling situations.

In determining absolute catch efficiencies we must know the true density of a given species and size class in the area or volume of water sampled. Once this value is known, the efficiency may be calculated by comparing it to the average density estimated from a standard unit of sampling effort. Such comparisons may be made in several ways: (1) a known number of marked fish (or unmarked if other fish are not present) may be released in the path of the gear just prior to sampling, (2) a known number of marked fish may be restricted in an enclosed site and then sampled, and (3) unmarked fish may be enclosed within the sample site, sampled, and their density later estimated by some other technique. Because a fish's avoidance to the gear may be affected by previous capture, handling, and marking, the ideal method utilizes an unmarked, and previously unhandled, fish population. A fourth method makes use of fish marked with acoustical transponding tags and a sonar receiving system to continuously observe the location of individual tagged fish prior to, during, and after the sample is taken. This highly refined and expensive technique has been successfully used by Jones (1974) on adult plaice in the North Sea, and it may be applicable for juvenile forms provided the acoustical tags are small enough not to interfere with the natural behavior of the fish.

We have used the first three methods discussed in our estimates of absolute catch efficiency for several gears on varied estuarine fishes during daylight (Table 4). In all cases, a crew ranging from two to five individuals was sufficient to conduct each efficiency experiment, and an experiment was completed within one-half day to three days, even when a tagging and recapture technique was necessary (i.e., method 2). Equipment used in our studies was typical of that available for most fish sampling programs, although the initial cost of a fine mesh (13 mm) blocking net, 100 m in length, to restrict fish within the sample site was relatively high (1977 cost is approximately $50 per meter). Locating a suitable study site in southeastern estuarine nursery areas for methods 2 and 3 is relatively simple since much of the habitat is composed of natural "pockets" in the form of small tidal creeks and embayments that can be easily enclosed.

The results presented in Table 4 indicate that there is some variation in the average efficiency of these four gears for the species and size classes tested. However, with the exception of the 16-m^2 drop-net efficiency for menhaden (5%), all other efficiencies are within the range of 32 to 75%. These efficiencies indicate that estimates of fish densities for the test species sampled by the respective haul seines and otter trawls are low by a factor of approximately 1.3 to 3.

By conducting comparative gear tests, using for a "standard" one gear whose absolute efficiency is known, the absolute efficiencies of the other gears may be obtained. We have done this with three otter trawls of different size (3.0, 4.6, and 6.1 m). Relative catch efficiencies for the three trawls for juvenile pinfish, for example, were 16:91:100 respectively. We observed an absolute catch efficiency of 52% for the 6.1-m trawl for juvenile pinfish, *Lagodon rhomboides*, indicating that the absolute catch efficiency should be 8% for the 3.0-m trawl, and 47% for the 4.6-m trawl.

It is my belief that strong emphasis should be given additional efforts to quantify the efficiency of our most standardized fish sampling gears for the more abundant species present in estuarine, freshwater, and marine environments. The costs of such efforts often will be high, but the rewards will be of significant value in improving the accuracy of our estimates of fish population size in varied aquatic systems. If each investigator would obtain some information relative to his specific gear and species of interest, we would be well on the way to a solution. The experience we have gained in our research on catch efficiency of the 6.1-m otter trawl for two size classes of both pinfish and spot has convinced us that usable catch efficiency values can be achieved, often at relatively low costs and within the bugetary restrictions of many

Table 4. Daylight catch efficiencies for several estuarine fish sampling gears. See text for description of method

Gear	Method	Species and age class	Average efficiency (%)
6.1-m otter trawl	2	Juvenile pinfish	52
		Yearling pinfish	54
		Juvenile spot	32
		Yearling spot	35
21-m haul seine	1	Juvenile pinfish	60
		Juvenile spot	72
		Adult rough silversides	56
		Adult mummichog	59
350-m long haul seine	1	Pelagic fishes	41
		Semi-demersal fishes	47
16-m² portable drop net	3	Juvenile Atlantic menhaden	5
		Juvenile striped mullet	35
		Juvenile pinfish	56
		Juvenile spot	75

research organizations. For certain species, habitats, and gear types, it may require a cooperative effort between several interested organizations. An excellent example of a large-scale cooperative field effort used to evaluate fish sampling methodology is provided by Hayne, Hall, and Nichols (1967).

SUMMARY

The goal of this paper was to provide information with which to plan, improve, and evaluate sampling programs designed to achieve accurate and precise estimates of juvenile fish abundance. Summary statements and recommendations follow:

1. A thorough knowledge of the life history of the species studied is required to design an efficient sampling program.

2. Quantitative sampling techniques are required to achieve estimates of absolute fish density.

3. Choice of appropriate sampling gear is dependent upon the species of fish studied, estuarine habitat involved, and budgetary limitations.

4. Seines, trawls, and other towed samplers provide the most economical sampling gear, yet their use as "quantitative" gear requires that the area they sample be evaluated during their operation.

5. Improvement of our estimates of fish abundance will depend largely upon the degree to which sampling error and sampling bias problems have been controlled.

6. Documentation of sample variation is important in determining the number of samples required to gain a given level of precision. In addition, information on the sources of variation will enable the investigator to best allocate his sampling effort in a stratified sampling program designed to improve the precision of his estimates.

7. Investigators should document the bias of their fish abundance estimates by quantifying the absolute and relative catch efficiencies of their standard sampling gears.

8. Basic research is needed to improve the art of sampling. Particular emphasis is needed to improve sampling technology, to better describe the distribution of fish in time and space, and to understand the response of fish to sampling gear.

9. Estimates of natural mortality rates of juvenile estuarine fishes appear to be most easily obtained when the fish are inhabiting well-defined estuarine nursery grounds. Although such restricted efforts will yield information only on a small portion of the survivorship curve, the resulting knowledge should provide a means of checking the remaining portions of the curve.

10. Finally, an objective, realistic view of the state-of-the-art of fish sampling methodologies must continue to be taken if we hope to correctly interpret the usefulness of our resulting estimates of fish abundance.

ACKNOWLEDGMENTS

This research was supported through a cooperative agreement between the National Marine Fisheries Service and the Energy Research and Development Administration.

LITERATURE CITED

1. Adams, S. M. 1976. Ecology of eelgrass, *Zostera marina* (L.), fish communities. I. J. Exp. Mar. Biol. Ecol. 22:293–311.

2. Allen, G. H., A. D. DeLacy, and D. W. Gotshall. 1960. Quantitative sampling of marine fishes – a problem in fish behavior and fishing gear, p. 448–611. *In* E. A. Pearson [ed.] Waste disposal in the marine environment. Pergamon Press, New York.

3. Alverson, D. L., and W. T. Pereyra. 1969. Demersal fish explorations in the Northeastern Pacific Ocean – an evaluation of exploratory fishing methods and analytical approaches to stock size and yield forecasts. J. Fish. Res. Board Can. 26(8):1985–2001.

4. Cain, R. L., and J. M. Dean. 1976. The annual occurrence, abundance and diversity of fish in South Carolina intertidal creeks. J. Mar. Biol. 36(4):369–379.

5. Cassie, R. M. 1971. Sampling and statistics, p. 174–209. *In* W. T. Edmondson and G. G. Winberg [ed.] A manual on methods for the assessment of secondary productivity in fresh water. Blackwell Scientific Publ., Oxford.

6. Clark, S. H. 1974. A study of variation in trawl data collected in Everglades National Park, Florida. Trans. Am. Fish. Soc. **103**(4):777–785.

7. Clutter, R. I., and M. Anraku. 1968. Avoidance of samplers, p. 57–76. *In* Zooplankton sampling. UNESCO, Monographs on Oceanographic Methodology, Geneva, Switzerland.

8. Cole, C. F. 1973. Can mortality rates be determined for estuarine ichthyoplankton populations, p. 163–192. *In* A. L. Pacheco [ed.] Proceedings of a workshop on egg, larval and juvenile stages of fish in Atlantic coast estuaries. Tech. Publ. No. 1, Middle Atlantic Coastal Fisheries Center, Highlands, New Jersey.

9. Copeland, B. J., and T. J. Bechtel. 1974. Some environmental limits of six Gulf coast estuarine organisms. J. Mar. Sci. **18**:169–204.

10. Derickson, W. K., and K. S. Price, Jr. 1973. The fishes of the shore zone of Rehoboth and Indian River Bays, Delaware. Trans. Am. Fish. Soc. **102**(3):552–562.

11. Eberhardt, L. L. 1975. Some quantitative issues in ecological impact evaluation, p. 309–315. *In* R. K. Sharma, J. D. Buffington, and J. T. McFadden [eds.] Proc. Workshop Biol. Significance Environ. Impacts, NR-CONF-002. U.S. Nuclear Regulatory Commission, Washington, D.C.

12. Eberhardt, L. L., and R. O. Gilbert. 1975. Biostatistical aspects, p. 783–917. *In* Source book of monitoring methods, Vol. 2. AIF/NESP-004, Atomic Industrial Forum.

13. Edwards, R. L. 1968. Fishery resources of the North Atlantic area, p. 52–60. *In* D. Gilbert [ed.] The future of the fishing industry in the U.S. College of Fish., Univ. Wash. Publ., Seattle, Washington.

14. Edwards, R., and J. H. Steele. 1968. The ecology of 0-group plaice and common dabs at Loch Ewe. I. Population and food. J. Exp. Mar. Biol. Ecol. **2**:215–238.

15. Edwards, R. R. C. 1973. Production ecology of two Caribbean marine ecosystems. I. Physical environment and fauna. Estuarine Coastal Mar. Sci. **1**:303–318.

16. Grosslein, M. D. 1971. Some observations on accuracy of abundance indices derived from research vessel surveys, p. 249–265. *In* Int. Comm. Northwest Atl. Fish. Redbook. Part III.

17. Gunter, G. 1967. Some relationships of estuaries to the fisheries of the Gulf of Mexico, p. 621–638. *In* George H. Lauff [ed.] Estuaries. Am. Assoc. Adv. Sci., Publ. No. 83. Horn-Shafer Co., Baltimore.

18. Hansen, D. J. 1969. Food, growth, migration, reproduction and abundance of pinfish, *Lagodon rhomboides*, and Atlantic croaker, *Micropogon undulatus*, near Pensacola, Florida, 1963–1965. U.S. Fish. Wildl. Serv. Fish. Bull. **68**:135–146.

19. Hayne, H. W., G. E. Hall, and H. M. Nichols. 1967. An evaluation of cove sampling of fish populations in Douglas Reservoir, Tennessee, p. 224–297. *In* Research Fisheries Resources Symposium. Am. Fish. Soc., Washington.

20. Hellier, T. R. 1959. The drop-net quadrat, a new population sampling device. Publ. Inst. Mar. Sci., Univ. Tex. **5**:165–168.

21. Herke, W. H. 1969. A boat-mounted surface push-trawl for sampling juveniles in tidal marshes. Prog. Fish-Cult. **31**:177–179.

22. ———. 1971. Use of natural, and semi-impounded, Louisiana tidal marshes as nurseries for fishes and crustaceans. Ph.D. Thesis, Louisiana State Univ., Baton Rouge. 264 p. Univ. Microfilms, Ann Arbor, Mich. (Diss. Abstr. **32**:2654-B.).

23. Hoese, H. D., and R. S. Jones. 1963. Seasonality of larger animals in a Texas turtle grass community. Publ. Inst. Mar. Sci., Univ. Tex. **9**:37–47.

24. Hoese, H. D., B. J. Copeland, F. N. Moseley, and E. D. Lane. 1968. Fauna of the Aransas Pass Inlet, Texas. III. Diel and seasonal variations in trawlable organisms of the adjacent area. Tex. J. Sci. **20**:33–60.

25. Hoss, D. E. 1974. Energy requirements of a population of pinfish, *Lagodon rhomboides* (Linnaeus). Ecology **55**(4):848–855.

26. Hunter, J. R., D. C. Aasted, and C. T. Mitchell. 1966. Design and use of a miniature purse seine. Prog. Fish-Cult. **28**:56–59.

27. Johnsen, R. C., and C. W. Sims. 1973. Purse seining for juvenile salmon and trout in the Columbia River Estuary. Trans. Am. Fish. Soc. **102**(3):341–345.

28. Jones, F. R. H. 1974. Objectives and problems related to research into fish behavior, p. 261–275. *In* F. R. H. Jones [ed.] Sea fisheries research. John Wiley Sons, New York.

29. Jones, R. 1976. The use of marking data in fish population analysis. FAO Fish. Tech. Paper No. 153. 42 p.

30. Jones, R. S. 1965. Fish stocks from a helicopter-borne purse net sampling of Corpus Christi Bay, Texas, 1962–1963. Publ. Inst. Mar. Sci., Univ. Tex. **10**:68–75.

31. Joseph, E. B. 1973. Analysis of a nursery ground, p. 118–121. *In* A. L. Pacheco [ed.] Proceedings of a workshop on egg, larval and juvenile stages of fish in Atlantic coast estuaries. Tech. Publ. No. 1, Middle Atlantic Coastal Fisheries Center, Highlands, New Jersey.

32. Kjelson, M. A., and D. R. Colby. 1977. The evaluation and use of gear efficiencies in the estimation of estuarine fish abundance, p. 416–424. *In* Proc. 3rd Int. Estuarine Res. Fed. Conf., Galveston, Texas, October 7–9, 1975.

33. Kjelson, M. A., and G. N. Johnson. 1975a. Description and evaluation of a portable drop-net for sampling nekton populations, p. 653–662. *In* Proc. 27th Annu. Conf. Southeastern Assoc. Game Fish Comm.

34. Kjelson, M. A., and G. N. Johnson. 1975b. Description and evaluation of long-haul seine for sampling fish populations in offshore estuarine habitats, p. 171–177. Proc. 28th Annu. Conf. Southeastern Assoc. Game Fish. Comm.

35. Kjelson, M. A., W. R. Turner, and G. N. Johnson. 1975. Description of a stationary drop-net for sampling nekton abundance in shallow waters. Trans. Am. Fish. Soc. **104**(1):46–49.

36. Kuipers, B. 1975. On the efficiency of a two-metre beam trawl for juvenile plaice (*Pleuronectes platessa*). Neth. J. Sea Res. **9**(1):69–85.

37. Kushland, J. A. 1974. Quantitative sampling of fish populations in shallow freshwater environments. Trans. Am. Fish. Soc. **103**(27):348–352.

38. Lagler, K. F. 1959. Freshwater fishery biology. W. C. Brown Co., Dubuque Iowa. 421 p.

39. Lippson, A. J. 1973. The Chesapeake Bay in Maryland: an atlas of natural resources. John Hopkins Press., Baltimore. 55 p.

40. Livingston, R. J. 1975. Diurnal and seasonal fluctuations of organisms in a north Florida estuary. Estuarine Coastal Mar. Sci. **4**:373–400.

41. Macketts, D. J. 1973. Manual of methods for fisheries resource survey and appraisal. FAO Fish. Tech. Paper No. 124. 40 p.

42. Marshall, H. L. 1976. Effects of mosquito control ditching on *Juncus* marsh and utilization of mosquito control ditches by estuarine fish and invertebrates. Unpublished Ph.D. Thesis, Univ. North Carolina, Chapel Hill, North Carolina. 190 p.

43. Massman, W. H., E. C. Ladd, and H. N. McCutcheon. 1952. A surface trawl for sampling young fishes in tidal rivers, p. 386–392. *In* Trans. 17th North Am. Wildl. Conf.

44. McErlean, A. J., S. G. O'Conner, J. A. Mihursky, and C. I. Gibson. 1973. Abundance, diversity and seasonal patterns of estuarine fish populations. Estuarine Coastal Mar. Sci. **1**:19–36.

45. McFadden, J. T. 1975. Environmental impact assessment for fish populations, p. 89–137. *In* R. K. Sharma, J. D. Buffington, and J. T. McFadden [eds.] Proc. Workshop Biol. Significance Environ. Impacts, NR-CONF-002. U.S. Nuclear Regulatory Commission, Washington, D.C.

46. McHugh, J. L. 1967. Estuarine nekton, p. 581–620. *In* George H. Lauff [ed.] Estuaries. Am. Assoc. Adv. Sci., Publ. No. 83. Horn-Shafer Co., Baltimore. 757 p.

47. Moseley, F. N., and B. J. Copeland. 1969. A portable drop-net for representative sampling of nekton. J. Mar. Sci. **14**:37–45.

48. Murphy, G. I., and R. I. Clutter. 1972. Sampling anchovy larvae with a plankton purse seine. U.S. Natl. Mar. Fish. Serv. Fish. Bull. **70**(3):789–798.

49. Odum, W. E. 1971. Pathways of energy flow in a south Florida estuary. Sea Grant – Tech. Bull. 7. Univ. Miami, Coral Gables, Florida. 162 p.

50. Oviatt, C. A., and S. W. Nixon. 1973. The demersal fish of Narragansett Bay: an analysis of community structure, distribution and abundance. Estuarine Coastal Mar. Sci. 1:361–378.

51. Perret, W. S., and C. W. Caillouet, Jr. 1974. Abundance and size of fishes taken by trawling in Vermilion Bay, Louisiana. Bull. Mar. Sci. 24:52–75.

52. Reintjes, J. W., and A. L. Pacheco. 1966. The relation of menhaden to estuaries, p. 50–58. In A symposium on estuaries and estuarine animals. Am. Fish. Soc., Spec. Publ. 3.

53. Roessler, M. 1965. An analysis of the variability of fish populations taken by otter trawl in Biscayne Bay, Florida. Trans. Am. Fish. Soc. 94:311–318.

54. Saila, S. B. 1961. The contribution of estuaries to the offshore winter flounder fishery in Rhode Island. Proc. Gulf Carib. Fish. Inst. 14th Annu. Sess. p. 95–109.

55. Seber, G. A. F. 1973. The estimation of animal abundance and related parameters. Hafner Press, New York. 499 p.

56. Trent, W. L. 1967. Attachment of hydrofoils to otter boards for taking surface samples of juvenile fish and shrimp. Chesapeake Sci. 8(2):130–133.

57. Turner, W. R. 1973. Estimating year-class strength of juvenile menhaden, p. 37–47. In A. L. Pacheco [ed.] Proceedings of a workshop on egg, larval and juvenile stages of fish in Atlantic coast estuaries. Tech. Publ. No. 1, Middle Atlantic Coastal Fisheries Center. Highlands, New Jersey.

58. Turner, W. R., and G. N. Johnson. 1973. Distribution and relative abundance of fishes in Newport River, N.C. NOAA Technical Report NMFS SSR-F, National Marine Fisheries Service, Department of Commerce, Seattle, Washington. 23 p.

59. Turner, W. R., and G. N. Johnson. 1974. Standing crops of aquatic organisms in tidal streams of the lower Cooper River system, South Carolina, p. 13–20. In Frank P. Nelson [ed.] The Cooper River environmental study. South Carolina Water Resources Comm. Report 117.

60. Watt, K. E. F. 1968. Ecology and Resource Management. McGraw-Hill, New York. 450 p.

61. Zar, J. H. 1975. Statistical significance and biological significance of environmental impacts, p. 285–293. In R. K. Sharma, J. D. Buffington, and J. T. McFadden [eds.] Proc. Workshop Biol. Significance Environ. Impacts, NR-CONF-002. U.S. Nuclear Regulatory Commission, Washington, D.C.

62. Zippin, C. 1958. The removal method of population estimation. J. Wildl. Manage. 22(1):82–90.

Striped Bass (*Morone saxatilis*) Monitoring Techniques in the Sacramento–San Joaquin Estuary*

Donald E. Stevens

Bay-Delta Fishery Project
California Department of Fish and Game
Stockton, California

ABSTRACT

Various methods have been used to monitor the striped bass population in the Sacramento–San Joaquin Estuary. Sampling in the spring with towed plankton nets has provided an adequate description of spawning time and area, but this sampling has not adequately measured egg standing crops and larva and post-larva mortality rates. Tow-net sampling effectively measures the abundance of young in midsummer. A midwater-trawl survey is satisfactory for measuring the abundance of young in the fall but not in the winter. Techniques have not been fully evaluated for monitoring one-year-old bass. Catch-per-unit-effort data from sportfishing party boats were useful for monitoring two-year-olds, until a change in angling regulations increased recruitment age. The Petersen method and indices developed from party-boat catches are the best methods for monitoring bass that are three years old and older. Long-term trends in catch can be monitored through postcard surveys and party-boat catches.

Key words: abundance indices, angler surveys, egg and larva surveys, fishing gear, mark-recapture, *Morone saxatilis,* mortality, population monitoring, striped bass

INTRODUCTION

The striped bass population in the Sacramento–San Joaquin Estuary is affected by environmental changes resulting from development of federal and state water projects. Knowledge of impacts of these changes is necessary to develop operating criteria for water projects that will minimize damage to the

---

*Funds for this work have been provided by Dingell-Johnson project, California, F-9-R, "A Study of Sturgeon and Striped Bass" supported by Federal Aid to Fish Restoration funds, the California Department of Water Resources, and the U.S. Bureau of Reclamation.

bass population. Assessing such impacts requires population monitoring. This paper describes techniques that the California Department of Fish and Game has employed during the past 40 years for monitoring striped bass at several life stages.

STUDY AREA

The Sacramento and San Joaquin river systems form a tidal estuary with a salinity gradient about 80 km long which extends from San Pablo Bay to the western delta (Fig. 1). River flows into the delta are quite variable and are partially controlled by upstream reservoirs. Flows peak in winter and spring. Kelley (1966) describes this region in detail.

Striped bass utilize the entire estuary and adjacent coastal area. Adult bass spend most of the year in saltwater, but in winter they begin migrating to the delta and rivers upstream for spawning in the spring (Chadwick 1967). The nursery area for young bass is the delta downstream to San Pablo Bay (Turner and Chadwick 1972).

MONITORING TECHNIQUES

Egg and Post-Larva Sampling in Spring

Pump Sampling

From 1967 to 1969, 0.5 HP Moyno utility pumps with a synthetic rubber helical rotor were used to continuously sample eggs and larvae drifting past sites in the Sacramento and San Joaquin rivers (Turner 1976). River water was pumped into containers where eggs and larvae were screened out. Pumping rate was about 35–55 liters per minute. Catches ranged up to 370 eggs and 450 larvae <5 mm standard length (SL) per 24-hr period. Pump sampling was implemented to obtain continuous samples that required only a few minutes of labor daily for sample collection. This method of sampling was terminated, however, because detritus often clogged the screens causing samples to overflow.

Plankton Net Sampling

Since 1963 several plankton nets have been used to capture bass eggs and larvae. L. W. Miller (*personal communication*) has described and evaluated some of these nets in detail.

From 1963 to 1966 the nets most frequently used had a 46-cm-diam mouth and a 102-cm-long cone which was made of 9-mesh-per-centimeter

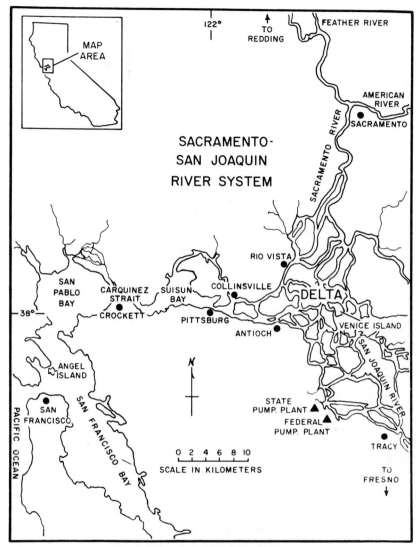

Fig. 1. Sacramento—San Joaquin estuary.

bolting cloth (Farley 1966). Mesh openings varied from about 600—800 μm. A screened metal bucket was attached to the cod end. Pygmy* flow meters often were mounted in the net mouth. Generally two nets were towed, one at the surface and the other about 4.5 to 7.5 m below the surface. A few surface samples were collected from highway bridges by setting nets in the current.

---

*Model 005—WA—130, Kahl Scientific Instrument Co. El Cajon, California.

About 30 stations were surveyed 2 or 3 days per week during the spawning season (April or May through June). Egg catches indicated when and where bass spawned (Farley 1966, Turner 1976).

From 1967 to 1975, egg and larva sampling was intensified to monitor spawning (Turner 1976) and to measure abundance of eggs and young at various stages for mortality rate estimation (D. H. Allen, *personal communication*). The nets from 1967 to 1974 had a 76-cm-diam mouth and a 3.2-m cone of 7.9-mesh-per-centimeter nylon marquisette (Turner 1976). Mesh openings were about 930 μm. A screened plastic quart jar was attached to the cod end. The nets were mounted on skis and were towed diagonally from bottom to surface to obtain depth integrated samples. The volume of water strained was measured with Pygmy or General Oceanics* meters. Surveys were made every other day during the spawning season. Ten-minute samples generally were taken at 3.2-km intervals from the western end of Suisun Bay to Rio Vista on the Sacramento River and to Venice Island on the San Joaquin River. Catch analyses revealed that the net mesh retained eggs but allowed many bass smaller than 7 mm SL to pass through (Turner 1976).

A more efficient net was used in 1975. This net has a 76-cm-diam mouth and is 2.9 m long. The mesh is 505-μm Nitex netting. This net's cylindrical cone design (Smith et al. 1968) filters water more effectively than the original design, and the mesh retains virtually all larvae. Catches of larvae $\leq$5 mm SL increased more than 50 times with the new net. The new and old net are almost equally efficient on young striped bass of 12 to 16 mm fork length (FL) (L. W. Miller, *personal communication*).

These surveys provide indices of total abundance of eggs and larvae. At each sampling station catches of eggs and larvae per cubic meter of water strained are multiplied by the water volume in the section of river represented by the sample. These products are summed to index the standing crop each sampling day.

These surveys have proven inadequate in two important respects.

1. Standing crop measurements for eggs are much too low in the delta. For example, in 1972 the total bass egg index was 7.5 million (Turner 1976), but as many as 170 billion eggs actually may have been spawned [as estimated from a length-fecundity relationship (Morgan and Gerlach 1950), abundance estimates for spawning females (*unpublished data*), and assuming 45% of total spawning occurred in the sampling area (Turner 1976)]. The magnitude of this discrepancy suggests either huge egg mortality and decomposition shortly after spawning or inefficient sampling. The latter is more likely, considering the lack of evidence for the former in laboratory and hatchery culture.

---

*Model 2030 General Oceanics Inc. Miami, Florida.

2. Surveys through 1974 do not demonstrate significant annual differences in larva and post-larva mortality rates calculated from abundance declines over time (D. H. Allen, *personal communication*). Differences in abundance indices measured by a summer survey, however, indicate egg–post-larva mortality rates probably varied from year to year (California Department of Fish and Game et al. 1974). Mortality rate differences may be masked by some combination of the following: (*a*) net efficiency problems, (*b*) sampling error, which may vary between areas of the estuary causing abundance and mortality estimates to vary with annual differences in bass distribution within the sampling area, and (*c*) abundance measurements that are biased by annual differences in migrations to and from the sampling area.

Summer Tow-Net Survey

Various tow nets have measured summer abundance of young bass since 1947 (Calhoun and Woodhull 1948, Erkkila et al. 1950, Calhoun 1953, Chadwick 1964), but a standard survey design was not adopted until 1962 (Chadwick 1964). This survey has been conducted annually (except 1966) ever since. It measures the abundance of bass of 17 to 50 mm FL.

Each annual survey consists of three to five subsurveys that take five days to complete. These subsurveys begin in mid-June and consist of sampling at 30 stations scattered from eastern San Pablo Bay to the eastern delta. They are conducted at two-week intervals until the mean length of the bass in the catch exceeds 38 mm FL. This length is attained between mid-July and mid-August.

Three 10-min, depth-integrated tows are made at each sampling station. These tows are diagonal from bottom to surface to reduce bias caused by variations in the vertical distribution of bass. The net (Calhoun 1953) is mounted on skis. The net's mouth area is about 1.5 m^2 and the net is 5.5 m long. It tapers to 39 cm in diameter at the cod end. The first 3.05 m of the cone is #6 thread webbing of 1.27-cm stretch mesh. The rest of the net is bobbinet with 3.1 holes per centimeter. These holes are about 2.5 mm in diameter. The bobbinet is sewn to the #6 thread webbing so that the webbing forms a fyke 60 cm long inside the bobbinet.

Total bass abundance during each subsurvey is indexed by multiplying the number caught at each sampling station by the volume of water represented by the station and summing these products. Indices of annual abundance are developed by plotting abundance against mean length for each subsurvey and interpolating abundance for selected lengths (Turner and Chadwick 1972). Abundance when the mean length of the population is 38 mm is the primary measure of survival between spawning and midsummer. Annual differences in length frequency distributions and in growth rates bias these indices somewhat (Turner and Chadwick 1972).

The abundance indices are highly correlated with river flows and water diversion rates (Turner and Chadwick 1972, California Department of Fish and Game et al. 1973-1976). These correlations are extremely helpful in evaluating water development plans and in developing water management recommendations (Chadwick et al. 1977), which are the primary objectives of the survey.

This survey, however, apparently underestimates bass abundance when high flows transport young bass to San Pablo Bay (Stevens 1977). Only one station in San Pablo Bay is sampled routinely. Coverage is not more intensive because bass catches always have been small at this station and few bass were caught elsewhere in San Pablo Bay in the 1950s (Calhoun 1953, Chadwick 1964). Although the catches of bass are generally low in San Pablo Bay, significant numbers of bass could inhabit that bay because its volume is large (Kelly 1966).

Fall and Winter Midwater Trawl Survey

Except for 1974, young bass abundance has been sampled during fall and winter each year since 1967. The primary objective is to determine if mortality during this period might vary enough between years to be a major factor causing variations in adult bass abundance.

A midwater trawl with a 3.6- by 3.6-m mouth is the sampling gear. The 17.7-m-long net is constructed of nine tapered panels of netting decreasing from 20-cm stretch mesh (sm) at the mouth to 1.3-cm sm at the cod end (L. W. Miller and D. Drake, *unpublished data*). It is towed by a 9.8-m boat. The trawl is released to the river bottom, and then it is pulled to the surface as it is towed, resulting in depth-integrated samples. Towing speed is about 1.5 knots. Tow duration is 12 min. VonGeldern (1972) describes in detail similar nets and fishing methods.

Each monthly survey (generally August or September to March) takes about six days to complete. One tow is made at each of 87 sampling sites scattered from San Pablo Bay through the delta. Single catches have ranged up to 445 bass, and most bass are 6 to 14 cm FL.

The method of tabulating the catch data is similar to that for the spring and summer surveys. All tows are assumed to strain equal amounts of water. For each of 17 groups of stations, mean catches are multiplied by the total volume of water represented by the stations in the grouping. These products are summed to obtain monthly abundance indices.

This survey has been partly successful. Fall abundance indices are highly correlated with the index from the summer tow-net survey [$r = 0.95$ for an index of abundance 60 days after the summer index is set vs the summer index (L. W. Miller and D. Drake, *unpublished data*)], suggesting that annual differences in mortality between summer and fall are minimal. However, after winter rains the monthly abundance indices are highly variable. They often produce survival rate estimates (calculated from the changes in monthly indices) that exceed 100% between months, an obvious impossibility for which there is no satisfactory explanation. These survival estimates were not improved by increasing sampling effort or adjusting catches for biases associated with water clarity (L. W. Miller and D. Drake, *unpublished data*).

Monitoring of Juvenile Bass

One- and two-year-old bass occasionally are taken in the midwater trawl designed to catch young bass. Catches of two-year-olds are too low to establish meaningful population indices, and the index for one-year-olds has not been evaluated. Attempts to take significantly larger numbers of these bass with a larger trawl (6.1- by 6.1-m mouth) have been unsuccessful.

Monitoring of Adult Bass Population and Catch

Effective methods of monitoring adult bass abundance are necessary to determine if conditions affecting the survival of young bass control the number of bass recruited to the fishery. Several approaches are being examined.

Petersen Mark-Recapture Estimates

To estimate the abundance of adult bass (fish $\geq$40.6 cm in total length), we have conducted a mark-recapture experiment since 1969. The modified Petersen method (Bailey 1951) is used: $N = M(C + 1)/(R + 1)$, where N = bass abundance, M = number of marked fish released, C = number of fish subsequently censused, and R = number of recaptured marks in the sample.

Gill nets and fyke traps (Hallock et al. 1957) are used to capture bass during their spring migration to the delta. The fish are tagged with numbered disc-dangler tags (Chadwick 1963) and released. The population is sampled through an annual summer-fall census of angler catches in San Francisco and San Pablo Bays and during subsequent spring tagging operations.

About 9000 to 18,400 tags have been applied annually. Seasonal employees stationed at six fishing ports in the Bay Area, Wednesday through Sunday each week from July through November, have observed annually from 18,000 to 39,000 bass and from 128 to 352 of the tags applied the preceding spring. In 1976, they observed 347 tags, including releases from all years, which is about 1.6% of the 22,000 bass censused

The abundance estimation procedures are complicated by sex- and age-sampling biases. Males spend more time on the spawning grounds than females (Chadwick 1967), so roughly twice as many males are tagged. In contrast, censused females slightly outnumber censused males. Three- and four-year-old bass are underrepresented in the tagged sample because many of those fish are not mature and they have not taken up adult migratory patterns (Chadwick 1967). Moreover, larger bass are more susceptible to capture in the gill nets (Chadwick 1967). Hence, all tagging and recapture samples are stratified by sex and age.

Sex is determined for each fish tagged. If milt is extruded, fish are classified male and if not they are classified female. About 75 to 90% of the censused fish are sexed by dissection. The remainder of the censused fish are assumed to have the same sex ratio as this sample.

In order to stratify by age, scales are sampled and lengths are measured on nearly all tagged bass. Scales are obtained from 75 to 90% of the censused bass. Length is measured for virtually all censused bass. A computer program (AGECOM, Abramson 1971) utilizes age-length relationships and length frequencies to apportion unaged fish into the appropriate year classes.

These procedures enable us to estimate abundance of individual year classes and to increase sample sizes for estimates of each year class with each successive sampling period (Figs. 2 and 3).

Two additional problems must be solved in estimating three-year-old bass abundance:

1. Only half of these fish are legal size during the tagging period, and recruitment is not complete until about six months later. Therefore, the marked:unmarked ratio observed during the first census after tagging would grossly underestimate the abundance. Our solution is to estimate abundance starting with the tagging sample taken the following spring (Figs. 2 and 3). This solution results in estimates that are biased slightly high (see evaluation of Ricker's condition 2 later in this section).

2. Few three-year-old females are tagged (maximum number was 230 in 1972). Therefore, this abundance is estimated indirectly by assuming that it is equal to that of the three-year-old males.

Due to the sampling biases, the most accurate annual estimates of total population abundance appear to be sums of the year-class estimates for both sexes, except that the age-3 estimates are first divided by 2 to eliminate fish recruited after the tagging period. These total abundance estimates have varied from 1.6 to 1.9 million bass annually. Estimates unstratified by sex and age (recruitment is estimated from growth rates based on monthly length frequencies and subtracted from the recapture samples) are about 20% lower. They range from 1.2 to 1.7 million bass annually.

Although sums of year-class estimates may yield the best total abundance estimates, relatively small samples result in rather wide confidence intervals on the individual estimates (Figs. 2 and 3). Low correlations for abundance estimates of three- and four-year-old bass vs young-of-the-year indices ($r <$ 0.31, 3 d.f.) and vs flows ($r <$ 0.58, 4 d.f.) in the first summer of life may partly reflect this lack of precision.

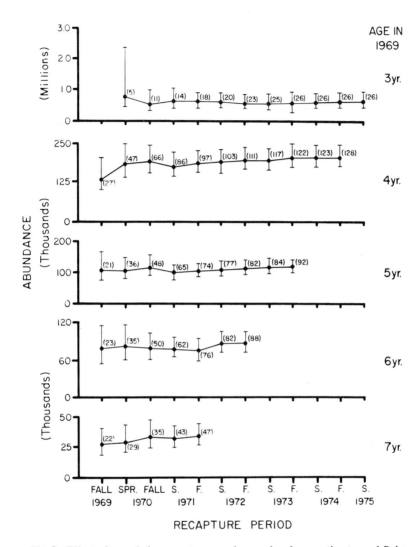

Fig. 2. Effect of cumulative recapture samples on abundance estimates and Poisson confidence intervals for male striped bass tagged in 1969. Abundance of three-year-olds was not estimated in fall 1969 because they were not fully recruited. Samples are not included after fish attained nine years of age (i.e., 1972 recaptures for fish seven years old in 1969) because ages cannot be estimated accurately for older bass. Numbers of tags recovered are in parentheses.

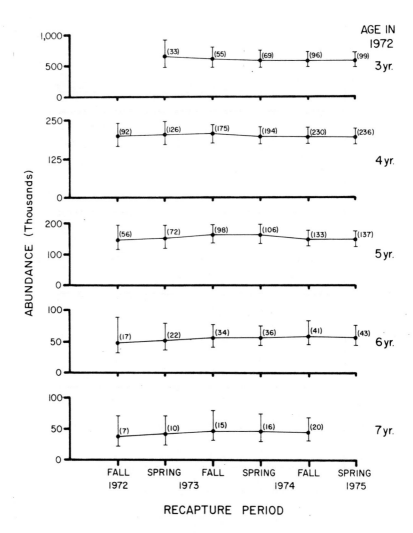

Fig. 3. Effect of cumulative recapture samples on abundance estimates and Poisson confidence intervals for male striped bass tagged in 1972. The abundance of three-year-olds was not estimated in fall 1972 because they were not fully recruited. The abundance of seven-year-olds was not estimated in spring 1975 because that year class was too difficult to classify accurately. Numbers of tags recovered are in parenthesis.

Ricker (1975) lists six conditions required for valid Petersen estimates. These conditions and applicable findings from the striped bass mark-recapture study are reviewed here to improve understanding of the limitations of the study.

1. Marked fish suffer the same natural mortality as the unmarked — mortality probably is somewhat higher for tagged bass than for untagged bass. A few bass die shortly after release, presumably due to tagging and handling. Since 1971, anglers and the tagging crew have recovered 341 dead bass within two weeks or so after tagging. This total is only 0.4% of the roughly 85,000 bass tagged. Although the dead fish that are recovered are removed from the release files, tagged bass that die and remain un-recovered cannot be accounted for.

   Delayed mortality also may occur due to irritation, which is sometimes caused by hydroids attaching to the tags. There is no way of measuring mortality from this source, but fish have been seen with large scars near the tag indicating healing following considerable tissue damage (Chadwick 1963, 1964; *unpublished data*).

   Hence, some unmeasured mortality probably occurs from the above sources. This greater mortality of marked fish would cause overestimates of abundance.

2. Marked and unmarked fish are equally vulnerable to fishing — tagging and handling bass reduces their vulnerability to angling by about one-third to one-half for about one month (Chadwick 1963). Essentially all resampling occurs later. Some of those fish saved by the low vulnerability during the first month probably are caught later and bias abundance estimates down-ward very slightly.

   Except for Department of Fish and Game gill nets, there is no fishing gear to bias returns by selectively snagging tagged fish. R/C ratios vary for both netted and angler-caught bass (Fig. 4). The ratios for netted bass tend to be somewhat higher, but they are not consistently higher than those for angler-caught bass. The higher ratios for netted bass may result from the nets snagging tags and/or the tendency for bass to return to the area where they were tagged (see condition 4).

   Initially, tagged three-year-old bass are more vulnerable than untagged three-year-olds, because only legal fish are tagged, and recruitment of three-year-olds is not complete until about six months later. Abundance is slightly overestimated (<3.5%, *unpublished data*) from the sampling that starts the following spring, because the higher early catch rates of tagged three-year-olds reduce R/C ratios.

3. Marked fish do not lose their marks — evidence of tag shedding consists of scars and broken tag wires, which have been detected for 297 bass since 1969. During the same period 5281 bass were observed with tags intact. In

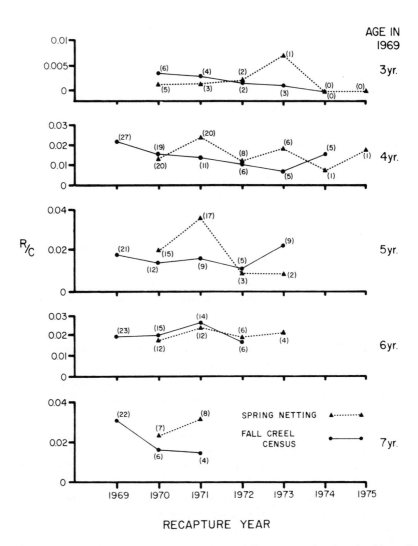

Fig. 4. **Ratios of male striped bass tagged in 1969, *R*, to total male striped bass, *C*, observed in various recapture samples.** Ratio was not calculated for three-year-olds in fall 1969 because that age group was not fully recruited. Ratios were not calculated after fish attained nine years of age (i.e., 1972 for fish seven years old in 1969) because ages cannot be estimated accurately for older bass. Numbers of tags recovered, *R*, are in parentheses.

1969, 1162 bass were double-tagged. Estimated tag loss was about 5% per year for four years after application. Subsequently, this loss rate appeared to decrease (G. E. Smith, *personal communication*). The effects of shedding on estimates of abundance are probably minimal, since consistent increases in abundance estimates (Figs. 2 and 3) or consistent decreases in R/C ratios (Fig. 4) are not apparent as successive years of recapture data are added.

4. Marked fish become randomly mixed with the unmarked fish — Chadwick (1967) found that summer and fall migrations were similar for tagged bass captured in different areas of the estuary, suggesting that population components are well mixed during the angler census. However, the following spring, individual bass tended to return to where they were tagged. This behavior should increase the probability of capture for tagged bass and may cause spring R/C ratios to be somewhat higher than those in the fall (Fig. 4).

5. All marks are recognized and reported on recovery — there is little likelihood of significant error from this source. Trained observers examine the entire catch used to estimate abundance.

6. Recruitment is negligible during the recovery period — all samples are stratified by year class, and except for three-year-olds (see condition 2), all year classes are fully recruited at the time of tagging.

Population Estimates Derived From Catch and Exploitation Estimates for Part of the San Francisco Bay Fishery

Adult striped bass abundance, $\hat{N}$, was estimated annually from 1969 to 1971 using the equation $\hat{N} = \hat{C}/\hat{u}$, where $\hat{C}$ is the estimated striped bass catch in north San Francisco Bay during the two- to four-month study period in late summer and fall, and $\hat{u}$ is an estimate of the fraction of the population represented by the catch (R. D. Rogers and D. E. Stevens, *unpublished data*). In addition, $\hat{C} = \hat{E} \times \hat{c}$, where $\hat{E}$ is an estimate of the number of fishing hours and $\hat{c}$ is the estimated catch per hour. $\hat{E}$ was estimated from hourly counts of boats fishing during approximately 20 half-day counting periods selected randomly each month. Counting was done from vantage points on Angel Island in the central bay. Census clerks stationed at major fishing ports obtained data with which to estimate $\hat{c}$. Returns of tags applied in the delta during the spring were used to estimate $\hat{u}$.

Catches and abundance should have been overestimated by this technique in 1969 and 1970 because catch-per-hour data, $\hat{c}$, were obtained only from anglers that fished for bass, whereas boats fishing for all species $\hat{E}$ were counted. In 1971 catch-per-hour data were collected from all anglers.

Despite the above bias, abundance appears to be underestimated in all three years. The estimates (after subtracting recruitment estimated from growth rates based on monthly length frequencies) for 1969, 1970, and 1971 are 1.3 million, 1.0 million, and 1.0 million, respectively. These estimates are 19, 38, and 38% less than sums of the year-class Petersen estimates.

Factors causing abundance to be underestimated include the following: (1) fractions of the population represented by the catches, $\hat{u}$, probably were overestimated because the tagged sample is biased toward the older fish, which tend to be harvested at higher rates than the younger fish, which are more abundant (Chadwick 1968, Miller 1974, *unpublished data*), and (2) effort, $\hat{E}$, would be underestimated if all fishing boats were not visible from the counting site.

In conclusion, this method appears suitable only for measuring the general magnitude of abundance.

Recruitment and Population Indices Derived from Sportfishing, Party-boat Catch Records

Party boats are boats taking parties of anglers fishing for a fee. Since 1938, party-boat operators have been required to report angling effort and catches to the Department of Fish and Game. These records have been used (Stevens 1977) to develop two sets of recruitment indices which are highly correlated with delta outflow during the first summer of life.

Before the minimum legal length increased from 30.5 to 40.6 cm in 1956, most bass were two years old when recruited. These recruits dominated the fishery in Suisun Bay, and catch-per-angler-day was used to index their abundance. Catch-per-day was affected by other factors affecting fishing success (e.g., water turbidity and temperature), but multiple regression analysis can be used to minimize this bias. After 1956, catch-per-day on party boats is not an appropriate recruitment index, because few party boats have fished in Suisun Bay and fish size varies too much elsewhere.

From 1958 to the present, total abundance can be estimated annually using party-boat catches in conjunction with tag return data. The concept is $N_t = C_t/u_t$, where N_t = abundance of legal ($\geq$40.6 cm) bass at the start of year t, C_t = total catch during year t of those bass that were legal at the start of the year, and u_t = fraction of the legal population harvested during year t.

Total catches C_t, are unknown, so reported catches on party boats, K_t, are used instead and are assumed to be a constant fraction of the total catches. Annual estimates of the fraction of the population harvested, $\hat{u}_t$, are available from tag returns (Chadwick 1968, Miller 1974, *unpublished data*). The abundance of legal bass, N_t, is then measured by the population index P_t as $P_t = K_t/u_t$.

Recruit abundance, R_t, is calculated assuming all recruitment occurs on May 1, which is approximately the first day used for annual tabulation of catch- and tag-return data. The formula is $R_t = P_t - P_{t-1} \cdot \hat{s}_{t-1}$, where $\hat{s}_{t-1}$ = estimated survival rate of legal bass based on tag returns in year $t - 1$. Since

1956, most recruits have been three years old (Robinson 1960, Miller and Orsi 1969); therefore, this index is assumed to measure abundance of that age class.

These population and recruitment indices are imprecise because (1) the party-boat catch does not form the same fraction of the total catch each year, (2) the fraction of the party-boat catch that is reported varies annually, and (3) recruitment does not consist entirely of three-year-old fish and does not occur instantaneously on May 1. These biases are discussed by Stevens (1977). Their overall effect was evaluated qualitatively by comparing trends in the recruitment index, R_t, with trends in annual mean weights of bass in the party-boat catch. The recruitment index tended to fluctuate inversely from the weights, suggesting that the recruitment index is a reliable indication of recruitment.

Except for 1970, from 1959 to 1974 R_t is related linearly to the logarithm of the mean June-July delta outflow three years earlier ($r = 0.82, P <$ 0.005), indicating that river flows early in life affect recruitment.

Recruitment Indices From Creel-Census, Catch-Per-Unit-Effort Data

A roving creel census was conducted during the spring in the delta from 1959 to 1965 (Miller and McKechnie 1969). One objective was to obtain recruitment indices from data for catch-per-unit-effort and age composition of the catch. This goal was not met because the vulnerability of bass to angling apparently varied between years. This technique has yet to be evaluated with data from the San Francisco—San Pablo Bay census used for Petersen estimates.

Catch Monitoring — Postcard Surveys and Party-boat
Catch-Per-Unit-Effort Data

Postcard questionnaires designed to measure angler success have been sent intermittently to random samples of buyers of fishing licenses since 1936 (Chadwick 1962, Seeley et al. 1963, McKechnie 1966, Emig 1971, Pelzman 1973) and each year since 1969. A comparison between total catch estimates derived from the post-1969 surveys (California Department of Fish and Game et al. 1975) and total catch estimates derived by multiplying the Petersen population estimates by estimates of the fraction of the population harvested (from tag returns) indicates that postcard respondents exaggerate their catches by a factor of about 6. Catch trends derived from the two methods show only fair agreement (Fig. 5), indicating that the postcard survey is insensitive to relatively small catch fluctuations. There is reasonably good agreement, however, between catch trends derived from postcard surveys and catch-per-angler-day on party boats since 1938 (Fig. 6), suggesting that the

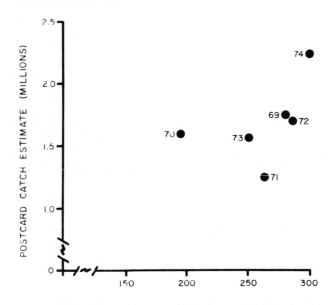

MARK–RECAPTURE CATCH ESTIMATE (THOUSANDS)

Fig. 5. Comparison between striped bass catch estimates derived from postcard questionnaires sent to anglers and catch estimates derived from mark-recapture studies. Mark-recapture catch estimates equal exploitation rate times the population estimate. Numbers adjacent to points indicate the year.

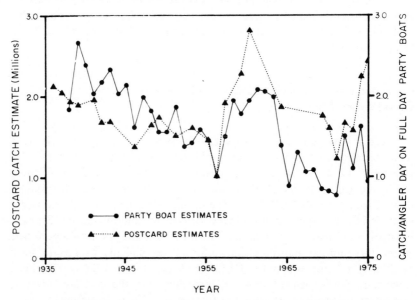

YEAR

Fig. 6. Relationship between angler success estimates from postcard questionnaires and sportfishing party-boat reports.

postcard survey and the party-boat catch reports both generally depict true long-term catch trends.

Interestingly, the postcard indices have increased relative to the party-boat catch indices over the years. This increase probably reflects party boats catching a reduced fraction of the total catch due to increased ownership of private boats.

SUMMARY

The California Department of Fish and Game has tried several techniques, with varying success, for monitoring striped bass at life stages from egg to adult.

Egg and larva surveys with plankton nets in the spring provide the best descriptions of where and when bass spawn. However, these surveys apparently underestimate standing crops of eggs, and estimates of larva mortality rates do not explain annual abundance differences measured by the summer tow-net survey. These deficiencies probably are partly due to inefficient net designs.

Rather precise abundance-flow correlations indicate that when spring and summer flows are low or moderate, young bass abundance is indexed effectively by an annual, summer tow-net survey. When high flows transport young bass to San Pablo Bay, this survey apparently underestimates bass abundance due to inadequate sampling in the bay.

Sampling with a midwater trawl satisfactorily measures abundance of young bass during the fall. In winter, catches are too variable to provide meaningful indices. The reason for this variability is unknown.

Techniques have not been fully evaluated for monitoring one-year-old bass. Catch-per-unit-effort data from sportfishing party boats effectively monitored two-year-olds until 1956, when a change in angling regulations increased recruitment age from two to three years.

The Petersen method and indices from party-boat catches appear to be the best monitoring techniques for bass three years old and older. The Petersen method has been used since 1969. At the present stage of data compilation, Petersen estimates for individual year classes do not correlate well with abundance indices for young bass of the same year classes or with delta outflows in the first summer of life. This lack of correlation may be partly due to error in the individual estimates associated with small recapture samples, coupled with the small number (six) of data points for each adult year class. Variations in mortality between the first and third year of life also could weaken the correlation. However, a correlation coefficient of 0.82 for the less precise party-boat catch index vs flow over a 16-year period suggests such variations are small relative to mortality variations for younger stages.

Long-term catch trends apparently can be monitored for the entire fishery through postcard surveys and party-boat catch-per-unit-effort data.

LITERATURE CITED

Abramson, N. J. (compiler) 1971. Computer programs for fish stock assessment. FAO Fish. Tech. Pap. **101**. p. var.

Bailey, N. J. J. 1951. On estimating the size of mobile populations from recapture data. Biometrika **38**:293–306.

Calhoun, A. J. 1953. Distribution of striped bass fry in relation to major water diversions. Calif. Fish Game **39**(3):279–299.

Calhoun, A. J., and C. A. Woodhull. 1948. Progress report on studies of striped bass reproduction in relation to the Central Valley Project. Calif. Fish Game **34**(1):171–188.

California Department of Fish and Game, California Department of Water Resources, U.S. Fish and Wildlife Service, and U.S. Bureau of Reclamation. 1973. Interagency ecological study program for the Sacramento–San Joaquin Estuary. Annu. Rep. No. 2. 59 p.

———. 1974. Interagency ecological study program for the Sacramento–San Joaquin Estuary. Annu. Rep. No. 3. 81 p.

———. 1975. Interagency ecological study program for the Sacramento–San Joaquin Estuary. Annu. Rep. No. 4. 81 p.

———. 1976. Interagency ecological study program for the Sacramento–San Joaquin Estuary. Annu. Rep. No. 5. 115 p.

Chadwick, H. K. 1962. Catch records from the striped bass sport fishery in California. Calif. Fish Game **48**(3):153–177.

———. 1963. An evaluation of five tag types used in a striped bass mortality rate and migration study. Calif. Fish Game **49**(2):64–83.

———. 1964. Annual abundance of young striped bass (*Roccus saxatilis*) in the Sacramento–San Joaquin Delta, California, Calif. Fish Game **50**(2):69–99.

———. 1967. Recent migrations of the Sacramento–San Joaquin River striped bass population. Trans. Am. Fish. Soc. **96**(3):327–342.

———. 1968. Mortality rates in the California striped bass population. Calif. Fish Game **54**(4):228–246.

Chadwick, H. K., Donald E. Stevens, and Lee W. Miller. 1977. Some factors regulating the striped bass population in the Sacramento–San Joaquin Estuary, California. (*In this volume*)

Emig, J. W. 1971. California inland angling survey for 1969, with corrections for the 1964 survey. Calif. Fish Game **57**(2):99–106.

Erkkila, L. F., J. W. Moffett, O. B. Cope, B. F. Smith, and R. S. Nielson. 1950. Sacramento–San Joaquin Delta fishery resources: Effects of Tracy pumping plant and delta cross channel. U.S. Fish and Wildl. Serv., Spec. Sci. Rept.: Fish., Vol. **56**. 109 p.

Farley, T. C. 1966. Striped bass, *Roccus saxatilis*, spawning in the Sacramento–San Joaquin river systems during 1963 and 1964. Calif. Dep. Fish Game Fish, Bull. **136**: 28–43.

Hallock, R. J., D. H. Fry, Jr., and D. A. LaFaunce. 1957. The use of wire fyke traps to estimate runs of adult salmon and steelhead in the Sacramento River. Calif. Fish Game **43**(4):271–298.

Kelley, D. W. 1966. Description of the Sacramento–San Joaquin Estuary. Calif. Dep. Fish Game Fish, Bull. **133**:8–17.

McKechnie, R. J. 1966. California inland angling survey for 1964. Calif. Fish Game **52**(4):293–299.

Miller, L. W. 1974. Mortality rates for California striped bass (*Morone saxatilis*) from 1965–1971. Calif. Fish Game **60**(4):157–171.

Miller, L. W., and R. J. McKechnie. 1969. Trends in the striped bass fishery in the Sacramento–San Joaquin Delta from 1959 to 1965. Calif. Fish Game, Anadromous Fish. Branch, Admin. Rep. No. 69–5. 26 p.

Miller, L. W., and J. J. Orsi. 1969. Growth of striped bass, *Morone saxatilis*, in the Sacramento–San Joaquin Estuary, 1961–1965. Calif. Fish Game, Anadromous Fish. Branch, Admin. Rep. No. 69–6. 7 p.

Morgan, A. R., and A. R. Gerlach. 1950. Striped bass studies on Coos Bay, Oregon, in 1949 and 1950. Oregon Fish. Comm. and Oregon Game Comm., Report 246. 31 p.

Pelzman, R. J. 1973. California inland angling survey for 1970. Calif. Fish Game **59**(2):100–106.

Ricker, W. E. 1975. Computation and interpretation of biological statistics of fish populations. Fish. Res. Board Can., Bull. 191. 382 p.

Robinson, J. B. 1960. The age and growth of striped bass (*Roccus saxatilis*) in California. Calif. Fish Game **46**(3):279–290.

Seeley, C. M., R. C. Tharratt, and R. L. Johnson. 1963. California inland angling surveys for 1959 and 1960. Calif. Fish Game **49**(3):183–190.

Smith, P. E., R. C. Counts, and R. I. Clutter. 1968. Changes in filtering efficiency of plankton nets due to clogging under tow. J. Cons. Perm. Int. Explor. Mer. **32**(2):232–248.

Stevens, D. E. 1977. Striped bass (*Morone saxatilis*) year class strength in relation to river flow in the Sacramento–San Joaquin Estuary, California. Trans. Am. Fish. Soc. **106**(1):34–42.

Turner, J. L. 1976. Striped bass spawning in the Sacramento and San Joaquin rivers in Central California from 1963 to 1972. Calif. Fish Game **62**(2):106–118.

Turner, J. L., and H. K. Chadwick. 1972. Distribution and abundance of young-of-the-year striped bass, *Morone saxatilis*, in relation to river flow in the Sacramento–San Joaquin Estuary. Trans. Am. Fish. Soc. **101**(3):442–452.

VonGeldern, C. E., Jr. 1972. A midwater trawl for threadfin shad, *Dorosoma petenense*. Calif. Fish Game **58**(4):268–276.

Striped Bass Ichthyoplankton Abundance, Mortality, and Production Estimation for the Potomac River Population

Tibor T. Polgar

Martin Marietta Corporation
Environmental Technology Center
Baltimore, Maryland

ABSTRACT

Methods are developed for estimating, from field survey data, the mortality rate and production for each successive ichthyoplanktonic stage. The abundance estimators used in the computation of these quantities are also derived. An age-dependent, ichthyoplankton population model is developed assuming either a uniform age distribution or an exponential age distribution within each stage. Striped bass egg and larval data from a 1974 ichthyoplankton survey in the Potomac River are used in model computations. The various model estimates are evaluated qualitatively, and the usefulness and limitations of the models are discussed.

Key words: abundance, entrainment loss, estimation procedures, ichthyoplankton, mortality, population dynamics, Potomac River, production, striped bass

INTRODUCTION

The potential of any fish stock for maintaining its size depends on its age structure, fecundity, natural mortality rate, and exploitation rate. Recruitment into the exploitable phase of the stock is generally related to the characteristics of the parent stock. In many instances, however, recruitment seems independent of the reproductive potential of the parent stock (Bannister et al. 1974), and for striped bass stocks, an inverse relationship may even exist between spawning-stock size and year-class success (Koo 1970). Yearly variation in mortalities during hatching and larval development are especially important in producing large fluctuations in the adult stock (Gulland 1965). Mortalities during the early life history, therefore, are major factors in determining stock stability and in assessing the importance of additional mortalities imposed by environmental disturbance.

110

The purpose of this paper is to present methods for estimating from field survey data the mortality rate and production for each successive ichthyoplanktonic stage. Clearly, field sampling yields information on densities of various ichthyoplanktonic stages at discrete times during spawning and development and at specific locations in the estuary. The problem is how to use this information to estimate standing stocks in each life stage and to derive mortality and production estimates for each stage. The four striped bass life stages being considered are eggs, yolk-sac larvae, finfold larvae, and post-finfold larvae. Surviving individuals in these stages have development times of approximately 2, 12, 11, and 30 days respectively. There is consensus among investigators that year-class success is determined by the end of the fourth stage, at which point the organisms are free-swimming, and they are referred to as fry, juveniles, or young-of-the-year.

Population parameters for ichthyoplankton have several direct uses. If mortality rates can be estimated, then it is possible to examine environmental conditions causing differences in these rates from year to year and from stage to stage. Specifically, questions regarding the "critical period" concept (May 1974) and compensation in early development may be resolved. Two of these questions are whether year-class success depends on the availability of food during any phase of development and whether transport into environments unfavorable for survival may cause catastrophic early mortalities. Both of these questions are particularly relevant in estuarine environments, where conditions such as flow and productivity vary radically on both the time and spatial scales.

Another use of these population parameters is to obtain better estimates of the cropping of ichthyoplankton due to power plant entrainment or any other point disturbance in the spawning and nursery area. Often, such estimates are simply based on standing crops of all ichthyoplanktonic stages near a specific location relative to standing crops in the entire system. It is clear, however, that if large mortalities are characteristic as development proceeds, the quantity of interest is not the total number of ichthyoplankton entrained during the entire period of development, but the fraction of those organisms lost (through all stages) that would have survived past the period of vulnerability.

ICHTHYOPLANKTON SAMPLING

The Maryland Power Plant Siting Program has been actively involved in sponsoring research to identify proper siting and operational practices for power plants to ensure the protection of aquatic habitats and to promote the survival of striped bass. As part of this program, much effort has been expended in studies of the striped bass that spawns in the Potomac River. Ichthyoplankton surveys have been carried out in the Potomac during the spawning and developmental periods in 1974, 1975, and 1976. For the numerical calculations in this paper, only data from the 1974 survey will be used; the survey design is described below. The surveys for 1975 and 1976 had similar designs, with minor modifications.

The Potomac River sampling grid in 1974 consisted of 12 cross-stream transects, separated by 6 to 12 km longitudinally. Geographically, the survey area spanned the region from 16 km below the Morgantown, Maryland, steam electric station (PEPCO) upstream to Washington, D.C. The transects farthest downstream (near Morgantown) were located in the mesohaline boundary region of the estuary. No striped bass spawning has been observed in this region, and major spawning activity has usually been seen 35 to 40 km upstream of the lowest transect. The transects were spaced to have an equal water volume between each transect pair (200×10^6 m^3 at MLW), and they were linearly separated by a distance of at least two tidal excursions. A single boat sampled all transects weekly during two days. The chosen distances between transects avoided multiple sampling of the same water masses, although at the cost of lower spatial resolution longitudinally.

Several tows were taken perpendicular to each cross-stream transect. The number of such tows at a transect was constrained by requiring the sampling of common-sized sub-areas of the river cross-sectional area of all transects. The 38 tow stations (total for all 12 transects) were sampled weekly throughout the 17-week duration of the spawning and developmental periods.

Each station location was sampled by an oblique tow, using a 1-m-diam, 0.505-mm-mesh ichthyoplankton net. The sampling procedure consisted of starting the tow as close to the bottom as possible, and straining 50 m^3 of water per 3-m depth interval until the net reached the surface. The strained volume was monitored and computed continuously, based on velocity data from net-mounted flow meters with on-board readouts. Since the volume sampled per 3-m depth interval was held constant, total volume per tow varied with river depth. Equal-volume sampling per depth interval and oblique sampling were chosen to compensate for possible diurnal fluctuations in distribution of ichthyoplankton with depth and for generally unknown vertical distribution properties. Further stratification of the sampling scheme (i.e., with depth) over such an extensive geographical area was cost-prohibitive. Similarly, replication at each tow station was not possible due to logistical and cost constraints. However, selected stations were sampled in triplicate on several occasions, bringing the total number of samples to 696 for the entire sampling period.

Ichthyoplankton were identified to species (where possible) and were counted for each tow. Striped bass ichthyoplankton were further classified into four life stages: eggs, yolk-sac larvae, finfold larvae, and post-finfold larvae. Length distributions of larvae were also determined. A large number of other measurements were taken concurrently with ichthyoplankton sampling, including temperature, salinity, dissolved oxygen, turbidity, and incident solar radiation. Polgar (1975), Polgar et al. (1975), and Mihursky et al. (1974) give further details on this and other elements of the striped bass program.

ESTIMATION OF ABUNDANCES

Collections on the sampling grid in any one week yield density estimates for each developmental stage for each tow at each transect. If density properties along the river axis are desired, however, it is necessary to obtain an estimator for transect-averaged densities. Pyne (1976) has shown that the minimum-variance unbiased estimator of average density, D, for n individual samples from the same distribution of organisms, collected in varying tow volumes, is

$$\bar{D} = \sum_{i=1}^{n} N_i \bigg/ \sum_{i=1}^{n} V_i \,, \tag{1}$$

where N_i is the number of ichthyoplankton found in V_i m^3 of water strained in the ith tow.

The equation for the average density of organisms in a river segment (rather than at a cross-stream transect) must have a form similar to Eq. (1), provided that all samples included in the calculation are assumed to originate from the same distribution of organisms. However, if a river segment is defined as a volume bounded by two adjacent, fixed transects, the organisms collected at each transect must be given weight in calculating an average density for a river segment. When tows are taken between fixed transects (whether these be fixed or randomized station locations), they must be assigned entirely to the river segment in question. If it is assumed that organisms collected at a transect are assigned equally to river segments on either side of the transect, an estimator for river segment density can be found. The following definitions will be used:

$N_{j,l',q}^{(k)} \equiv$ the number of ichthyoplankton on life stage k collected in the qth tow at transect l' on the jth week;

$k \equiv \{0, 1, 2, 3\}$, the index designating the life stage: 0 = eggs, 1 = yolk-sac larvae, 2 = finfold larvae, and 3 = post-finfold larvae;

$j \equiv \{1, ..., \tau_{yr}\}$, the index number for weekly collections, which may vary from year to year;

$l' \equiv \{1, ..., 12\}$, the index for fixed cross-stream transects, with l' = 1 being the furthest downstream transect and l' = 12 being the furthest upstream transect;

$l \equiv \{1, ..., 11\}$, the index number for the river segment enclosed by transects l' and l' + 1;

$q \equiv \{1, ..., Q_{j,l'}\}$, the index number for tows taken on transect l' in week j, with the total number of tows on that transect at that time being $Q_{j,l'}$; variable from week to week because occasional replicates are included in estimates.

Accordingly, the average "instantaneous" density of life stage k in river segment l (enclosed by transects l' and $l' + 1$) on weekly sampling occasion j is

$$D_{j,l}^{(k)} = \frac{\displaystyle\sum_{q=1}^{Q_{j,l'}} N_{j,l',q}^{(k)} + \sum_{q=1}^{Q_{j,l'+1}} N_{j,l'+1,q}^{(k)}}{\displaystyle\sum_{q=1}^{Q_{j,l'}} V_{j,l',q} + \sum_{q=1}^{Q_{j,l'+1}} V_{j,l'+1,q}} , \qquad (2)$$

where $V_{j,l',q}$ is the strained volume in the qth tow on transect l' on weekly sampling occasion j. When only fixed transect sampling is carried out, as in 1974, Eq. (2) represents the average "instantaneous" river segment density.

If fixed transect sampling is augmented by tows at a number of fixed and random stations within a river segment, as in the 1975 survey, the collected ichthyoplankton in these tows are assigned entirely to that segment. Thus, if the randomized sampling effort is fairly uniform among river segments and if $l' + \frac{1}{2}$ denotes the location of tows taken within a river segment, the average "instantaneous" density for life stage k in river segment l is

$$D_{j,l}^{(k)} = \frac{\left[\frac{1}{2}\displaystyle\sum_{q=1}^{Q_{j,l'}} N_{j,l'q}^{(k)} + \sum_{q=1}^{Q_{j,l'+\frac{1}{2}}} N_{j,l'+\frac{1}{2},q}^{(k)} + \frac{1}{2}\sum_{q=1}^{Q_{j,l'+1}} N_{j,l'+1,q}^{(k)} \right]}{\left[\frac{1}{2}\displaystyle\sum_{q=1}^{Q_{j,l'}} V_{j,l',q} + \sum_{q=1}^{Q_{j,l'+\frac{1}{2}}} V_{j,l'+\frac{1}{2},q} + \frac{1}{2}\sum_{q=1}^{Q_{j,l'+1}} V_{j,l'+1,q} \right]} . \qquad (3)$$

Densities obtained using either Eq. (2) or Eq. (3) may be converted into abundances, $A_{j,l}^{(k)}$ for river segment l on weekly sampling occasion j as follows:

$$A_{j,l}^{(k)} = R_l D_{j,l}^{(k)} , \qquad (4)$$

where R_l is the volume of river segment l. These abundance values are "snapshot" representations of the estimated ichthyoplankton standing crop in any one river segment at the time of measurement. Later in the mathematical development, it will become necessary to extrapolate these relatively shcrt-term representations of abundance to the periods between weekly sampling intervals. It should also be noted that unless resulting egg production estimates agree well with the estimated fecundity of the parent stock, these abundance estimates are only relative. The accuracy of population parameter estimates obviously depends on the extent of variations in sampling efficiency with developmental stage. No such variations are considered in this paper. However, derived egg production estimates are compared with the estimated reproductive potential of the spawning stock.

In the ichthyoplankton population models that follow, the entire spawning and developmental period is considered as an event for a single cohort, even though the process is time- and space-dependent. The problem is reduced to variation in the age variable, with zero age being the time of spawning. Nonetheless, abundances are calculated on a weekly and segmental basis at first (Eq. 4), because the estimation procedure requires that tows sample the same distribution of organisms. Later, abundance values are summed over time and space for each life stage.

A MODEL WITH UNIFORM AGE DISTRIBUTION WITHIN EACH STAGE

The simplest but most unrealistic approach to calculating stage-to-stage productions from observed abundances requires estimates only of development times in each stage. The immediate drawback of this method is easily recognized from its major assumption. This assumption postulates for each stage that the age-frequency distribution for individuals in ichthyoplankton collections is a uniform distribution (i.e., equal frequencies for all ages). The uniform age assumption, by necessity, implies that all death occurs at the instant of development into the next stage. Thus, mortality rates are not obtainable; only productions into each stage and survival probabilities through each stage can be derived.

It is easily shown from the first moment of the uniform age distribution that the mean age in each stage, $\bar{a}^{(k)}$, is

$$\bar{a}^{(k)} = \frac{1}{2} t^{(k)} , \tag{5}$$

where $t^{(k)}$ is the development time or duration for stage k. Clearly, since mortality really occurs continuously throughout the entire development time within each stage, we would expect mean ages of less than one-half the development time in the samples. A more realistic age structure is included in the formulation in the next section.

If we assume that each weekly abundance observation in each river segment represents a constant production rate for that week (a quite realistic assumption with the given weekly sampling frequency except perhaps for eggs), then production for that week may be computed. The production rate into stage k is simply the rate of appearance of stage k ichthyoplankton of zero age. For the uniform age distribution this production rate, of course, is also the rate of appearance into all subsequent age groups (or cohorts) within stage k.

The production into stage k in segment l for week j can be expressed as

$$P_{j,l}^{(k)} = R_l D_{j,l}^{(k)} \left[t^*/t_{j,l}^{(k)} \right] = t^* A_{j,l}^{(k)}/t_{j,l}^{(k)} , \tag{6}$$

where the ratio $t^*/t_{j,l}^{(k)}$ specifies the average number of days between sampling periods, t^* (taken to be seven days in the present case), as a fraction of the development time for stage k. For eggs this ratio is greater than 1.0 and production is greater than abundance, while for the following three stages this ratio is less than 1.0 and production is less than abundance.

The time of development from stage k to $k + 1$, $t_{j,l}^{(k)}$, is written as a function of time and location in Eq. (6), because it depends explicitly on water temperature. A regression was developed for the dependence of egg development time in hours, $t_{j,l}^{(0)}$, on temperature (Polgar et al. 1975):

$$t_{j,l}^{(0)} = -4.7\overline{T}_{j,l} + 131.6 . \tag{7}$$

Here $\overline{T}_{j,l}$ denotes the average water temperature (°C) in river segment l on sampling occasion j. Other developmental times have not been investigated sufficiently for inclusion of temperature dependence, and thus they are taken as constants (Doroshev 1970, Mansueti 1958): $t^{(1)} = 12$ days (yolk-sac-larval development time); $t^{(2)} = 11$ days (finfold-larval development time); $t^{(3)} = 30$ days (post-finfold-larval development time).

The total production into stage k, $N^{(k)}(0)$, therefore, can be expressed as the sum (or "integral") of $P_{j,l}^{(k)}$ values over all sampling periods and river segments as follows:

$$N^{(k)}(0) = \sum_j \sum_l P_{j,l}^{(k)} . \tag{8}$$

The survival probability through stage k is

$$S^{(k)} = \frac{N^{(k+1)}(0)}{N^{(k)}(0)} . \tag{9}$$

It should be noted that if temperature dependence of development times is ignored, total production can be expressed directly by the sum of all observed weekly abundances — that is,

$$N^{(k)}(0) = \frac{t^*}{t^{(k)}} \sum_j \sum_l A_{j,l}^{(k)} = \frac{t^*}{t^{(k)}} A^{(k)} , \tag{10}$$

where $A^{(k)}$ denotes the sum of observed standing-crop values for stage k for all weeks and all segments. Equation (10) implies that the entire spawning process can be regarded as the development of a single cohort, with age as the only independent variable. Consequently, the derived probability of survival for each stage represents an average over weeks and segments. Since no temperature effects will be considered in subsequent models, the notation shown in Eq. (10) will be followed.

Table 1 summarizes the results from the model assuming uniform age distribution and using 1974 data. If the temperature effect on egg-hatching time is included in the calculations (Eqs. 6 through 8), only egg production and survival probability are affected; the new values for these two variables are 4.15×10^9 and 0.0150 respectively. It should be noted that for the last three stages, production values are less than abundance values. This result is expected since $t^*/t^{(k)}$ is less than 1.0 for these three stages (Table 1), indicating that development times for $k = 2, 3$, and 4 are long enough so that a cohort passing through any of these stages may be sampled during more than a single weekly sampling period.

Although unrealistic, the model computations in Table 1 have been carried out for use in approximations when an exponential age distribution within each stage is assumed.

A MODEL WITH EXPONENTIAL AGE
DISTRIBUTION WITHIN EACH STAGE

As a consequence of the assumption of a uniform age distribution, all production estimates are likely to be underestimated. It is not clear how stage-to-stage survival probabilities are affected. A more realistic estimation procedure would have to take cognizance of the rapid decrease in the size of a cohort as it passes through a stage.

The following assumptions will be made for each ichthyoplanktonic stage in constructing a mortality model:

1. Mortality rate is constant over the development time. However, mortality rate may vary from stage to stage.

2. Production is constant over a period equivalent to the time interval between sampling. Duration of this period is denoted as t^*, and the times of specific standing-crop observations are considered centered within these periods.

Table 1. Uniform-age-distribution model results

Stage	k	Development time (days), $t^{(k)}$	Sum of instantaneous abundances[a] (1974 data), $A^{(k)}$	Production,[b] $N^{(k)}(0)$	Survival probability to next stage,[c] $S(k)$	Mean age[d] (days), $\bar{a}^{(k)}$
Eggs	0	≈2	1.63×10^9	5.71×10^9	0.0193	1.0
Yolk-sac larvae	1	12	1.07×10^8	6.24×10^7	0.0454	6.0
Finfold larvae	2	11	4.44×10^6	2.83×10^6	0.1205	5.5
Post-finfold larvae	3	30	1.46×10^6	3.41×10^5		15.0

[a]Calculated using Eq. (4) summed over all weeks and all segments.
[b]Calculated using Eq. (10).
[c]Calculated using Eq. (9).
[d]Calculated using Eq. (5).

3. Mortality occurs continuously and is proportional to the number of individuals in the stage at any time. Removing time as an independent variable (with only age remaining) implies that the entire ichthyoplankton population originates as a single cohort of eggs. Therefore, a mortality rate for stage k represents an average over time and space. Temperature dependence of developmental times is ignored.

Under these assumptions, an exponential age distribution of individuals within each stage is governed by

$$\frac{d[N^{(k)}(a)]}{da} = -\lambda^{(k)}N^{(k)}(a) , \tag{11}$$

where $\lambda^{(k)}$ is the mean instantaneous mortality rate of ichthyoplankton in stage k and a is age. The solution to Eq. (11) for the total number of individuals of age a in stage k produced in the entire system may be written as

$$N^{(k)}(a) = N^{(k)}(0) \exp[-\lambda^{(k)}a] \quad \text{for } 0 \leq a \leq t^{(k)}. \tag{12}$$

$N^{(k)}(a)$ may be related to the observed abundances as follows:

$$t^* \sum_j \sum_l A_{j,l}^{(k)} = t^* A^{(k)} = \int_0^{t^{(k)}} N^{(k)}(a)\, da . \tag{13}$$

In other words, the sum of the observed abundances of stage k, $A^{(k)}$, times the number of days between sampling periods, t^*, equals the age distribution function integrated from 0 to the development time for that stage, $t^{(k)}$.

After substituting Eq. (12) into Eq. (13) for $N^{(k)}(a)$, integrating the right side of Eq. (13), and solving for $N^{(k)}(0)$ one obtains

$$N^{(k)}(0) = \frac{t^* A^{(k)}\lambda^{(k)}}{1 - \exp[-\lambda^{(k)}t^{(k)}]} , \tag{14}$$

and thus, Eq. (12) becomes

$$N^{(k)}(a) = \frac{t^* A^{(k)}\lambda^{(k)} \exp[-\lambda^{(k)}a]}{1 - \exp[-\lambda^{(k)}t^{(k)}]} \quad \text{for } 0 \leq a \leq t^{(k)}. \tag{15}$$

Eqs. (14) and (15) hold for each stage k, with stage-specific death rates $\lambda^{(k)}$ and development times $t^{(k)}$. Since the sole independent variable is age, the number of ichthyoplankton in stage k with age equal to the development time for that stage, $t^{(k)}$, must represent the number of ichthyoplankton entering stage $k + 1$ — that is,

$$N^{(k)}\left(t^{(k)}\right) = N^{(k+1)}(0), \; k = \left\{0, 1, 2\right\} \; . \tag{16}$$

With this condition, a system of equations is obtained of the form

$$\frac{A^{(k)}\lambda^{(k)} \exp[-\lambda^{(k)}t^{(k)}]}{1 - \exp[-\lambda^{(k)}t^{(k)}]} = \frac{A^{(k+1)}\lambda^{(k+1)}}{1 - \exp[-\lambda^{(k+1)}t^{(k+1)}]} \; . \tag{17}$$

Clearly, three relationships of the above form exist, with four unknown mortality rates, $\lambda^{(k)}$. To estimate the values of any three of the mortality rates, and therefore of all production values, $N^{(k)}(0)$, an independent estimate of one of the mortality rates must be obtained. Two methods for specifying the system of Eq. (17) will be attempted.

Method I: Estimates Derived from the Results of the Assumption of Uniform Age Distribution

The probability of survival from one stage to the next is

$$S^{(k)} = \frac{N^{(k+1)}(0)}{N^{(k)}(0)} \; . \tag{18}$$

Although this equation appeared in developing the model of ichthyoplankton age structure assuming uniform age distribution (Eq. 9), this expression is valid for any model and assumption concerning age distribution.

In the exponential age distribution model, this probability is also expressed as

$$S^{(k)} = \exp\left[-\lambda^{(k)}t^{(k)}\right] \; . \tag{19}$$

Equating these two results and substituting Eq. (10), which is based on the assumption of a uniform age distribution, in the numerator and denominator of Eq. (18), we obtain the basis for an approximation:

$$\frac{t*A^{(k+1)}/t^{(k+1)}}{t*A^{(k)}/t^{(k)}} \approx \exp[-\lambda^{(k)}t^{(k)}] \tag{20}$$

or

$$\lambda^{(k)} \approx \frac{1}{t^{(k)}} \left\{ \ln \left[\frac{t^{(k+1)}}{t^{(k)}} \right] + \ln \left[\frac{A^{(k)}}{A^{(k+1)}} \right] \right\} . \tag{21}$$

This approximation is only reasonable if the mortality rate in stage k is small compared to the rate of production, that is, if $\lambda^{(k)} \ll 1$. If we hypothesize that the mortality rate is lower in later stages, it is logical to approximate $\lambda^{(2)}$ based on the assumption of uniform age distribution, since Eq. (21) involves both stages k and $k + 1$. The approximated value of $\lambda^{(2)}$ by application of Eq. (21) is

$$\lambda^{(2)} \approx 0.1922 \text{ day}^{-1} .$$

With this value, the system of three Eqs. (17) may be solved by successive application of Newton's method to give estimates of the mortality rates for the other three stages, that is, $\lambda^{(0)}$, $\lambda^{(1)}$, and $\lambda^{(3)}$. Then Eq. (14) is solved for each stage to give estimates of production, that is, $N^{(k)}(0)$. Table 2 summarizes the numerical results.

Method II: Equal Mortality Rates in Two Stages

Assuming equal mortality rates for the last two stages seems reasonable in view of the expected leveling off of mortality with development and the attainment of some degree of motility. Thus, if finfold and post-finfold larvae are assumed to have equal mortality rates, then for each of these two stages

$$\frac{A^{(k)}\lambda^{(k)} \exp[-\lambda^{(k)}t^{(k)}]}{1 - \exp[-\lambda^{(k)}t^{(k)}]} = \frac{A^{(k+1)}\lambda^{(k)}}{1 - \exp[-\lambda^{(k)}t^{(k+1)}]} \tag{22}$$

from Eq. (17) with $\lambda^{(k+1)} = \lambda^{(k)}$. Mortality rates for other stages are obtained directly from Eq. (21), and production for each stage is obtained as in Method I. Table 3 summarizes the numerical results obtained using this method.

DISCUSSION OF MODEL RESULTS

Clearly, the assumption of an exponential age distribution within each stage leads to estimates of production for all stages that are higher than those obtained using the assumption of a uniform age distribution (compare Tables 2 and 3 with Table 1). This result follows from the constraint imposed by

Table 2. Exponential-age-distribution model results

(Method I)

Stage	k	Development time (days), $t^{(k)}$	Sum of instantaneous abundances[a] (1974 data), $A^{(k)}$	Mortality rate[b] (day^{-1}), $\lambda^{(k)}$	Production,[c] $N^{(k)}(0)$	Survival probability to next stage,[d] $S^{(k)}$	Mean age[e] (days), $\bar{a}^{(k)}$
Eggs	0	≈2	1.63×10^9	2.348	2.692×10^{10}	0.0091	0.41
Yolk-sac larvae	1	12	1.07×10^8	0.321	2.458×10^8	0.0213	2.86
Finfold larvae	2	11	4.44×10^6	0.192[f]	6.787×10^6	0.1207	3.69
Post-finfold larvae	3	30	1.46×10^6	0.071	8.196×10^5	0.1206	10.04

[a]Calculated using Eq. (4) summed over all weeks and all segments.

[b]Calculated using a system of three Eqs. (17).

[c]Calculated using Eq. (14)

[d]Calculated using Eq. (18) or (19).

[e]Calculated as follows:

$$\bar{a}^{(k)} = \int_0^{t^{(k)}} a N^{(k)}(a)\, da \Big/ \int_0^{t^{(k)}} N^{(k)}(a)\, da = \frac{1}{\lambda^{(k)}} \left\{ \frac{1 - \exp[-\lambda^{(k)} t^{(k)}] - \lambda^{(k)} t^{(k)} \exp[-\lambda^{(k)} t^{(k)}]}{1 - \exp[-\lambda^{(k)} t^{(k)}]} \right\} = \frac{1}{\lambda^{(k)}} - \frac{t^{(k)}}{\exp[\lambda^{(k)} t^{(k)}] - 1}.$$

[f]Calculated using Eq. (21).

Table 3. Exponential-age-distribution model results

(Method II)

Stage	k	Development time (days), $t^{(k)}$	Sum of instantaneous abundances[a] (1974 data), $A^{(k)}$	Mortality rate[b] (day^{-1}), $\lambda^{(k)}$	Production,[c] $N^{(k)}(0)$	Survival probability to next stage,[d] $S^{(k)}$	Mean age[e] (days), $\bar{a}^{(k)}$
Eggs	0	≈2	1.63×10^{9}	2.346	2.690×10^{10}	0.0092	0.42
Yolk-sac larvae	1	12	1.07×10^{8}	0.322	2.465×10^{8}	0.0211	2.85
Finfold larvae	2	11	4.44×10^{6}	0.125f	5.200×10^{6}	0.2523	4.28
Post-finfold larvae	3	30	1.46×10^{6}	0.125f	1.311×10^{6}	0.0234	7.28

[a]Calculated using Eq. (4) summed over all weeks and all segments.

[b]Calculated using a system of two Eqs. (17).

[c]Calculated using Eq. (14).

[d]Calculated using Eq. (18) or (19).

[e]Calculated as follows:

$$a^{(k)} = \int_0^{t^{(k)}} aN^{(k)}(a)\,da \bigg/ \int_0^{t^{(k)}} N^{(k)}(a)\,da = \frac{1}{\lambda^{(k)}} \left\{ \frac{1 - \exp[-\lambda^{(k)}t^{(k)}] - \lambda^{(k)}t^{(k)}\exp[-\lambda^{(k)}t^{(k)}]}{1 - \exp[-\lambda^{(k)}t^{(k)}]} \right\} = \frac{1}{\lambda^{(k)}} - \frac{t^{(k)}}{\exp[\lambda^{(k)}t^{(k)}] - 1}.$$

[f]The mortality rates $\lambda^{(2)}$ and $\lambda^{(3)}$ are assumed to be equal, and their value is calculated from Eq. (22).

death occurring only at the "instant" of stage transition under the assumption of uniform age distribution. Thus, the mean age within each stage is exactly one-half of the corresponding development time (Table 1, last column), and the number of individuals in the zero age group in each stage is equal to the number of individuals in any other age group in that stage. For the exponential model, however, mean age is always less than one-half the development time (Tables 2 and 3, last column) and the number of individuals in the zero age group always exceeds the number of individuals in any other age group within that stage. Thus, production rates are higher. Although the exponential model requires an independent estimate of one of the four mortality rates, it is more realistic.

Estimates of mortality rate and production (and thus of survival probability and mean age) for the first two stages (eggs and yolk-sac larvae) computed by the two exponential methods agree well (compare Table 2 with Table 3). Conceptually, Method I seems more reasonable because $\lambda^{(k)} > \lambda^{(k+1)}$ and $S^{(k)} < S^{(k+1)}$ as k increases (Table 2), which corresponds to what is thought to occur in the field. Because Method I yields more reasonable results than Method II, the uniform-age-distribution model becomes a necessary part of the estimation process, because it is used in obtaining an independent estimate for one of the mortality rates (Eqs. 20 or 21).

The sensitivity of model results to consistent sampling errors in any one stage differs in the two types of models. Under the assumption of uniform age distribution, undersampling of any one stage affects the population parameters of that stage and of only the immediately preceding stage. Under the assumption of exponential age distribution, the undersampling of a particular stage will affect the population parameters of that stage and of all preceding stages.

Adult sampling, concurrent with the ichthyoplankton surveys (Morgan and Wilson 1974) and with acoustic fish surveys (Zankel et al. 1975), showed that the size of the spawning population was 1×10^6 striped bass during peak spawning activity. Based on fecundity and sex-ratio measurements, this population should have produced 73×10^9 eggs, as compared to $4-27 \times 10^9$ eggs in Tables 1–3. There are a great number of biological, sampling-design, and sampling-efficiency factors that could account for the difference between the model estimates of egg production and the potential production inferred from adult population characteristics. I will address this discrepancy only from the point of view of model formulation.

It is possible to refine the ichthyoplankton population model to include both time and age as independent variables, thus allowing mortality rates to vary in time, as they no doubt do. It is also possible, but more ambitious, to include physical transport processes that continually redistribute ichthyoplankton. Just accounting more carefully for changes in abundance from week to week and from river segment to river segment (rather than simply assuming production and abundance constant for each week) could possibly yield higher estimates of egg production. More important, however, is the effect of possible diurnal-spawning activity on the egg production estimate. Consideration of diurnal-spawning activity is important because all sampling is usually carried out during the same part of a day, and egg mortality is so high that, if sampling is out of phase with spawning, radical underestimation of egg production may result. Some analyses have already been carried out on this problem (Polgar and Fanzone 1977), and calculations have shown that values as high as 90.5×10^9 could be obtained for the 1974 egg production.

ACKNOWLEDGMENTS

This work was supported by the Power Plant Siting Program of the State of Maryland, under the Department of Natural Resources contract 18-74-04(75). I am grateful to J. F. Fanzone and G. M. Krainak of the Martin Marietta Corporation for their contributions to model development and data management respectively. I am deeply indebted to J. Mihursky and his staff, of the Chesapeake Biological Laboratory, including J. Cooper, K. Wood, R. Prince, E. Gordon, and R. Block, who were responsible for ichthyoplankton collection, sorting, and identification.

LITERATURE CITED

Bannister, R. C. A., D. Harding, and S. J. Lockwood. 1974. Larval mortality and subsequent year-class strength in the plaice (*Pleuronectes platessa* L.), p. 21–37. *In* J. H. S. Blaxter [ed.] The early life history of fish. Springer-Verlag, New York.

Doroshev, S. I. 1970. Biological features of eggs, larvae and young of striped bass (*Roccus saxatilis,* Walbaum) in connection with the problem of its acclimatization in the U.S.S.R. J. Ichthyol. **10**(2):235–248.

Gulland, J. A.. 1965. Survival of the youngest stages of fish, and its relation to year-class strength. Spec. Publ. ICNAF **6**:363–371.

Koo, T. S. Y. 1970. The striped bass fishery in the Atlantic states. Ches. Sci. **11**(2):73–93.

Mansueti, R. 1958. Eggs, larvae and young of the striped bass, *Roccus saxatilis.* Maryland Dept. Res. and Ed., Contr. No. 112.

May, R. C. 1974. Larval mortality in marine fishes and the critical period concept, p. 3–19. *In* J. H. S. Blaxter [ed.] The early life history of fish. Springer-Verlag, New York.

Mihursky, J. A., R. M. Block, J. E. Cooper, K. Wood, R. Prince, and E. W. Gordon. 1974. Interim report on 1974 Potomac Estuary horizontal ichthyoplankton distributions. Chesapeake Biological Laboratory, University of Maryland, UMCEES Ref. No. 74-170.

Morgan, R. P., and J. Wilson. 1974. Interim report on 1974 spawning stock assessment, Potomac River Fisheries Program. Chesapeake Biological Laboratory, University of Maryland.

Polgar, T. T. 1975. Impact of Potomac River power plants on early life stages of striped bass — preliminary results. Rec. Maryland Power Plant Siting Act 4(3):1—6.

Polgar, T. T., R. E. Ulanowicz, D. A. Pyne, and G. M. Krainak. 1975. Investigations of the role of physical transport processes in determining ichthyoplankton distributions in the Potomac River: interim report for 1974 spawning season data. Maryland Power Plant Siting Program, Ref. No. PPRP 11/PPMP 14.

Polgar, T. T., and J. F. Fanzone. 1977. Diurnal periodicity of spawning activity, and its effect on ichthyoplankton population estimates (*in preparation*).

Pyne, D. A. 1976. On the comparison of two binomial or Poisson populations when sampling units are non-homogeneous. Dept. Math. Sci., The Johns Hopkins University, Ref. No. 246.

Zankel, K. L., L. H. Bongers, T. T. Polgar, W. A. Richkus, and R. E. Thorne. 1975. Size and distribution of the 1974 striped bass spawning stock in the Potomac River. Martin Marietta Corp., Ref. No. PRFP 75-1.

Methods for Calculating Natural Mortality Rate, Biomass Production, and Proportion Entrained of Lacusterine Ichthyoplankton

P. A. Hackney

Tennessee Valley Authority
Norris, Tennessee

ABSTRACT

The impact of ichthyoplankton entrainment in lacusterine environments has been difficult to ascertain because the total number produced is usually unknown. Additionally, larvae in lakes are vulnerable to entrainment for varying periods of time, not just instantaneously as when transported past cooling water intakes located on rivers.

A method for determining the number of larvae produced in lakes is given. Also, methods of calculating the natural mortality rate and biomass production of lacusterine ichthyoplankton are presented.

This paper was presented orally at the conference but was not available for publication in the *Proceedings*.

Confidence Intervals on Mortality Rates Based on the Leslie Matrix

Douglas S. Vaughan

Marine Experiment Station
Graduate School of Oceanography
University of Rhode Island
Kingston, Rhode Island

ABSTRACT

The variability of an estimator of the natural instantaneous mortality rate of age class 0 (M_0) is studied by Monte Carlo simulations. Certain input parameters are more important than others in affecting this variability. The deterministic estimator is assumed to be the result of nondeterministic events in the real world. The proposed estimator is found to be biased, but it can be corrected using the difference between the deterministic and nondeterministic estimators of M_0.

Key words: Atlantic menhaden, confidence intervals, Leslie matrix, Monte Carlo simulation, mortality rates

INTRODUCTION

Since many fish species are characterized by high fecundity and high mortality in age class 0, it is unreasonable to extrapolate the juvenile natural mortality rate from estimates of the adult natural mortality rate. Accurate estimates of mortality rates for early life history stages are often difficult to obtain directly. An indirect method for estimating the instantaneous natural mortality rate for age class 0 has been derived based on the Leslie matrix algorithm (Vaughan and Saila 1976, Van Winkle et al. 1974).

Both an invariant and a stable age structure were assumed in developing this indirect estimate, which can be expressed mathematically as follows:

$$n_x(t) = n_0(t-x) \prod_{j=0}^{x-1} p_j \, ,$$

$$\lambda = n_x(t)/n_x(t-1) \, , \quad \text{for all ages } x \, ,$$

(1)

128

where $n_x(t)$ is the number of individuals of age x at time t, p_j is the probability that an individual survives through age class j independent of time t, and λ is assumed constant for all ages x. Significant variations in fishing or natural mortality rates over time will affect the first assumption, and the presence of a dominant year class impairs the latter assumption.

Schreiber (1959) describes the Monte Carlo method as "the application of random output from a given distribution of numbers as input to functions or processes under study in order to determine the distribution of outputs of the processes in question." This method can be used to assess the effects of variations in the input parameters to the Leslie matrix algorithm on the estimate of the natural instantaneous mortality rate for age class 0, $\hat{M}_0$. The cumulative frequency distribution of $\hat{M}_0$ can then be used directly to determine empirical confidence intervals, or the resultant distribution may be approximated by the normal distribution with its concommitant confidence intervals. This empirical study is necessary because the actual underlying distribution of $\hat{M}_0$ would be extremely difficult to obtain theoretically due to the nonlinearity of the formula for obtaining $\hat{M}_0$:

$$\hat{M}_0 = -\ln\left\{ \frac{\lambda}{m_1 + \sum_{x=1}^{k-1}\left[\left(m_{x+1}/\lambda^x\right)\left(\prod_{j=1}^{x} p_j\right)\right]} \right\}, \tag{2}$$

where m_x is the fecundity of a fish (both sexes combined) in age class x, and k is the number of age classes (starting with 0).

METHODOLOGY

Leslie's (1945, 1948) deterministic, discrete-time model incorporates age-specific fecundity and mortality. The population matrix described by Leslie transforms a population's age structure from one unit of time to the next. This matrix has nonzero elements on the first row and the subdiagonal. The elements of the top row consist of the product of the age-specific survival rate and the fecundity per individual of the following year class, $p_x m_{x+1}$, while the subdiagonal contains the age-specific survival rate, p_x.

The data requirements for indirectly estimating M_0 from Eq. (2) are the age-specific fecundity, sex ratio, and probability of survival for age classes 1 through k. Atlantic menhaden (*Brevoortia tyrannus*) is used as an example in this paper, and the data requirements for this species are given in Tables 1 through 6. A linear least-squares regression (Fig. 1) was developed for the natural logarithm (ln) of fecundity per female, m', vs the natural logarithm of fish length (L in millimeters; $\hat{\rho} = 0.937$) (Higham and Nicholson 1964).

Table 1. Length (L), proportion of females (q), and fecundity (m' and m) of the Atlantic menhaden, *Brevoortia tyrannus*, by age class

Age class (x)	Length[a] (females) (L_x)	Proportion[b] of females (q_x)	Fecundity[c] Female (m_x')	Fecundity[c] Individual (m_x)
0	75.5	0.4949		
1	175.5	0.5074		
2	238.4	0.4975	75,429[d]	37,526
3	278.0	0.5192	196,969	102,266
4	302.9	0.5370	297,177	159,584
5	318.5	0.5434	378,407	205,626
6	328.3	0.5763	437,861	252,340
7	334.5	0.6552	478,874	313,422
8+	339.8	0.6403	516,042	330,422

[a]See Reintjes (1969, p. 10):
$$L \text{ (in millimeters)} = 345 \left\{ 1 - \exp[-0.464(x + 0.032)] \right\}, x = 0.5, 1.5, 2.5,$$
... , 7.5, 9.0 years.

[b]See Reintjes (1969, p. 12).

[c]A least-squares, fecundity-length regression for females (see Fig. 1) was developed from data in Higham and Nicholson (1964, Table 8):
$$m' = (3.66 \times 10^{-7})L^{4.7995}.$$
The fecundity per individual, m, was obtained from the fecundity per female, m', by multiplying by the proportion of females, q, in that age class.

[d]The fecundity per female for age class 2 obtained from Fig. 1 was multiplied by 0.8, the proportion of mature females in that age class (Reintjes (1969), to give 75,429 eggs per female.

Table 2. Catch, $C_{x,t}$, of Atlantic menhaden (in millions of fish) of age x in year t in landings by purse seiners, 1955–1971

Year (t)	$C_{x,t}$ for ages (x) 0 through 8+								
	0	1	2	3	4	5	6	7	8+
1955	761.01	674.15	1057.68	267.31	307.21	38.07	10.53	1.84	0.64
1956	36.37	2073.26	902.60	319.60	44.78	150.68	28.70	6.72	1.99
1957	299.58	1599.98	1361.77	96.73	70.80	40.52	36.93	4.26	1.10
1958	106.06	858.16	1635.35	72.05	17.25	15.94	9.09	4.88	0.43
1959	11.40	4038.72	851.29	388.27	33.41	11.87	12.36	4.55	1.77
1960	72.17	281.01	2208.63	76.37	102.20	23.77	7.95	2.36	0.65
1961	0.25	832.42	503.60	1209.57	19.18	29.38	2.86	0.81	0.24
1962	51.58	514.11	834.52	217.25	423.37	30.75	24.60	2.98	0.70
1963	96.89	724.23	709.20	122.53	44.97	52.38	10.42	3.33	0.56
1964	302.59	703.95	604.98	83.50	17.94	7.85	6.62	1.31	0.32
1965	294.12	739.28	417.55	77.75	12.17	1.81	1.22	0.74	0.07
1966	349.46	550.83	404.11	31.70	3.88	0.37	0.11	0.11	0.04
1967	6.95	633.20	265.68	72.76	5.09	0.49	0.01		
1968	154.61	376.28	535.52	65.68	10.67	0.98	0.06		
1969	158.08	372.37	284.31	47.81	5.44	0.14	0.01		
1970	24.19	861.67	468.58	38.14	6.77	0.52			
1971	73.97	258.80	525.82	89.33	16.77	2.70			

Source: Nicholson 1975.

Table 3. Annual age-specific exploitation rates, $u_{x,t}$, and maximum instantaneous fishing mortality rates, F_t, 1955–1967

Year (t)	$u_{x,t}$ for ages (x) 2 through 7						$\bar{u}_t^a$	F_t^b
	2	3	4	5	6	7		
1955	0.715	0.729	0.613	0.533	0.574	0.480	0.607	0.935
1956	0.870	0.759	0.450	0.776	0.860	0.859	0.762	1.437
1957	0.909	0.715	0.696	0.739	0.847	0.908	0.802	1.621
1958	0.749	0.531	0.447	0.516	0.636	0.734	0.602	0.922
1959	0.860	0.708	0.525	0.556	0.826	0.875	0.725	1.291
1960	0.566	0.553	0.639	0.788	0.840	0.908	0.716	1.258
1961	0.650	0.715	0.311	0.510	0.447	0.536	0.528	0.751
1962	0.854	0.800	0.876	0.723	0.871	0.842	0.828	1.758
1963	0.881	0.861	0.830	0.876	0.883	0.912	0.874	2.070
1964	0.880	0.869	0.903	0.855	0.895	0.949	0.892	2.224
1965	0.917	0.946	0.970	0.943	0.917	0.949	0.940	2.819
1966	0.829	0.839	0.876	0.974	1.000	1.000	0.920	2.522
1967	0.788	0.871	0.837	0.891	1.000	1.000	0.877	2.099

$^a \bar{u}_t$ is the average of the $u_{x,t}$ values for the given year.

$^b F_t = -\ln(1 - \bar{u}_t)$.

Table 4. Annual catch and effort, adjustment of effort using catchability coefficients (Q), and the catch-per-unit-of-effective-effort (CPUE)

Year (t)	Weight landed (1000 metric tons) (C_t)	Observed vessel weeks (f_t^*)	Q_t^a	Relative Q_t^b	Effective effort $(f_t)^c$	CPUE$_t^d$ ($\times 10^3$)
1955	641.4	2492	0.000375	0.239	595.6	1.077
1956	712.1	2573	0.000558	0.356	916.0	0.777
1957	602.8	2461	0.000659	0.421	1036.1	0.591
1958	510.0	2254	0.000409	0.261	588.3	0.867
1959	659.1	2652	0.000487	0.311	824.8	0.799
1960	529.8	1999	0.000629	0.402	803.6	0.659
1961	575.9	2296	0.000327	0.209	480.0	1.200
1962	537.7	2260	0.000778	0.497	1123.2	0.479
1963	346.9	2277	0.000909	0.580	1320.7	0.263
1964	269.2	1839	0.001209	0.772	1419.7	0.190
1965	273.4	1800	0.001566	1.000	1800.0	0.152
1966	219.6	1435	0.001757	1.122	1610.1	0.136
1967	193.5	1346	0.001559	0.996	1340.6	0.144
1968	234.8	1228				0.191[e]
1969	161.6	1010				0.160[e]
1970	259.3	834				0.311[e]
1971	250.3	865				0.289[e]

[a] $Q_t = F_t/f_t^*$; F_t values are from Table 3.

[b] Relative $Q_t = Q_t/Q_{1965}$.

[c] $f_t = f_t^* \cdot Q_t/Q_{1965}$.

[d] CPUE$_t = C_t/f_t$.

[e] Catch-per-unit-effort (unadjusted) for 1968–1971.

Table 5. Age-specific total instantaneous mortality rate for year t, $Z_{x,t}$, and weighted total instantaneous mortality rate for year t, Z_t.[a]

Year (t)	$Z_{x,t}$ for ages (x) 1 through 7							Weighted Z_t
	1	2	3	4	5	6	7	
1955	1.128	1.627	2.217	1.143	0.713	0.880	0.352	1.209
1956	1.533	2.357	1.630	0.223	1.529	2.031	1.933	1.676
1957	0.402	1.192	1.158	0.925	0.929	1.458	1.727	1.103
1958	1.335	1.776	1.106	0.712	0.592	1.030	1.352	1.169
1959	1.567	2.385	1.309	0.314	0.375	1.630	1.920	1.445
1960	0^b	0.087	0.866	0.731	1.602	1.769	1.770	0.930
1961	1.837	1.691	1.900	0.378	1.028	0.809	0.996	1.317
1962	0.830	2.080	1.737	2.252	1.244	2.162	1.834	1.707
1963	1.242	2.212	1.994	1.818	2.141	2.146	2.415	1.969
1964	1.749	2.289	2.163	2.531	2.099	2.429	3.167	2.307
1965	1.482	2.467	2.886	3.382	2.689	2.295	2.807	2.491
1966	1.535	1.531	1.646	1.886	3.428			1.900
Average age-specific mortality (Z_x)	1.220	1.808	1.718	1.358	1.531	1.694	1.843	1.602
Inverse of standard deviation ($1/\sigma_{Z_x}$)	1.804	1.484	1.770	0.993	1.092	1.718	1.267	

$^a Z_{x,t}$ is calculated using Eq. (7) and values for f_t from Table 4 and values of $C_{x,t}$ from Table 2. Z_t is calculated as

$$\left(\sum_{x=1}^{7} Z_x/\sigma_{Z_x}\right) \bigg/ \sum_{x=1}^{7} 1/\sigma_{Z_x}.$$

$^b Z_{1,1960} = -0.109$, which implies a probability of survival greater than 1; $Z_{1,1960}$ was set equal to 0 and was not included in the calculation of the average age-specific mortality, the inverse of the standard deviation, or Z_{1960}.

Table 6. Effective effort $(\bar{f}_t)$, Paloheimo's (1961) weighted effective effort $(\bar{f}_t)$, adult total instantaneous mortality rate (Z_t), and adult probability of survival (p_t) by year

The mean $(\hat{\mu})$ and standard deviation $(\hat{\sigma})$ for each variable and the slope (v), y-intercept (u), and correlation coefficient $(\hat{\rho})$ for the functional regression of Z_t on $\bar{f}_t$ are given below the double line.

Year (t)	$f_t{}^a$	$\bar{f}_t{}^a$	$Z_t{}^b$	$p_t{}^c$
1955	595.6	629.0	1.209	0.298
1956	916.0	859.2	1.676	0.187
1957	1036.1	802.6	1.103	0.332
1958	588.3	598.1	1.169	0.311
1959	824.8	736.2	1.445	0.236
1960	803.6	629.4	0.930	0.395
1961	480.0	618.5	1.317	0.268
1962	1123.2	1068.2	1.707	0.181
1963	1320.7	1217.3	1.969	0.140
1964	1419.7	1388.0	2.307	0.100
1965	1800.0	1564.9	2.491	0.083
1966	1610.1	1370.9	1.900	0.150
1967	1340.6			
$\hat{\mu}$	1043.2	956.9	1.602	(0.201)
$\hat{\sigma}$	422.7	350.0	0.493	
v			0.0014087	
u			0.25404	
$\hat{\rho}$			0.9376	

$^a f_t$ and $\bar{f}_t$ are measured in units of 1965 vessel-weeks; f_t values are from Table 4; $\bar{f}_t$ values are calculated from Eq. (9).

$^b Z_t$ values are from Table 5.

$^c p_t = \exp(-Z_t)$.

Age-specific sex ratios, q_x, for the Atlantic menhaden can be found in Reintjes (1969, p. 12). The length-age relationship based on von Bertalanffy's growth function can also be found in Reintjes (1969, p. 10) for both females (Fig. 2) and males.

Total instantaneous mortality rates, Z, as a function of fishing effort can be determined from catch, effort, and cohort data. The probability of survival can be directly computed from the total instantaneous mortality rate by the relationship,

$$p = \exp(-Z). \tag{3}$$

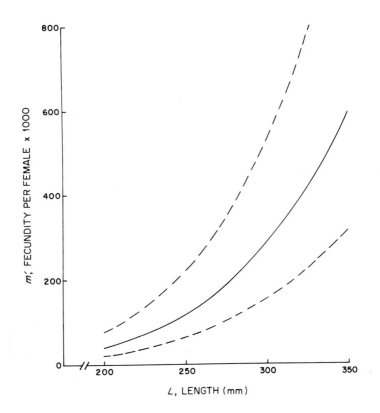

Fig. 1. Fecundity per female is presented as a function of length (mm) based on the least-squares regression: $m' = (3.666 \times 10^{-7})L^{4.7995}$. The 95% confidence interval is indicated by the dashed lines.

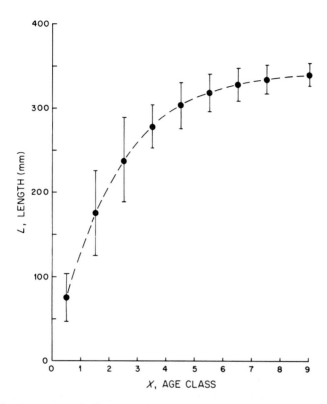

Fig. 2. The von Bertalanffy growth function for female Atlantic menhaden is shown, as well as the approximated 95% confidence interval for each age class (at the midpoints). The equation for the von Bertalanffy growth function is $L = 345\left\{1 - \exp[-0.464(x + 0.032)]\right\}$.

A recent technical report from NOAA (Nicholson 1975) updates the catch statistics in Schaaf and Huntsman (1972). Table 2 gives the number of fish caught by purse seines by age from 1955 through 1971. Using these data after the manner of Schaaf and Huntsman (1972), the maximum age-specific exploitation rates in year t, $u_{x,t}$, and maximum instantaneous fishing mortality rates, F_t, were determined for the years 1955 through 1967 (see Table 3); that is,

$$u_{x,t} = C_{x,t}\bigg/\sum_{j=0}^{k-x} C_{x+j,\,t+j} \, , \qquad (4)$$

where $C_{x,t}$ is the catch of age-x fish in year t, k equals 8 (the maximum age class in the landings), and the denominator of the right-hand expression represents the virtual population of an age class in year t. Note also that

$$F_t = -\ln(1 - \bar{u}_t) . \qquad (5)$$

To account for the variability in effort, a relative catchability coefficient for year t, Q_t, was calculated from

$$Q_t = F_t / f_t^* , \qquad (6)$$

where F_t is again the maximum instantaneous fishing mortality rate and f_t^* is the actual effort in vessel-weeks expended in year t. Effective effort in year t, f_t, is equal to the actual effort, f_t^*, times Q_t divided by Q for 1965, thus converting effective effort to 1965 vessel-weeks. Effective effort in year t, f_t, and catch-per-unit-effective-effort in year t, CPUE_t, are presented in Table 4.

The age-specific total instantaneous mortality rate in year t, $Z_{x,t}$, was calculated from

$$Z_{x,t} = \ln\left[(C_{x,t}/f_t)/(C_{x+1,t+1}/f_{t+1})\right] , \qquad (7)$$

where again f_t is the effective effort for year t (values given in Table 4) and $C_{x,t}$ is the catch of age class x in year t (values given in Table 2). Since age class 1 was not fully recruited into the catchable stock, and since Schaaf and Huntsman (1972) suggest that the Atlantic menhaden are recruited at age 1.5, an inflation factor was determined for the catch of age class 1. Assuming the landings by age are proportional to the actual age structure, then the proportion of age class 1 caught, PC, is

$$\text{PC} = \int_{1.5}^{2.0} \exp(-F_t)dt \Big/ \int_{1.0}^{2.0} \exp(-F_t)\, dt = \frac{1 - \exp(0.5\,F)}{1 - \exp(F)}. \qquad (8)$$

Assuming an instantaneous fishing mortality rate, F, equal to 1.05, which is the average of the estimates by Dryfoos et al. (1973; $F = 0.95$) and Schaaf and Huntsman (1972; $F = 1.15$), PC is then equal to 0.372. The inflation factor for age class 1 is equal to the inverse of the proportion caught or 2.690.

An analysis of variance was performed on the age-specific total instantaneous mortality rates, $Z_{x,t}$, with the result that significant differences ($\alpha = 0.001$) were found between years, but not between age classes 1 through 7 ($\alpha = 0.05$). In calculating the average total instantaneous mortality rate for year t, Z_t, each age-specific total instantaneous mortality rate in year t was weighted by the inverse of the standard deviation of the average age-specific total instantaneous mortality rate, Z_x (Table 5). This procedure gives the most weight to that age class having the smallest year-to-year fluctuation.

A review of the literature was made regarding the appropriate regression for the yearly total instantaneous mortality rate, Z_t, on the yearly effective effort, f_t. Paloheimo (1961) describes a new linear formula which he found preferable to the Beverton and Holt (1956) iterative procedure. In this method, Z_t is regressed against a weighted effective effort (or just "weighted effort"), $\bar{f}_t$, where

$$\bar{f}_t = (1 - a)f_t + af_{t+1} ,$$

$$a = \frac{-\sum \ln[1 - \exp(-Z_j)]/Z_j}{\sum Z_j} .$$

(9)

The summations used in calculating a are over the range of estimated Z_t's in increments of 0.1, thus giving a weighted average of the effective effort between the years t and $t + 1$. Table 6 presents $\bar{f}_t$ values based on Nicholson's (1975) data where $a = 0.29$.

The functional regression (Teissier 1948), which minimizes the sum of the products of the vertical and horizontal distances of each point from the regression line, was used with Paloheimo's linear formula, due to the significant error in the independent variable $\bar{f}_t$; the correlation coefficient was 0.938. The y-intercept (0.25404) yields an estimate of the natural instantaneous mortality rate for adult menhaden (Fig. 3; Table 2). The total instantaneous mortality rate, Z_t, as a function of weighted effective fishing effort is

$$Z_t = 0.25404 + 0.0014087\bar{f}_t .$$

(10)

In order to apply Monte Carlo simulation techniques, it is necessary to have estimates of the standard deviation for each of the age-specific input variables (see Table 7). The standard deviation in estimating the sex ratio for each age class, q_x, which is binomially distributed, can be approximated from

$$\hat{\sigma}(q_x) = \sqrt{q_x(1 - q_x)/n_x} ,$$

(11)

where n_x is the number of fish sexed in age class x. The standard deviation involved in predicting an individual value of the dependent variable, Y, of a least-squares regression based on n observations for a given value of the independent variable, X, is

$$\hat{\sigma}(\hat{Y}) = \hat{\sigma}_{y/x}\sqrt{1 + \frac{1}{n} + \frac{(X - x)^2}{(n - 1)\hat{\sigma}_x^2}} ,$$

(12)

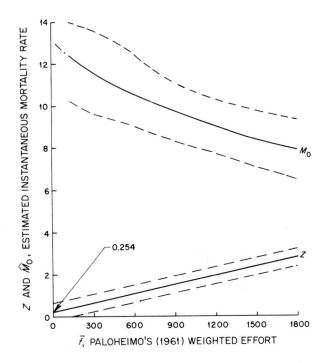

Fig. 3. The adult total instantaneous mortality rate, Z, and its 95% confidence interval (functional regression) is shown as a function of Paloheimo's (1961) weighted effort, $\bar{f}$. Also presented are the estimates of instantaneous natural mortality rate for age class 0, $\hat{M}_0$, and their empirically derived confidence interval as a function of $\bar{f}$. The functional regression for the adult total instantaneous mortality rate vs Paloheimo's (1961) weighted effective effort is $Z = 0.25404 + 0.0014087 f$.

Table 7. Standard deviation for input parameters in Monte Carlo simulation study

Age class (x)	Standard deviation of q ($\times 10^{-3}$)	Standard deviation of L (mm)	Standard deviation of $\ln(m)$
1	2.182	25.8	
2	1.941	25.8	0.306
3	2.822	12.5	0.303
4	3.991	14.2	0.304
5	5.389	11.7	0.306
6	8.090	10.0	0.307
7	15.68	8.3	0.309
8+	31.04	6.7	0.309

Effort (f)	Standard deviation of effort ($\hat{\sigma}_f$)	Standard deviation of mortality ($\hat{\sigma}_Z$)
0	306.433	0.190
300	264.311	0.180
600	233.999	0.174
900	220.431	0.171
1200	226.636	0.172
1500	251.152	0.177
1800	289.362	0.186
804.190	222.681	0.171
646.776	230.663	0.173

where

$$\hat{\sigma}_{y/x} = \sqrt{\frac{\sum(\hat{y}_i - y_i)^2}{n-2}} \; ,$$

$$\hat{\sigma}_x^{\;2} = \frac{\sum(x_i - \bar{x})^2}{n-1} \; .$$

The corresponding estimated standard deviation, $\hat{\sigma}(\hat{Z})$, for the functional regression (Teissier 1948, Ricker 1973) of yearly total instantaneous mortality rate, Z_t, on yearly weighted effective effort, $\bar{f}_t$, is

$$\hat{\sigma}(\hat{Z}) = \sqrt{\hat{\sigma}_Z^{\;2}(1 - \hat{\rho}^2) + v^2(1 - \hat{\rho})^2(\bar{f}_t - \hat{\sigma}_{\bar{f}})^2} \; , \qquad (13)$$

where v is the slope of the functional regression line, $\hat{\rho}$ is the correlation coefficient for the regression, and $\hat{\sigma}_Z$ and $\hat{\sigma}_{\bar{f}}$ are the standard deviations for Z_t and $\bar{f}_t$, respectively.

The estimated standard deviation for the length-age class relationship was approximated from data given by Nicholson (*personal communication*). Assuming normality, the range in length (in millimeters) for each age class is approximately six standard deviation units wide. Thus dividing the range in length for each age class by six gives an estimate of the standard deviation. Finally, the standard deviation for the weighted effective effort, $\hat{\sigma}_{\bar{f}}$, was approximated by applying Eq. (6) to the least-square regression ($\hat{\rho} = 0.817$) of Paloheimo's (1961) weighted effective effort on year, t.

The resultant input formulations, assuming Gaussian distributed error terms, used in the Monte Carlo simulations were

$$\bar{f}^* = \bar{f} + \epsilon_1 , \tag{14}$$

$$Z = 0.25404 + 0.0014087\bar{f}^* + \epsilon_2 , \tag{15}$$

$$q_x^* = q_x + \epsilon_3 , \tag{16}$$

$$L = 345 \left\{ 1 - \exp\left[-0.464(x + 0.032) \right] \right\} + \epsilon_4 , \tag{17}$$

$$m' = (3.666 \times 10^{-7})L^{4.7995} \exp \epsilon_5 . \tag{18}$$

For each simulation experiment all age classes were treated independently. Each error term (ϵ_i) was assumed normally distributed with mean zero and standard deviation as described in the previous paragraph. GGNOR from the International Mathematical and Statistical Languages (IMSL) package was used to generate normal deviates from the standard normal distribution. The standard normal deviate was then multiplied by the estimated standard deviation and added to the expected value of the dependent variable. Where the weighted effort, $\bar{f}$, or total instantaneous mortality rate, Z, might become negative, they were set to zero, thus skewing the underlying distribution for low values of effort and total instantaneous mortality rate.

Simulation runs, of 5000 experiments each, were made, varying weighted effort, $\bar{f}$, and its corresponding standard deviation, $\hat{\sigma}_{\bar{f}}$, the intrinsic growth rate of the population, r, the standard deviation of the intrinsic growth rate, $\hat{\sigma}_r$, and a factor (denoted FACTOR) multiplying all inputted standard deviations for length, fecundity, and mortality. An additional simulation run of 25,000 experiments was made for a particular set of conditions.

EMPIRICAL RESULTS

An initial series of runs was made varying weighted effort from 0 to 1800 in increments of 300. A stable intrinsic growth rate ($r = 0$) was assumed, and runs were made with $\hat{\sigma}_r$ equal to 0.0 and 0.1. The value of $\hat{\sigma}_r = 0.1$ was arbitrarily determined. Table 8 presents the deterministic ($M_0{}'$) and

Table 8. The deterministic (M_0') and nondeterministic ($\hat{M}_0$) estimates of the natural instantaneous mortality rate for age class 0 (M_0), the lower (M_L) and upper (M_U) bounds of the empirically derived 95% confidence intervals, the bias ($\hat{M}_0 - M_0'$), and the standard deviation of M_0 ($\hat{\sigma}_M$) are given for various inputted weighted efforts (f)

The intrinsic growth rate of the population (r) is zero and its corresponding standard deviation ($\hat{\sigma}_r$) is 0.0 or 0.1. The mean simulated effort ($\bar{\bar{f}}$) is also given.

$\bar{f}$	$\bar{\bar{f}}$	M_0'	M_L	$\hat{M}_0$	M_U	BIAS	$\hat{\sigma}_M$
			$\hat{\sigma}_r = 0.0$				
0	127	12.281	10.387	12.381	13.488	0.100	0.855
300	321	11.396	9.750	11.551	13.280	0.155	0.980
600	604	10.419	9.090	10.533	12.290	0.114	0.799
900	901	9.637	8.497	9.706	10.987	0.069	0.631
1200	1194	9.004	7.846	9.056	10.273	0.052	0.618
1500	1495	8.435	7.227	8.468	9.656	0.033	0.626
1800	1801	7.910	6.533	7.922	9.222	0.012	0.682
			$\hat{\sigma}_r = 0.1$				
0	124	12.264	10.260	12.412	14.028	0.148	0.965
300	313	11.414	9.625	11.598	13.620	0.184	1.064
600	600	10.418	9.060	10.557	12.475	0.139	0.856
900	906	9.614	8.349	9.692	11.057	0.078	0.695
1200	1196	8.991	7.764	9.050	10.361	0.059	0.659
1500	1506	8.404	7.107	8.445	9.743	0.041	0.672
1800	1801	7.899	6.480	7.926	9.318	0.027	0.724

nondeterministic ($\hat{M}_0$) estimates of the natural instantaneous mortality rate for age class 0, M_0, as well as the lower, M_L, and upper, M_U, bounds of the 95% confidence interval read off the cumulative frequency distribution of $\hat{M}_0$ values from 5000 simulation experiments. The third place beyond the decimal point was determined by linear interpolation. The bias ($\hat{M}_0 - M_0'$) and the standard deviation of $\hat{M}_0$, $\hat{\sigma}_M$, are also given. Note that for weighted effort of zero, the actual mean of the simulated efforts, $\bar{f}$, is about 125, which results from setting negatively simulated efforts equal to zero.

A functional fit (Teissier 1948) of catch (metric tons) to Paloheimo's (1961) weighted effort was calculated using the quadratic regression model forced through the origin ($\hat{\rho} = -0.969$):

$$C = 1530.068\bar{f} - 0.95131\bar{f}^2. \tag{19}$$

A value of $\bar{f}$ equal to 804.190 maximizes the resultant catch for this relationship. Using a $\hat{\sigma}_{\bar{f}}$ equal to 222.681, the above weighted effort was used in a series of runs, which are presented in Table 9. Initially, the error in the intrinsic growth rate, $\hat{\sigma}_r$, was changed. Next, the intrinsic growth rate, r, itself

Table 9. The deterministic ($M_0{}'$) and nondeterministic ($\hat{M}_0$) estimates of the natural instantaneous mortality rate for age class 0 (M_0), the lower (M_L) and upper (M_U) bounds of the empirically derived 95% confidence interval, the bias ($\hat{M}_0 - M_0{}'$), and the standard deviation of M_0 ($\hat{\sigma}_M$) for that effort which maximizes the quadratic fit of catch vs weighted effort (functional regression)

$\bar{\bar{f}}$		$M_0{}'$	M_L	$\hat{M}_0$	M_U	BIAS	$\hat{\sigma}_M$
	$\hat{\sigma}_r$			$r = 0$			
799	0.01	9.885	8.722	9.961	11.347	0.076	0.675
804	0.05	9.872	8.634	9.955	11.392	0.083	0.693
800	0.10	9.873	8.593	9.970	11.540	0.097	0.750
806	0.50	9.533	7.123	10.105	14.110	0.572	1.747
	r			$\hat{\sigma}_r = 0.1$			
806	−1.0	13.411	11.438	13.631	16.685	0.220	1.345
800	−0.5	11.405	9.960	11.566	13.703	0.161	0.951
800	0.0	9.873	8.593	9.970	11.540	0.097	0.750
805	+0.5	8.554	7.281	8.610	9.908	0.056	0.663
805	+1.0	7.392	6.134	7.423	8.635	0.031	0.641
	FACTOR			$r = 0$ and $\hat{\sigma}_r = 0.1$			
802	2.00	9.860	8.203	10.051	11.874	0.191	0.929
800	1.00	9.873	8.593	9.970	11.540	0.097	0.750
805	0.50	9.860	8.708	9.939	11.395	0.079	0.684
805	0.25	9.860	8.752	9.932	11.355	0.072	0.667
808	0.10	9.849	8.782	9.917	11.317	0.068	0.646

was varied. Finally, a multiplicative factor, FACTOR, ranging from 0.10 to 2.00 was introduced. The standard deviations of fecundity, length, and mortality were multiplied by FACTOR.

A similar sequence of runs was made based on the functional fit (Teissier 1948) of catch on effort using the Gompertz regression model ($\hat{\rho} = -0.963$):

$$C = 6932.394\bar{f} - 927.777\bar{f}(\ln \bar{f}),\qquad(20)$$

where the maximizing effort is $\bar{f} = 646.776$. The results of these runs are presented in Table 10 using $\hat{\sigma}_{\bar{f}} = 230.663$.

The confidence intervals given in Tables 8 through 10 are based on *one* real world observation where the standard deviation of $\hat{M}_0$, $\hat{\sigma}_M$, estimated via simulation runs is assumed to be the population standard deviation for the given sets of conditions. Given the effort which maximizes the quadratic curve for catch vs effort (Eq. 13) and given the data in Table 9, the width of the confidence interval from the lower bound to the predicted mortality rate

Table 10. The deterministic (M_0') and nondeterministic ($\hat{M}_0$) estimates of the natural instantaneous mortality rate for age class 0 (M_0), the lower (M_L) and upper (M_U) bounds of the empirically derived 95% confidence interval, the bias ($\hat{M}_0 - M_0'$), and the standard deviation of M_0 ($\hat{\sigma}_M$) for that effort which maximizes the Gompertz fit of catch vs weighted effort (functional regression)

$\bar{\bar{f}}$		M_0'	M_L	$\hat{M}_0$	M_U	BIAS	$\hat{\sigma}_M$
	$\hat{\sigma}_r$			$r = 0$			
647	0.01	10.293	9.055	10.395	12.015	0.102	0.752
644	0.05	10.298	8.982	10.407	12.103	0.109	0.783
645	0.10	10.293	8.983	10.425	12.220	0.132	0.821
648	0.50	9.934	7.490	10.580	15.050	0.646	1.884
	r			$\hat{\sigma}_r = 0.1$			
648	−1.0	14.343	11.941	14.562	18.015	0.219	1.548
648	−0.5	11.987	10.273	12.188	14.805	0.201	1.154
645	0.0	10.293	8.983	10.425	12.220	0.132	0.821
649	+0.5	8.889	7.598	8.972	10.420	0.083	0.703
648	+1.0	7.671	6.374	7.713	8.995	0.042	0.656
	FACTOR			$r = 0$ and $\hat{\sigma}_r = 0.1$			
645	2.00	10.283	8.595	10.495	12.590	0.212	0.992
645	1.00	10.293	8.983	10.425	12.220	0.132	0.821
652	0.50	10.267	9.043	10.369	12.100	0.102	0.778
647	0.25	10.285	9.071	10.380	12.147	0.095	0.779
651	0.10	10.267	9.073	10.360	12.042	0.093	0.756

($M_{0,L} = \hat{M}_0 - M_L$) and the width of the confidence interval from the predicted mortality rate to the upper bound ($M_{0,U} = \hat{M}_U - M_0$) were calculated (Table 11). Thus, if a series of estimates by year, or by some other time interval, can be obtained for age-specific sex ratios, length, fecundity, and mortality, then the respective half-interval widths can be reduced by dividing by $\sqrt{n}$, where n is the length of the series of estimates. The resultant reduced half-interval widths can then be subtracted and added to the predicted mortality rate, that is,

$$M'_{0,L} = \hat{M}_0 - M_{0,L}/\sqrt{n}, \tag{21}$$

$$M'_{0,U} = \hat{M}_0 + M_{0,U}/\sqrt{n}. \tag{22}$$

Two separate runs were made for the effort maximizing the quadratic curve of catch vs effort ($\bar{f} = 804.190$; Eq. 19) with $\hat{\sigma}_{\bar{f}}$ equal to 222.681, r equal to 0.0, and $\hat{\sigma}_r$ equal to 0.1. These simulation runs were for 5000 and

Table 11. The lower $(M_{0,L})$ and upper $(M_{0,U})$ half-interval widths derived empirically, and the half-interval width based on the normal approximation $(1.96\ \hat{\sigma}_M)$ for 95% confidence intervals. Simulation runs based on $\bar{f} = 804.190$ and $\hat{\sigma}_{\bar{f}} = 222.681$.

$\bar{f}$		$M_{0,L}$	$M_{0,U}$	$1.96\ \hat{\sigma}_M$
	$\hat{\sigma}_r$	$r = 0$		
799	0.01	1.239	1.386	1.323
804	0.05	1.321	1.437	1.359
800	0.10	1.377	1.570	1.471
806	0.50	2.982	4.005	3.425
	r	$\hat{\sigma}_r = 0.1$		
806	−1.0	2.193	3.054	2.636
800	−0.5	1.606	2.137	1.864
800	0.0	1.377	1.570	1.471
805	+0.5	1.329	1.298	1.300
805	+1.0	1.289	1.212	1.257
	FACTOR	$r = 0$ and $\hat{\sigma}_r = 0.1$		
802	2.00	1.848	1.823	1.821
800	1.00	1.377	1.570	1.471
805	0.50	1.231	1.456	1.340
805	0.25	1.180	1.423	1.307
808	0.10	1.135	1.400	1.265

25,000 experiments respectively. The half-interval widths were determined directly from the frequency distribution for a single estimate of M_0. Equations (21) and (22) were then used to calculate the confidence intervals for an increasing number of estimates of M_0. The results of these calculations are presented in Table 12 for the two simulation runs of 5000 and 25,000 experiments.

The bias $(\hat{M}_0 - M_0{}')$, a constant deflection of the estimator from the parameter to be estimated, increases with increasing $\hat{M}_0$ (Tables 8, 9, and 10). In addition, the bias decreases with increasing intrinsic growth rate, r, and increases both with increasing error in the intrinsic growth rate, $\hat{\sigma}_r$, and with increasing multiplicative factor, FACTOR (Tables 8, 9, and 10). A step-wise regression based on the data given in Tables 8, 9, and 10 resulted in the following multiple regression equation ($\hat{\rho} = 0.99$):

$$\text{BIAS} = -0.19248 + 0.02951\hat{M}_0 - 0.04650\lambda + 1.44997\hat{\sigma}_\lambda{}^2$$

$$+ 0.02942(\text{FACTOR})^2 , \tag{23}$$

Table 12. Half-interval widths derived empirically and based on a normal approximation with increasing estimates of M_0 for runs of 5000 and 25,000 simulated experiments

Simulation runs based on $f = 804.190$, $\hat{\sigma}_f = 222.681$, $r = 0$, and $\hat{\sigma}_r = 0.1$

Number of estimates of M_0 (n)	Number of simulation runs					
	N = 5000			N = 25,000		
	$M_{0,L}/\sqrt{n}$	$M_{0,U}/\sqrt{n}$	$1.96\,\hat{\sigma}_M/\sqrt{n}$	$M_{0,L}/\sqrt{n}$	$M_{0,U}/\sqrt{n}$	$1.96\,\hat{\sigma}_M/\sqrt{n}$
1	1.377	1.570	1.471	1.343	1.594	1.451
5	0.616	0.702	0.658	0.601	0.713	0.649
10	0.435	0.496	0.465	0.425	0.504	0.459
20	0.308	0.351	0.329	0.300	0.356	0.324
30	0.251	0.287	0.269	0.245	0.291	0.265
40	0.218	0.248	0.233	0.212	0.252	0.229
50	0.195	0.222	0.208	0.190	0.225	0.205
100	0.138	0.157	0.147	0.134	0.159	0.145
500	0.062	0.070	0.066	0.060	0.071	0.065
1,000	0.044	0.050	0.047	0.042	0.050	0.046
5,000	0.019	0.022	0.021	0.019	0.023	0.021
10,000	0.014	0.016	0.015	0.013	0.016	0.015
25,000	0.009	0.010	0.009	0.009	0.010	0.009
∞	0	0	0	0	0	0

where $\lambda = \exp r$, the maximal eigenvalue of the Leslie matrix, and $\hat{\sigma}_\lambda^2$ is the variance associated with λ.

Since a deterministic M_0' is actually the result of nondeterministic processes occurring in nature, M_0' may be a biased estimate of the true natural instantaneous mortality rate for age class $0, M_0$. Therefore, an improved estimator of M_0 would be $M_0^* = M_0' - \text{BIAS}$. The new lower (M_L^*) and upper (M_U^*) bounds of the 95% confidence interval for a single estimate of M_0^* would then be

$$M_L^* = M_0^* - M_{0,L} \,,$$

$$M_U^* = M_0^* + M_{0,U} \,. \tag{24}$$

These new confidence intervals for the data in Table 9 are summarized in Table 13.

NORMAL APPROXIMATION

As the number of estimates of M_0 increases, the frequency distribution of M_0 approaches in probability the normal distribution. The 95% confidence interval about the sample mean is

$$(\overline{M}_0 - 1.96\sigma_M/\sqrt{n} \,,\, \overline{M}_0 + 1.96\sigma_M/\sqrt{n}) \,, \tag{25}$$

Table 13. Empirical and normally approximated 95% confidence intervals corrected for bias

Simulation runs based on $\bar{f} = 804.190$, and $\hat{\sigma}_{\bar{f}} = 222.681$.

f		$M_0{}^*$	Empirical		Normal	
			$M_L{}^*$	$M_U{}^*$	$M_L{}^*$	$M_U{}^*$
	$\hat{\sigma}_r$		$r = 0$			
799	0.01	9.809	8.570	11.195	8.486	11.132
804	0.05	9.789	8.468	11.226	8.430	11.148
800	0.10	9.776	8.399	11.346	8.305	11.247
806	0.50	8.961	5.979	12.966	5.536	12.386
	r		$\hat{\sigma}_r = 0.1$			
806	−1.0	13.191	10.998	16.245	9.555	15.827
800	−0.5	11.244	9.638	13.381	9.380	13.108
800	0.0	9.776	8.399	11.346	8.305	11.247
805	+0.5	8.498	7.169	9.796	7.198	9.798
805	+1.0	7.361	6.072	8.573	6.104	8.618
	FACTOR		$r = 0$ and $\hat{\sigma}_r = 0.1$			
802	2.00	9.669	7.821	11.492	7.848	11.490
800	1.00	9.776	8.399	11.346	8.305	11.247
805	0.50	9.781	8.550	11.237	8.441	11.121
805	0.25	9.788	8.608	11.211	8.481	11.095
808	0.10	9.781	8.646	11.181	8.516	11.046

where $\overline{M}_0$ is the mean of the estimates of M_0, σ_M is the true population standard deviation (estimated here by $\hat{\sigma}_M$ based on 5000 simulation experiments), and n is the number of estimates of M_0. The number 1.96 is the appropriate multiple of the true population standard deviation of a normal distribution that results in a 95% confidence interval. Tables 8, 9, and 10 contain $\hat{\sigma}_M$ values for all of the simulation runs made. Table 11 contains a column headed by $1.96\hat{\sigma}_M$ which is one-half the width of the confidence interval for a single estimate of M_0 based on the normal distribution. Table 12 extends the results of Table 11 to an increasing number of estimates of M_0.

In Table 13 the confidence intervals based on the normal approximation have been corrected for bias in the same manner as the empirical confidence intervals. Note that for both $M_L{}^*$ and $M_U{}^*$ the normal approximation tends to result in lower values than does the empirical method. This result is due to the skewness in the underlying distribution of M_0.

As the number of estimates of M_0 increases (see Table 12), the frequency distribution of M_0 approaches normality, thus becoming more symmetrical.

Therefore if only a few estimates of M_0 are available, an empirical confidence interval would be preferable to a normal approximation.

SUMMARY

1. For a standard deviation of the intrinsic growth rate less than or equal to 0.1, the bias and width of the 95% confidence interval show only a slight corresponding increase, whereas increasing the standard error from 0.1 to 0.5 results in a considerable increase in the bias and width of the confidence interval.

2. Negative intrinsic growth rates ($r < 0$) result in a greater bias and wider confidence intervals for the estimator of M_0 as compared to a stable population ($r = 0$). On the other hand, positive intrinsic growth rates ($r > 0$) result in less bias and narrower confidence intervals.

3. Decreasing the multiplicative factor (i.e., decreasing the standard deviation of fecundity, length, and mortality of adults) does not greatly reduce the bias or width of the confidence intervals.

4. For a low number of estimates of M_0, the empirically derived confidence intervals are preferred over the normal approximation. In addition, the reduction in the width of the confidence intervals (empirical or normal approximation) is a function only of $\sqrt{n}$.

5. Estimates of M_0 should be "improved" by correcting for bias, which has been determined to be a function of the nondeterministic estimate of M_0 ($\hat{M}_0$), the natural logarithm of the intrinsic growth rate, the variance of the natural logarithm of the intrinsic growth rate, and the square of the multiplicative factor.

LITERATURE CITED

Beverton, R. J. H., and S. J. Holt. 1956. A review of methods for estimating mortality rates in exploited fish populations, with special reference to sources of bias in catch sampling. Rapp. Conseil Expl. Mer. **140**(1):67–83.

Dryfoos, R. L., R. P. Cheek, and R. L. Kroger. 1973. Preliminary analyses of Atlantic menhaden, *Brevoortia tyrannus*, migrations, population structure, survival and exploitation rates, and availability as indicated from tag returns. Fish. Bull. **71**(3):719–734.

Higham, J. R., and W. R. Nicholson. 1964. Sexual maturation and spawning of Atlantic menhaden. Fish. Bull. **63**(2):255–271.

Leslie, P. H. 1945. On the uses of matrices in certain population mathematics. Biometrika **33**:183–212.

————. 1948. Some further notes on the uses of matrices in population mathematics. Biometrika **35**:215–245.

Nicholson, W. R. 1975. Age and size composition of the Atlantic menhaden, *Brevoortia tyrannus*, purse seine catch, 1963–71. U.S. Fish Wildl. Serv. Rep., Fisheries No. 684. 28 p.

Paloheimo, J. R. 1961. Studies on estimation of mortalities. I. Comparison of a method described by Beverton and Holt and a new linear formula. J. Fish. Res. Board Can. 18(5):645–662.

Reintjes, J. W. 1969. Synopsis of biological data on the Atlantic menhaden, *Brevoortia tyrannus*. U.S. Fish Wildl. Serv. Circ. 320. 30 p.

Ricker, W. E. 1973. Linear regressions in fishery research. J. Fish. Res. Board Can. 30(3):409–434.

Schaaf, W. E., and G. R. Huntsman. 1972. Effects of fishing on the Atlantic menhaden stock: 1955–1969. Trans. Am. Fish. Soc. 101(2):290–297.

Schreiber, A. L. 1959. Monte Carlo methods as tools for system research. Office of Naval Research, Rept. No. 7, Contract Nonr. 2525-(00). 19 p.

Teissier, G. 1948. La relation d'allometrie sa signification statistique et biologique. Biometrics 4:14–48.

Van Winkle, W., B. W. Rust, C. P. Goodyear, S. R. Blum, and P. Thall. 1974. A striped-bass population model and computer programs. ORNL/TM-4578, ESD-643, Oak Ridge National Laboratory, Oak Ridge, Tennessee. 200 p.

Vaughan, D. S., and S. B. Saila. 1976. A method for determining mortality rates using the Leslie matrix. Trans. Am. Fish. Soc. 105(3):380–383.

Part III

Compensation and Stock-Recruitment Relationships

An Argument Supporting the Reality of Compensation in Fish Populations and a Plea to Let Them Exercise It

James T. McFadden

School of Natural Resources
University of Michigan
Ann Arbor, Michigan

> *Hope from him had almost drained;*
> *His children all had been entrained.*
> *The last striped bass in Hudson River*
> *Gave a pained, convulsive shiver.*
>
> *His civil rights had been infringed –*
> *On intake screen he hung impinged.*
> *But yet one chance to outwit fate:*
> *"I think I still can compensate!"*

Key words: compensation, density-dependent, density-independent, exploitation

INTRODUCTION

The concept of compensation is very important in assessing the impact of power plant operation on fish populations because compensation represents the capacity of a fish population to offset, in whole or in part, reductions in numbers caused by entrainment and impingement. Not everyone has rushed to embrace the concept, however. In fact, government regulatory agencies and their scientific staffs involved in power plant licensing proceedings have seemed reluctant to credit many fish populations with significant natural compensatory capacity and even more reluctant to make quantitative allowance for this capacity when evaluating power plant impact. An example, presented for its utility not as a criticism, is the original position of the U.S. Nuclear Regulatory Commission on Indian Point Unit 2 on the Hudson River in New York and later applied to the proposed Summit Power Station Units 1 and 2 in Delaware: "Examination of relevant information concerning striped bass populations has led the staff to conclude that density-dependent regulatory processes as described above are not operative in East Coast striped bass larval and juvenile populations, because the breeding stock is not sufficient to saturate the nursery areas to population levels at which such processes would be operative" (USAEC 1974, citing USAEC 1972). Some softening of this stand was reflected in the NRC staff's review of Indian Point Unit 3 (USNRC 1975) and its testimony for the Summit Power Station (Christensen et al.1975) for which it was maintained that the fishery operates in a compensatory manner but that natural compensatory processes should not be relied on to a major extent to offset power plant impact in heavily fished populations.

By contrast, I have always maintained, as a general proposition, that some natural compensatory capacity must be recognized as a logical necessity; that general observation of natural populations, a wide range of specific scientific studies, and the prevailing concepts of the field of fishery management all support this view; and that the operation of compensation must be taken into account in order to make realistic decisions in management of fish populations, whatever the source of disturbance or impact. Most people seem to intuitively expect that killing some fish will reduce the size of the population − a proposition which seems logical enough but is not always true. Upon a little reflection, I would argue that it should be at least equally obvious that a sustained removal of fish from a population will eventually result in reduced natural death rate or increased reproduction by the remainder.

This paper presents an argument in favor of the reality of compensation in fish populations − it is not a balanced examination of pros and cons. The ideas, at first developed in the relative tranquility of an academic environment, have been shaped through a succession of regulatory agency hearings in which too much open-mindedness has seemed the short road to intellectual extinction. They are called "adversary" proceedings for good reason. Hence, while trying to be fair and logical, my main goal has been to make converts rather than compromises. Success has been limited, but I remain optimistic.

HISTORICAL DEVELOPMENT OF THE GENERAL CONCEPT

The term "compensation" refers to the tendency of populations of living organisms to experience (a) an increase in death rate or decrease in birth rate as they grow in density, thus establishing some ultimate upper limit, and (b) a decrease in death rate or increase in birth rate as population density declines, thus leading to stabilization before extinction, or even to an eventual return to higher numbers.

The concept arises directly from recognition that living organisms depend on resources such as food and space, which are available in limited amounts. When density of organisms is low, competition for resources is slight; when density is high, competition becomes intense to the disadvantage of survival or reproduction. It is also possible that large populations will attract greater attention from predators, either because the latter congregate in the vicinity of a food source or reproduce more rapidly when prey is abundant or because predators become conditioned to seek the more abundant prey. Cannibalism also may contribute to compensation in some species because the large numbers of parents from which initially large broods arise also constitute a large pool of predators. The number of physical and biological forces that may play a role in compensation is very large, and the possible modes of operation are numerous and complex. As Cushing (1975, p. 237) states, "Any description of the phenomenon reveals the stability and leaves the possible mechanisms indistinct." This is not to say that no mechanisms of compensation can be identified; many indeed have been identified, and a number of examples will be cited in the following sections. However, of the number of mechanisms operating simultaneously (often with a high degree of inter-action) and sequentially through the various life-history stages (with intensity of operation at each stage determined in part by the population density established in the preceding stage) only a few may be statistically verifiable during a single period of observation. Undoubtedly too, the operation of one mechanism under certain environmental conditions may preempt the operation of other potentially effective mechanisms.

Thus a common state of affairs is to observe a population's stability as generated by compensation and to estimate the compensatory capacity while being unable to explain all or even very many of the mechanisms involved. This latter limitation has not prevented scientists from developing highly successful management programs grounded on the reality of compensation, as will be shown.

Because compensation involves changes in birth rates or death rates in response to changes in population density, the biological processes involved are often termed "density dependent." Thus, a density-dependent increase in mortality rate might be expected to result from a substantial increase in population density. Strictly speaking, only processes that reduce population growth at high densities and increase growth at low densities — "direct density-dependent processes"* — are compensatory. Nevertheless, the term "density dependence" is often used synonymously with "compensation."

The concept of compensation has been long and widely accepted. In 1798 Thomas Malthus published his famous *Essay On The Principle Of Population* (Malthus 1798) in which he maintained that, when unchecked, a population increases in a geometrical ratio; that the increase is always up to the

---

*Mortality may also take an *inverse* density-dependent form; that is, the mortality rate decreases with an increase in population density. This could happen where very abundant prey "saturate" the capacity of a predator to capture them, thereby causing a smaller fraction of the prey to be caught when they are more abundant. Other types of "protection in numbers" would bring about the same result. Inverse density-dependent mortality is not synonymous with compensation.

limits of the means of subsistence; and that further increase is prevented by war, famine, pestilence, etc. He termed the processes involved in regulation of population size the "struggle for existence." His observations, based on human populations, clearly encompassed the compensatory concept that population growth declines as population size increases. The ideas presented by Malthus were not entirely new; for example, Machiavelli had realized 275 years earlier that human populations in some areas might increase beyond the limits of subsistence and be checked by want and disease. In 1835, Quetelet, a Belgian statistician, concluded that a population's resistance to growth increases in proportion to the square of the rate of population growth (Quetelet 1835). In 1838, Verhulst, a student of Quetelet, published a short essay developing an equation describing the course of population increases in proportion to population density; his equation generated the S-shaped population growth curve so familiar today, the logistic curve. Nearly 100 years later, this formulation of population growth was rediscovered by Pearl and Reed (1920) and was soon shown to describe the performance of such diverse organisms as yeast, protozoa, fruit flies, and man. Pearl recognized the importance of density-dependent mortality and reproduction: "In general there can be no question that this whole matter of influence of density of population, in all senses, upon biological phenomena, deserves a great deal more investigation than it has had. All indications are that it is one of the most significant elements in the biological, as distinguished from the physical, environment of organisms" (Pearl 1930, p. 145).

In the last four decades, the concept of compensation has been debated, refined, and amplified. A well-balanced review is presented by Krebs (1972, p. 269–288). Major contributions to our understanding have come from studies of insects, fishes, birds, and mammals. Despite the long history of the concept and its simple, almost intuitive basis, its explicit formulation and modern emphasis are usually attributed to the Australian entomologist Nicholson (1933). His inspiration was later referred to by another notable population ecologist (Haldane 1953) as "a blinding glimpse of the obvious."

Thus, the emergence of the concept of compensation can be traced over the past 450 years. It seems to have originated in observations of local human populations "struggling for existence" in the face of limited resources. The first formulations were crude and partially incorrect. It was Pearl's work in the 1930s that established the generality of compensation among diverse types of living organisms. Common to all these perceptions was the realization that populations possessed a potential to increase at a much more rapid pace than was realized, except possibly when they were so small that the resources upon which they depended were, for all practical purposes, infinite in extent. To appreciate the reality, pervasiveness, and imminence of the operation of compensation within a population, one should look first to this "biotic potential" as it is sometimes called and then envision as a population grows in size "the absorption of the potential increase by innumerable checks (in the shape of mortality) . . . . The survivors flourish by the deaths of their brothers and sisters, and the stable numbers must be the result of a fine control of mortality, perhaps a density-dependent one" (Cushing 1975).

Some feel for the magnitude of biotic potential can be gained by artificially simplifying the life history of a reasonably representative fish. The striped bass, which has inspired endless volumes of testimony in power plant hearings, serves as a useful example here. Assume that each fish spawns once in its lifetime at age 6; half are females, each producing 700,000 eggs; each fish reaches 457 mm in length; and all the eggs survive to become mature fish. These assumptions cause the biotic potential to be seriously underestimated since some females spawn at an earlier age and many live to spawn again at older ages. Beginning with one spawning pair, such a striped bass population would grow to astronomical numbers by the end of three generations (18 years). At the end of the third generation, all the fish laid end to end would encircle the earth at the equator 500,000,000 times or would stretch from the earth to the sun and back again 60,000 times. This is the biotic potential, which is increasingly suppressed by density-related mortality factors as a striped bass population grows from some minimal initial density toward an upper limit set by the carrying capacity of the environment. In 1879–1881, 435 striped bass were transplanted from the east coast to San Francisco Bay; within 20 years, annual catches were 500 tons (Merriman 1951). Under minimal densities, the population grew explosively; within a relatively short time, numbers became large enough that population growth leveled off. The innate capacity for growth is the foundation of the compensatory capacity possessed by striped bass.

Approaching population processes of fish in general from this perspective, one envisions an established population as one in which survival or reproduction (or both) have been vastly suppressed — precisely for the reason that the population has become large. The natural factors that operate to suppress the biotic potential are many and are complexly intertwined — availability of food, predators, disease, and physical factors such as temperature. Many of them (probably all of them under certain conditions) have greater suppressive effect when the population is large than when it is small. If some new effect that kills off part of the population is introduced, it reduces the suppressive effect of many factors in the population's environment. As a consequence, survival rate or reproductive rate becomes higher — the population *compensates* in part for the reduction in size. When something causes a population to either increase or decrease in size, there is a tendency for eventual return to average size when the perturbation is removed. "Populations do not usually become extinct or increase to infinity. This is what is loosely termed the 'balance of nature'" (Krebs 1972).

Rather than being a fragile living system then, the population typically is vigorous and resilient. This sense was captured admirably by that eminent interpreter of marine science, the late Rachel Carson, who, in describing the life of the seacoast, wrote, "Whenever the sea builds a new coast, waves of living creatures surge against it, seeking a foothold, establishing their colonies. And so we come to perceive life as a force as tangible as any of the physical realities of the sea, a force strong and purposeful, as incapable of being crushed or diverted from its ends as the rising tide" (Carson 1955).

The 10-year span from the mid-1960s to the mid-1970s has seen an impressive and timely increase in public awareness of environmental problems. With this awakening has come an accurate perception of the fragility of the ecosystem, which may seem at first to conflict with the description of life as a force "as incapable of being crushed or diverted from its ends as the rising tide." The apparent conflict is easily explained. Populations of most living organisms have little means of coping with wholesale destruction of environmental resources upon which they depend. If specific foods or narrow temperature ranges to which they are highly adapted are destroyed, the population perishes. Likewise, exposure to toxic substances such as the many organic compounds introduced into the environment by man, with which a natural population has no evolutionary experience, is likely to prove disastrous. Most of the current environmental awareness is built on public recognition of these two classes of problems — wholesale destruction of environmental resources and the release of exotic toxic substances. A third class of man-caused problems — the imposition on a population of increased mortality that takes a form similar to natural predation — has an entirely different effect on most species. This is the kind of impact to which the population has been adapted by thousands or millions of years of evolutionary experience. The agent of mortality — predatory fish, commercial or sport fishermen, or power plants — is an indifferent matter from the standpoint of population response. When the population is reduced in numbers, the survival rate or reproductive rate among the remaining members tends to increase; a compensating response is generated. This is the reality upon which successful management of agriculture, forestry, wildlife, and fisheries is carried on today. The population has a measurable and often impressive capacity to persist in a healthy state in the face of deliberate removals by man. Populations of most species, while fragile when deprived of basic life requirements or exposed to exotic toxicants, are robust in the face of this predation-type mortality.

A GENERAL CASE ARGUMENT

A formal argument for the general operation of compensation in animal populations can be developed as follows. It is generally observed that populations fluctuate within some more or less well-described bounds; that is, they neither increase without limit nor commonly decline to extinction during the normal time span of human observation.

A useful simplification is to represent a population as persisting at or near some average level of abundance or equilibrium level represented by K_0 (Fig. 1). As a generalization, the birth rate is expected to decline and the death rate to increase as a population becomes larger and larger. Linear relationships are used to simplify the illustration, even though the real relationships would almost certainly be nonlinear. Death rate would reflect the combination of natural deaths and any deaths imposed on the population by activities of man. In Fig. 1, the equilibrium population density, K_0, is maintained, on the average, by the balance between the death rate (I_0, representing the prevailing natural death rate plus a 0-level of man-caused deaths) and the birth-rate characteristic of the population at density K_0. If a low level of man-induced mortality is added to the baseline natural mortality, the overall death rate would increase to a level I_L (Fig. 1). The population would then decline and the birth rate would consequently increase until a new equilibrium density had been reached (K_L) at which the birth rate equaled the new death rate I_L. Imposition of a still higher man-induced mortality would increase the total death rate to I_H, and the population would equilibrate eventually at a still lower average density K_H. Thus, one can think of the population as fluctuating through time around some average level of abundance determined by the overall death rate (Fig. 2). For a 0-level man-induced environmental

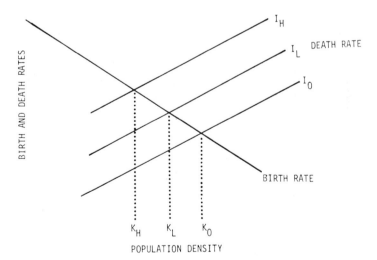

Fig. 1. Relationship of birth rates and death rates to population density. The death rate I_0 and the population density K_0 represent zero environmental impact; the death rate I_L and the corresponding population density K_L represent a low level of environmental impact superimposed on a fish population; the death rate I_H and the corresponding population density K_H represent a fish population subjected to a high level of environmental impact. In each case, the population equilibrates at that density at which the birth rate and death rate are equal.

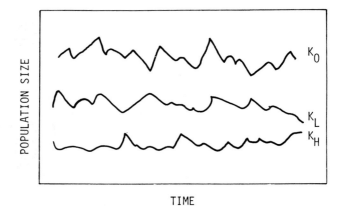

POPULATION SIZE

K_0

K_L

K_H

TIME

Fig. 2. Hypothetical population histories corresponding to zero, low, and high levels of environmental impact represented in Fig. 1.

impact, this would be a population density K_0; for a low level of environmental impact, a population density K_L; and for a high level of environmental impact, a still lower population density K_H.

This example was chosen to illustrate as simply as possible the relationship among death rate, birth rate, and average population density where compensation is operative. The increase in total mortality depicted in this example could have been offset at lower population levels by an increase in survival among members of the population not killed by the man-induced impact rather than by the increase in birth rate; the principle would be the same.

Referring again to the example in Fig. 1, the initial condition postulated is a population fluctuating around an average equilibrium level K_0; for this population, the birth rate (b) and death rate (d) on the average are equal. Thus, the rate of population increase (r) is 0:

$$r = b - d = 0 .$$

When the population is first subjected to additional mortality (no matter how slight) caused by man, the death rate is increased and exceeds the birth rate. The population's rate of growth (r) becomes negative — that is, the population declines in numbers:

$$b - d < 0 .$$

If the increment of mortality is sustained, the population eventually will dwindle to extinction unless the birth rate increases or the natural mortality rate decreases sufficiently to allow the overall birth and death rates to again become equal. This would be true if the removal is sustained over a long enough period, even if only a single organism per year is killed over and above the pre-impact natural mortality rate. Therefore, every population that is subjected to sustained additional mortality through man's activities and does not become extinct must possess some compensatory capacity. The list from everyday observation is impressively long. Species of birds and mammals commonly killed along highways — raccoons, squirrels, skunks, deer, pheasants, woodpeckers, sparrows — must have some compensatory capacity, or the new predator, the automobile, by now would have pursued them far down the trail toward extinction. The same can be said for pest insects attacked with weapons ranging from rolled newspapers to organic chemicals. Any wild plant or animal harvested by man for sport or subsistence — and many have been pursued since antiquity — must have some form of compensatory capacity to have survived. However, this compensatory capacity is not unlimited, as proven by the extinction or near extinction of a number of species caused by man's predatory activity. Ecologists now understand that compensatory capacity is most limited in species having relatively low maximum reproductive rates.

Simulation models are valuable aids in making decisions about management of complex systems, such as fish populations, and a variety of models have been used to estimate power plant impacts. The compensation argument just developed finds support in the realistic exercise of such simulation models. If a fish population model is operated from an initial state representing a population at a stable equilibrium or one fluctuating around an average equilibrium level and if removal of a single additional fish per year is simulated over a long period, the population will dwindle to extinction unless compensatory processes are simulated as well. The extinction case clearly is not a realistic simulation of commonly observed population performance. Further, because populations incorporate negative feedback processes under completely natural conditions, it does not suffice to confine compensatory mechanisms to those accounted for by the activities of man — such as density-dependent fishing. The general case argument requires that some *natural* compensatory process be operative to represent realistically and logically the performance of a population of living organisms.

COMPENSATION IN FISH POPULATIONS

Both the historical development of the general concept of compensation and the formal argument presented above support the view that compensation is operative in fish populations. This section supports this view with three additional lines of evidence:

- The historical development of the concept is traced through its applications in fishery management, and the present-day consensus held by the world's leading fishery scientists is summarized.

- A selection of compensatory mechanisms that have been convincingly demonstrated to operate in fish populations is reviewed.

- A large number of cases, in which substantial levels of exploitation by sport or commercial fisheries have been sustained by fish populations without serious depletion, are cited.

History of the Concept of Compensation in Fisheries

Historically, the formal foundation of modern fish population dynamics was laid down in the mid-1930s. Picking up the thread of historical development of the concept of compensation with the rediscovery of the logistic curve by Pearl and Reed (1920), the first explicit application of the concept in this form to fisheries seems to have been in 1933 when a parallel between the sigmoid population growth form in yeast and growth in numbers in a fish stock was noted (Hjort, Jahn, and Ottestad 1933). The first formal application of the logistic to management problems in a major fish stock was Graham's (1935) application to the plaice stock of the North Sea.

The logistic and its various modifications such as the "Schaefer model" (Schaefer 1954) have since been applied to such important fisheries as the Antarctic blue whale (Chapman 1964), Icelandic cod (Gulland 1961), Pacific halibut (Schaefer 1954), and Icelandic haddock (Gulland (1961). "The development of the logistic curve in fisheries and in other fields implied that loss of stock with increased mortality was compensated by increased recruitment" (Cushing 1974, p. 237). Thus, since the publication of Graham's work in 1935, the concept of compensation clearly has been a basic tenet of scientific management for the world's major fish stocks.

Interpretation and application of the logistic curve to fishery management is conveniently summarized in a recent book, *The Management of Marine Fisheries*, by J. A. Gulland, an internationally noted scientist with the Food and Agricultural Organization of the United Nations (Gulland 1974, p. 68–86). He makes the following points about compensation from the basis of the logistic model:

> Though the curves of Figures 3 and 10 and the model on which they are based are highly simplified descriptions of the changes in a fish stock under exploitation, they do illustrate most of the biological features important to fishery management. The first is that it is impossible to exploit a fish population without causing some change. This may seem obvious, but with the present day concern with the natural environment, and the desire to minimize ecological disturbance, there may be a feeling that a well-managed fishery should cause no changes, which is impossible . . . The second important point is almost the converse of this. That is, provided the catches are not too great, the decline in abundance is not continual. After a time

the population will reach a new equilibrium, at which the same catches can be maintained indefinitely year after year. Finally, if the stock is allowed to be depleted too far, though still without driving it to extinction, its productivity and the catches that can be taken will be reduced.

Ricker's commentary (1958, p. 250) on the same basic compensatory principle applied in fishery management is

> The principal reasons for lessened surplus production at higher stock densities are three:
>
> 1. Near maximum stock density efficiency of reproduction is reduced, and quite commonly the actual number of recruits is less than at smaller densities. In the latter event, reducing the stock will increase recruitment.
>
> 2. When food supply is limited, food is less efficiently converted to fish flesh by a large stock than by a smaller one. Each fish of the larger stock gets less food individually, hence a larger fraction is used merely to maintain life, and a smaller fraction is used for growth.
>
> 3. An unfished stock tends to contain more older individuals, relatively, than a fished stock. This makes for decreased production in at least two ways: (a) Larger fish tend to eat larger foods, so an extra step may be inserted in the food pyramid, with consequent loss of efficiency of utilization of the basic food production. (b) Older fish convert a smaller fraction of the food they eat into new flesh — partly, at least, because mature fish annually divert much substance to maturing eggs and milt.
>
> Under reasonably stable natural conditions, the net increase of an unfished stock is zero, at least on the average; its growth is balanced by natural deaths. Introducing a fishery increases production per unit of stock by one or more of the methods above and so creates a surplus which can be harvested. In these ways "a fishery, acting on a fish population, itself creates the production by which it is maintained" (Baranov). Notice that effects 1 and 3 above may often increase the *total* production of fish flesh by the population — it is not merely a question of diverting some of the existing production to the fishery, although that also occurs.

The logistic-type models have been used successfully in fishery management since their inception and are still used today, but they have long been recognized as embodying important basic principles in an over-simplified way. For example, they do not distinguish the contributions of growth of existing stock and recruitment of new individuals; they assume that the entire population is involved equally in compensatory response; they represent density-related changes in populations as occurring instantaneously and continuously.

A very important advance in conceptualizing fish population dynamics in a more realistic way is Ricker's (1954) exposition of the problem of stock and recruitment. Ricker emphasized the compensatory nature of the numerical relationship between parent fish and the progeny they produced and the importance of the earlier life history stages in compensation, and he formulated a mathematic model embodying these concepts, which has been applied to major fish stocks throughout the world. Ricker's model operates on the same basic principles given in the formal argument presented earlier in this paper and implicit in the logistic model — but in a much more refined and realistic way. It is explained here in graphical form because of its usefulness in clarifying the compensatory responses of fish populations to new increments of mortality (e.g., fisheries, power plant operations, and pollution).

Consider a parent stock of fish and the stock of progeny that it produces, expressing both the parents and progeny in the same units of measurement. If the *rate* of future replacement of the present population is independent of the *size* of the population, the relationship between parental stock and progeny will be described by a 45° diagonal line as shown in Fig. 3 (replacement reproduction), and this will be referred to as a *density-independent* relationship between parents and progeny. If environmental conditions permit survival of a very large parental stock, that stock will produce a generation of progeny equal in size to itself; by the same token, if unfavorable environmental conditions reduce the parental stock to some very low

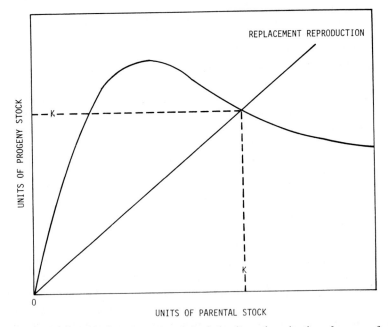

Fig. 3. Relationship between parental stock density and production of progeny for a hypothetical fish population.

density, it will produce again a generation of progeny equal to itself. Under these situations the size of the population could, by chance, increase without limit or dwindle to extinction; no compensatory process operates to increase the rate of population growth at low levels of density, thus deflecting it from decline to extinction, or to decrease the rate of population growth at very high levels of population density, thus deflecting it from unlimited expansion. To persist within some more or less well-defined limits of abundance, a fish stock must have some compensatory (density-dependent) processes.

The curve in Fig. 3 represents a density-dependent relationship between parental and progeny stocks. At very low levels of parental stock, the population tends to increase severalfold in the progeny generation. At point K, the parental stock is replaced by exactly the same size progeny stock (the reproduction curve intersects the 45° diagonal), and this density is the equilibrium point or replacement level of reproduction. If no environmental fluctuation deflects the stock from point K, it will remain perpetually at that density, exactly replacing itself over each succeeding generation. At densities above replacement reproduction, the parental stock will fail to replace itself and the population will decline back toward the equilibrium point. If stock density is deflected by environmental conditions below the equilibrium point, the parental stock will more than replace itself; that is, the population will tend to increase back toward replacement level over succeeding generations. At replacement level, the parental stock exactly replaces itself in the face of baseline natural mortality, producing no surplus progeny as a buffer against removal by an environmental impact such as power plant operation, a fishery, or pollution.

Figure 4 explores the situation in which an increment of mortality is imposed on the population of Fig. 3, thus deflecting it away from the replacement level of parental stock. Let us say, for example, that an amount of parental stock equal to the line segment $\overline{cK}$ is removed from the population before reproduction. The parental stock now consists of $\overline{Oc}$ units, and this parental stock produces $\overline{ca}$ units of progeny. At this stock density, the parents produce $\overline{cb}$ units of offspring (sufficient to replace themselves) plus a surplus $\overline{ab}$, which may be removed by the fishery or killed by power plant operations or pollutional inputs but which still leaves the population equilibrated at a density $\overline{Oc}$. For this situation, the removal, $\overline{ab}$, from the population is about 28% of the progeny stock, $\overline{ac}$.

If an additional increment of removal is imposed on the stock (e.g., a total of $\overline{de}$ units of progeny stock), the removal rate $(\overline{de}/\overline{df})$ will be 60% and the population will sustain this level of removal, equilibrating at a density $\overline{Of}$. To hold the stock at this reduced density, 60% average removal must be sustained. If this rate of removal is reduced, the parental stock will more than replace itself and succeeding generations will tend to increase until the population equilibrates once again at a higher level of density. A still higher percentage removal (70% $[\overline{gh}/\overline{gi}]$ for example), if sustained, will reduce the population to the density $\overline{Oi}$.

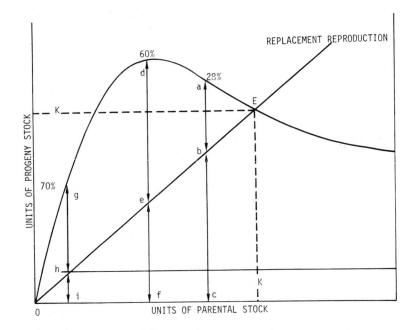

Fig. 4. Equilibrium exploitation rates for parent-progeny relationship of Fig. 3.
(Equilibrium at c units of parental stock is maintained with an exploitation rate of 28%; a level of f units of stock by an exploitation rate of 60%; and a level of i units of stock by an exploitation rate of 70%.)

Two important points emerge. First, an increment of removal imposed on a fish stock drives the stock to a lower average density at which the population once again equilibrates, but the increment of mortality imposed and sustained does not drive the population into a steady downward spiral leading to severe depletion or extinction. This point was advanced earlier in summarizing compensation concepts from Gulland (1974) and Ricker (1958). Second, the rate of removal must be sustained from generation to generation if the stock is to be held at a reduced level of abundance; increasingly higher percentage removals must be sustained if the stock is to be driven to successively lower levels of density.

Cushing (1975, p. 238) states:

The equation relating recruitment to parent stock used initially by Ricker and later by Beverton and Holt is a convenient summary of present opinion on the natural regulation of numbers in a fish population. Recruitment depends on stock modulated by density-dependent mortality.

Gulland (1974, p. 101) also reflects on the widespread acceptance of this type of formulation of compensation:

> Other curves could be derived with other assumptions, but so far for all stocks the observations of stock and recruitment, though often scattered, have been consistent with at least one of the theoretical curves.

To promote discussion of the present state of knowledge and of future research requirements in the face of the large post-war increase in global fishing intensity and problems in the world food supply, a symposium on fish stocks and recruitment was convened in 1970 by the International Council for the Exploration of the Sea, the International Commission for North Atlantic Fisheries, and the Food and Agricultural Organization of the United Nations. There were 82 participants from 20 countries. The published proceedings (Parrish 1973) recorded the following scientific opinions on the importance and prevalence of compensation or density dependence in fish populations:

> When J. Tanner (1966) concluded after the analysis of density dependence in 111 different populations representing 71 species, "It is significant that the processes known to regulate vertebrate populations affect either reproduction or the survival of juveniles," he could have been reciting from the current litany of fisheries biology (Paulik 1973, p. 302).

> The papers and discussions indicate that recruitment . . . in both marine and freshwater fish and shellfish populations is determined by a complex of density-dependent and density-independent factors. The former may act as the main source of control governing the form of the relationship between recruitment and spawning stock size (egg production), and the latter give rise to the well-known short-term, irregular fluctuations in recruitment characteristic of some teleost species having high fecundity. The papers and discussions indicate further that in most species for which detailed information is available these factors operate mainly during the early stages of development (i.e. between the egg and the end of the first year of life) so that year-class strength is determined and population control mechanisms for most fish stocks operate before the individuals enter the exploited phase" (Parrish 1973, p. 5).

> It was the consensus of the meeting that density dependence does occur in all fish stocks at some point, and this is itself a major transition from earlier dogma. As R. Jones and Bowers pointed out, the precise age at which density becomes of major significance varies from stock to stock so that, for example, the plaice may have properties that are different from haddock. Plaice growth may be

density-regulated while they are in the plankton, and a relatively constant number may transform to the demersal habitat.

Dr. LeCren pointed out that freshwater species such as trout and salmon are really not very different in their fundamental biological machinery from the marine ones and that the same kind of density dependencies probably occur in marine and fresh water. The general acceptance of density dependence as relevant to all marine stocks is probably of great value since it avoids the nasty theoretical problem, specifically, that the absence of any density-dependent feedback system implies a random walk process in stock size which would predict much greater temporal variations in population size than are actually observed (Slobodkin 1973, p. 10).

Compensatory Mechanisms

As discussed earlier for animal populations in general, a wide range of environmental agents and compensatory mechanisms may be involved in the overall compensatory response generated by a fish population. Different mechanisms may operate at different levels of population density or under different environmental conditions, and the effect of one compensatory mechanism may preempt the activation of alternate mechanisms that are potentially available. A wide range of population processes also may play a significant role in the compensatory response. Furthermore, compensation may be masked statistically either by its own effectiveness, which may so stabilize some fish populations that observations on population processes at very different levels of density are hard to obtain, or by the imposition of a large amount of random variation in population parameters caused by density-independent factors. Thus, it is usually not profitable to focus on a single mechanism of compensation or on even a small complex of mechanisms to test the reality or extent of the phenomenon. However, this is not to say that the existence of such mechanisms is a matter of theory, speculation, or mystery; on the contrary, the compensatory operation of many different mechanisms and population processes has been proven in fish populations. The 17 examples in Table 1 drawn from the scientific literature provide a concrete understanding of the remarkable resilience of fish populations in the face of increases in mortality caused by man.

Capacity of Fish Populations to Withstand Mortality

While it is clear that the concepts in fisheries science were developed through experience with stocks subjected to exploitation by man, nothing explicit has yet been stated here about the degree of resiliency possessed by these stocks. Remaining unanswered are such questions as how large an

Table 1. Examples of the resiliency of fish populations to man-induced mortality

1. A dense population of perch in Lake Windermere was subjected to an extensive experimental fishery for five years (LeCren 1958); reduction of the population to 3% of its original density resulted in a fourfold increase in both mean weight and fecundity of adults.

2. Brown trout populations in a small Swedish lake were compared for six years before and six years during exploitation with gill nets and sport fishing gear (Lindstrom, Fagerstrom, and Gustafson 1970). Although mean size decreased, the individual growth rates (lengths) increased an average of 10.5% after exploitation.

3. Bluegills in three large Michigan ponds were subjected to annual reductions of 0, 60, and 90% in young-of-the-year during a five-year experiment (Beyerle and Williams 1972). Survival averaged 0.5, 8.1, and 12.2% respectively. Growth of fish in all three ponds was similar, evidently as the result of compensatory survival rates.

4. Commercial exploitation of plaice off the coast of Scotland increased greatly after 1956 (Bagenel 1963). Fecundity increased from 137,000 eggs per female in 1956–1957 to 157,000–161,000 eggs per female in 1958–1961. The data suggest that heavy fishing reduced the population size but that survivors had proportionately more food, resulting in higher fecundity.

5. Large brook trout in a Canadian lake were subjected to 90% experimental exploitation by gill nets (Smith 1956). Survival of planted fingerlings was two times that existing before gill netting of large trout. Although growth data are not easily interpreted, the growth rate of fingerlings appeared to have decreased after exploitation.

6. Data from seven brook trout populations were studied to compare fished and unfished populations (Jensen 1971). Fishing resulted in more young and fewer old fish. Increased age-specific fecundity compensated for increased mortality from fishing. Since fecundity increased with size, the data suggested that growth may be the most important variable in a fish population's adjustment to exploitation.

7. Reductions in numbers of adult pike were observed in Lake Windermere (Kipling and Frost 1969). There was a significant increase in eggs per gram of fish after the population density had been falling for several years, suggesting a compensatory response in fecundity.

8. Ciscoes in three Canadian lakes were subjected to varying degrees of exploitation (Miller 1950). Total mortality averaged 61 and 60% in two mildly exploited lakes and 70.5% in an unfished lake. Under moderate and then heavy exploitation, total mortality in one lake averaged 80 and 94% respectively. The comparison suggests that fishing mortality reduces natural mortality, but that the overall effect of moderate to heavy exploitation is an increase in total mortality.

9. Brook trout in a Wisconsin stream were subjected to varying amounts of sport fishing (McFadden 1961). There was a significant regression of natural mortality on angling mortality so that exploitation reduced natural mortality. Total mortality increased with exploitation, but there was a broad range in the number of spawners and size of the egg complement, which would result in adequate numbers of progeny.

Table 1 (continued)

10. Rainbow trout populations maintained by stocking were compared in five New Zealand lakes (Fish 1968). In three of the lakes supporting 7–18 fish per acre, growth in weight was two to three times that in two lakes supporting 31–35 fish per acre. The inverse relationship between number of trout and their average weight suggests that the lakes would produce large trout if the population were kept small.

11. Rainbow trout in a New Zealand lake were subjected to a sport fishery of increasing intensity for 14 years (Percival and Burnet 1963). The growth of two- and three-year-old fish was negatively correlated with population size. The data suggest that an increase in exploitation increases the survival rate of juveniles. The larger number of surviving juveniles resulted in a reduction in growth rate and maximum size of the fish.

12. Catch and escapement of an exploited sockeye salmon population were studied in Bristol Bay, Alaska (Mathisen 1969). Growth in length of smolts could be expressed as a negative exponential function of population density. A similar relationship exists for maturing salmon during their migration toward fresh water. Thus, as exploitation increases, the growth in individual fish increases.

13. Records of 111 animal populations representing 71 species were analyzed to determine the relationship between the rate of increase of a population and the population density (Tanner 1966). Of the seven fish species examined, Atlantic salmon, yellow perch, walleye, and northern pike showed a statistically significant negative correlation of population growth rate with population density; freshwater drum and goldeye showed a negative correlation and lake trout a positive correlation, but these were not statistically significant. The data strongly support the concept of a compensatory increase in survival and/or fecundity following reduction in the size of a population.

14. Populations of rainbow and brook trout were compared among New York ponds (Eipper 1964). Growth rates of both trout species were inversely related to population density.

15. The number of spawners was compared with the number of progeny for haddock, Pacific herring, and coho, sockeye, and pink salmon (Ricker 1954). The data suggest that the survival rate of progeny increases as the number of spawners decreases. Within limits, a reduction in spawners can also result in increased numbers of surviving progeny.

16. Plaice, haddock, sole, turbot, and cod in the North Sea were subjected to varying degrees of exploitation (Beverton and Holt 1957). Survival, particularly of plaice, was strongly inversely correlated with the size of the adult population.

17. The size of young sockeye salmon in a British Columbia lake was compared with population density for 11 years (Foerster 1944). There was a statistically significant negative correlation ($r = 0.82$) between the density of the lake population and the mean weight of migrants.

annual removal can be sustained, at what level of added mortality will the population be drastically reduced in numbers, at what level of added mortality will it be threatened with extinction, and what would constitute an intolerable or irreversible reduction of a fish population. As a foundation for responding to these questions from a basis of empirical evidence, the principles unfolded through the arguments and historical accounts previously presented are summarized as follows:

• An undisturbed or unexploited population (one at the maximum equilibrium level on the average) produces enough new individuals to just replace natural losses; there is no surplus production.

• A population, if it is to be held at maximum size, cannot be exposed to any additional mortality either from natural causes or from man's activities because, at maximum stock size, no surplus production is available to absorb an increment of mortality.

• At the opposite extreme, as a population approaches 0 size, surplus production approaches 0. In very small populations, the rate of surplus production per individual is very high; however, so few individuals are present that the overall rate of increase for the population is very low.

• Should exploitation or some other form of environmental impact occur, population size will be reduced. At levels of density lower than equilibrium, surplus production will be available to absorb the environmental impact while maintaining the population at the new but reduced level.

• Maximum surplus production occurs at some intermediate level of stock density.

• If an added mortality is large enough, population size will be reduced to a level at which the fish may become undesirably scarce, and the population may be vulnerable to accidental or even inevitable extinction and may be able to generate only a small surplus production for the benefit of man or as a cushion against further decline.

In ecological terms, the significance of an impact imposed by man on a fish population is not a matter of "good" or "bad" but rather a matter of one or more states of reality defined by the average level of abundance of the impacted stock and the magnitude of its surplus production. Increasing overall impact on the fish population decreases standing stock, increases resource utilization, causes maximum surplus production at an intermediate level of exploitation, and varies the potential for disposition of the surplus production as, for example, between fisheries and kills resulting from power generation. Because the objective of fisheries management is to maximize some form of productivity on a sustained basis (often the total weight of the fish harvested), identification of the fishing rate that will produce the greatest yield on a long-term basis has usually been emphasized. There has been no particular value attached to building up the population to maximum size.

Unlike esthetically valued species such as brightly feathered birds, fish are rarely accessible to the admiring gaze of the public and none has a reputation for song.

The emphasis on maximizing harvest has led to a wide range of historical experience with initially unexploited fish populations that have been subjected to harvest by sport and commercial fisheries and have persisted in a healthy ecological state despite increased overall mortality. This experience constitutes unassailable proof of the general operation of the phenomenon of compensation in fish populations. If the natural death rate had not declined or the birth rate increased as population density was reduced, the large removals by fishing would quickly have driven these populations to extinction.

In reviewing a substantial number of published estimates of exploitation rates (Table 2), it becomes clear that cases in which $\geq 25\%$ of the exploitable age classes in a population have been removed annually are common. The figures generally represent situations in which substantial exploitation has been underway for fairly long periods (usually for decades) and is continuing. Clearly then, many populations possess compensatory reserve sufficient to offset very substantial increments of man-induced mortality. The repeated removal of 25–50% of a fish population and the sustained reduction of abundance of the fish to a level well below the pre-fishing stock size, with hopes for survival of the stock dangling by the seemingly precarious conceptual thread of compensation, may at first seem a drastic proposition to the interested layman or the scientist lacking a background in scientific management of fisheries. However, such treatment, rather than being an extreme to which populations can at great peril be pushed, is a normal and ecologically sound treatment that permits fish stocks to operate at maximum productivity. Earlier management models (e.g., the logistic model) predicted that maximum sustained yield would be obtained at an average population level of 50% of the virgin stock. Current more flexible and realistic models (e.g., that of Ricker) indicate that, for the stock recruitment relationship characteristic of many important fish, maximum sustained yield would be obtained under exploitation rates of 25–75% and that the average equilibrium level of the populations at maximum sustained yield would be less than half the virgin stock level (Ricker 1958, p. 239, 268). The broad experience sampled in Table 2 confirms these management tenets, and the tenets provide a basis for assessing the limits within which fish populations can be managed safely by man. It is not suggested here that harvest of fish by fishermen and power plants up to the level of maximum sustained yield is socially desirable, but simply that it is ecologically safe.

Yield statistics from commercial and sport fisheries provide another reflection of the capacity of fish stocks to sustain themselves in the face of substantial man-induced mortalities. The relatively unproductive Great Lakes of North America have produced yields from 1 to 7 lb per acre per year; a

Table 2. Summary of published estimates of exploitation rates in fish populations

Name of species		Exploitation rate, %	Location	Reference
Scientific	Common			
Alosa sapidissima	American shad	25	Connecticut River, Conn.	Walburg 1960
Ambloplites rupestris	Rock bass	16	Oliver Lake, Ind.	Gerking 1950
		21	Escanaba Lake, Wis., 1946—69	Kempinger et al. 1975
Aplodinotus grunniens	Freshwater drum	47	Upper Miss. River impoundments, 1944—48	Butler 1965
		58	Upper Miss. River impoundments, 1944—48	Butler 1965
		31	Upper Miss. River impoundments, 1944—48	Butler 1965
Clupea harengus	Atlantic herring	29	South coast, Ireland, 1906—36	Burd and Bracken 1965
		10	South coast, Ireland, 1951—55	Burd and Bracken 1965
		42	South coast, Ireland, 1956—60	Burd and Bracken 1965
		25	South coast, Ireland, 1961—63	Burd and Bracken 1965
Coregonus clupeaformis		40	Georgina Bay, Lake Huron	Cucin and Regier 1965
		21	Lake Superior	Dryer 1964
Cynoscian nebulosus	Spotted seatrout	19	Pine Island, Fla., 1961	Iversen and Moffett 1962
Esox lucius	Northern pike	38	Fletcher Floodwater, Mich.	Christensen and williams 1959
		14	Lake George, Minn.	Groebner 1964
		32—49	Grove Lake, Minn.	Groebner 1964
		23	Ball Club Lake, Minn.	Johnson and Peterson 1955
		46	Escanaba Lake, Wis., 1946—69	Kempinger et al. 1975
		32	Murphy Flowage, Wis.	Snow 1958
		50	Wisconsin waters	Threinen et al. 1966
		22—28	Grace Lake, Minn.	Wesloh and Olson 1962
Esox masquinongy	Muskellunge	27	Escanaba Lake, Wis., 1946—69	Kempinger et al. 1975
		14—70	Nogies Creek, Ont., 1952—60	Muir 1963
Gadus morhua	Atlantic cod	11	Gulf of St. Lawrence, 1949—52	Paloheimo and Kohler 1968
		25	Gulf of St. Lawrence, 1955—65	Paloheimo and Kohler 1968

Table 2 (continued)

| Name of species | | Exploitation | Location | Reference |
Scientific	Common	rate, %		
Hippoglossoides platessoides	American plaice	31	Gulf of St. Lawrence	Poweles 1969
Ictalurus nebulosus	Brown bullhead	25	Shoe Lake, Ind.	Ricker 1945
Ictalurus punctatus	Channel catfish	30	Sacramento Valley, Calif.	McCannon and LaFaunce 1961
Lepomis gibbosus	Pumpkinseed	29	Escanaba Lake, Wis., 1946–69	Kempinger et al. 1975
Lepomis macrochirus	Bluegill	25	Sugar Loaf Lake, Mich.	Cooper and Latta 1954
		35	Gordy Lake, Ind.	Gerking 1953
		42	Escanaba Lake, Wis., 1946–69	Kempinger et al. 1975
		15–20	Muskellunge Lake, Ind.	Ricker 1945
		36	Spear Lake, Ind.	Richer 1955
Lepomis microlophus	Redear sunfish	29	Gordy Lake, Ind.	Gerking 1953
		23	Muskellunge Lake, Ind.	Ricker 1945
Micropterus dolomieui	Smallmouth bass	05–18	Oneida Lake, N.Y.	Forney 1961
		22	Waugoshance Point, Lake Michigan	Latta 1963
Micropterus salmoides	Largemouth bass	17	Oliver Lake, Ind.	Gerking 1950
		36	Gordy Lake, Ind.	Gerking 1953
		20	Clear Lake, Calif.	Kimsey 1957
		20–48	Southerland Res., Calif.	LaFaunce et al. 1964
		14	Gladstone Lake, Minn.	Maloney et al. 1962
		11	Browns Lake, Wis., 1953	Mraz and Threinen 1957
		12	Shoe Lake, Ind.	Ricker 1945
Perca flavescens	Yellow perch	15	Escanaba Lake, Wis., 1946–69	Kempinger et al. 1975
Pleuronectes platessa	Plaice	49	North Sea, 1929–38	Beverton and Holt 1957
		33	North Sea, 1950–64	Gulland 1968

Table 2 (continued)

Name of species		Exploitation rate, %	Location	Reference
Scientific	Common			
Pomoxis nigromaculatus	Black crappie	11	Oliver Lake, Ind.	Gerking 1953
		26	Escanaba Lake, Wis., 1946–69	Kempinger et al. 1975
Pseudotolithus typus, P. senegalensis		40	Coast of Nigeria, 1961–62	Longhurst 1964
Salmo gairdneri	Rainbow trout	13–17	New York streams	Hartman 1959
		20–26	New York lakes	Hartman 1959
Salmo salar	Atlantic salmon	66	Little Codroy River, Newfoundland, 1955–63	Murray 1968
Salmo trutta	Brown trout	23	Sydenham River, Ont., 1966–67	Marshall and MacCrimmon 1970
Salvelinus fontinalis	Brook trout	59	Sydenham River, Ont., 1966–67	Marshall and MacCrimmon 1970
		19–75	Lawrence Creek, Wis.	McFadden 1961
Stizostedion vitreum	Walleye	29	Escanaba Lake, Wis., 1946–69	Kempinger et al. 1975
		27	Many Point Lake Minn.	Olson 1957
		20–40	Escanaba Lake, Wis.	Patterson 1953, Niemuth et al. 1959
		15–28	Spirit Lake, Iowa	Rose 1947, 1955
		07	Nipegon Bay, Lake Superior, 1955	Ryder 1968
		13	Nipegon Bay, Lake Superior, 1956	Ryder 1968
		34	Nipegon Bay, Lake Superior, 1957	Ryder 1968
		05	Fife Lake, Mich.	Schneider 1969
Tilapia esculenta		49	Lake Victoria, Africa, 1958–59	Garrod 1963
		42	Lake Victoria, Africa, 1959	Garrod 1963
		32	Lake Victoria, Africa, 1959–60	Garrod 1963
		34	Lake Victoria, Africa, 1960	Garrod 1963

productive North Sea fishery has yielded about 27 lb per acre, Escanaba Lake in Wisconsin has yielded 4–41 lb per acre over a 24-year period (the overall average being 20 lb per acre), the Gulf of Mexico's fin-fish catch attributable to estuarine production has averaged about 50 lb per acre in recent years, and commercial fishing in Chesapeake Bay has yielded about 155 lb per acre. Clearly, these substantial removals would deplete fish populations rapidly were those populations unable to compensate through increased survival or reproduction rates.

Experience with the world's major stocks has proven both that they have the capacity to withstand impressively high levels of exploitation and produce a substantial surplus at population levels well below their pre-exploitation abundance, and that they have the ability to rebound to higher levels of abundance after being heavily exploited if fishing intensity is reduced. This is entirely in keeping with the concepts of population dynamics, and there is important empirical proof of this recovery capacity. Fishing reduced the plaice stock in the southern North Sea to one-third or one-half its original abundance, but relaxation of fishing during wartime rapidly doubled its size (Cushing 1975, p. 135). The Pacific halibut population decreased by a factor of 7 during a 20-year period of fishing, then increased in size when protected by a closed season (Fukuda 1962). A North Sea herring stock was reduced to one-tenth its previous abundance during a 12-year period and subsequently began to increase as a result of reduced fishing pressure (Cushing 1975, p. 136). Sturgeon fishing in the Amur River basin was banned for 12 years to restore overfished stocks; by the end of the ban, the proportion of sexually mature fish had increased sixfold and the average size and age of fish had increased as well – evidence of restoration of the stock (Krykhtin 1972). Similar examples are reported for the whitefish of Lake Wabamun (Miller 1949) and for lake trout of Lake Opeongo (Fry 1949). Thus, reductions in fish stocks caused by exploitation have commonly been proven to be reversible.

APPLICABILITY OF FISHERY EXPLOITATION PRINCIPLES
TO POWER PLANT IMPACTS

Cushing (1975, p. 138–139) points out that "Many populations of wild animals are exploited but the commercial fish stocks are numerous . . . . No other wild populations have been so well documented for such long periods of time." This imposing body of accumulated experience with fish populations has provided the basis for the development of successful management principles in fishery science. However, are these principles and the experience from which they are derived, based as they are on fishery exploitation of the (usually) mature age groups of a stock, applicable to the situation in which some fraction of the earliest life stages (eggs, larvae, and juveniles) is cropped by entrainment and impingement? Can the contention that "The agent of mortality — predatory fish, commercial or sport fisherman, or power plant — is an indifferent matter from the standpoint of population response" be sustained? The questions are important ones. If answered affirmatively, any empirical evidence for compensation in a particular stock, in combination with one of the generally proven compensation models based upon the entire foundation of fishery-management science (e.g., Ricker or Beverton-Holt stock-recruitment function), can be used to provide estimates of power plant impact. This taking into account of compensation would greatly increase the degree of realism associated with assessing the effects of power-plant-induced mortality on fish populations.

It turns out that the question whether it is different (possibly worse) to kill young fish than to kill older fish was answered more than 20 years ago by Ricker (1954, p. 607):

> Exploitation that takes fish at an age when natural mortality is still compensatory means, for practical purposes, a fishery for young during the first year or two of their life — the earlier the better. The removal of such young is at least partly balanced by increased survival and/or growth of the remainder; in fact, the effects of removals at this stage are equivalent to reduction of the spawning stock which produced the brood in question. If the reproduction curve for the population is of any of the types 3–8, such reduction will at first increase net production of recruits, which will produce more eggs and permit a larger catch of young in future years. This ascending spiral of abundance may continue until the level of stock is reached which produces maximum recruits.

Ricker goes on to say ". . . it is clear that any general prejudice against exploiting young fish is unsound." Ricker's analysis of the situation has not been challenged. Insofar as population response is concerned, no different principles are involved. Killing some fish during the egg, larval, and juvenile stages (which is what power plants do) is no worse than killing the parents that would have produced these young (which is what fisheries do). The entire foundation of fishery management experience and principles, therefore, can be applied with confidence to problems of power plant impact. It can be reasoned that fish populations can readily sustain "exploitation" by power plants at levels comparable to those experienced in commercial and sport fisheries. In a well-managed fishery, an annual exploitation rate of 50% or more for some species might well be a goal eagerly sought by fishery managers and fishermen rather than a threat to be guarded against. Even for species with rather low compensatory capacity, an exploitation rate of approximately 30% and reduction of the stock to about 46% of its pre-exploitation abundance would be necessary to achieve the very conventional goal of maximum sustained yield (Ricker 1958, p. 268).

As has been pointed out, the workability of such management plans, based on current fishery concepts and models, has been directly proven by empirical data such as that summarized in Table 2. A 50% exploitation rate in a fishery, however, is not likely to be comparable to a 50% exploitation rate caused by a power plant. Once a particular year class of fish becomes vulnerable to a fishery, it is likely to be exploited during each remaining year of its life, although the rate may vary with age. For example, assuming recruitment to the fishery at age 4, a 50% exploitation rate could well mean that half the members of age class 4 are captured, half of the survivors of age class 4 are captured at age 5, and so on for older age classes.

Compared with this annually repeated fishery mortality, a 50% exploitation rate caused by a power plant, which would affect each year class only once during its lifetime (during the first year when the young are entrained or impinged), would represent a very much smaller mortality. What relative impacts, measured in terms of stock reductions, the two mortalities would be translated into depends on the characteristics of the fishery. Clearly, the power plant mortality, falling as it does on the very young, precedes the age of first reproduction. The fishery mortality, on the other hand, could begin several years after the fish first reproduce, and some part of the population's reproductive potential would then be protected from reduction. However, most fish reach a size desirable in sport or commercial fisheries by the age at which they make significant reproductive contributions. Therefore, the common case will be for a given exploitation rate caused by a fishery, and hence repeatedly imposed in successive years on each year class of fish, to translate into a greater impact than would the same exploitation rate caused by a power plant, and hence imposed only once on each year class.

In most real-world situations, power plant impact would be added to a pre-existing fishery exploitation rate rather than applied to an unexploited stock. Even in this case, however, the addition of an exploitation rate greater than 25% to a pre-existing fishery exploitation rate of 25% would not endanger the stock of many species.

LITERATURE CITED

Bagenal, T. B. 1963. Variations in plaice fecundity in the Clyde area. J. Mar. Biol. Assoc. U.K. 43(2):391–399.

Beverton, R. J. H., and S. H. Holt. 1957. On the dynamics of exploited fish populations. Fish Invest. London Ser. 2, 19. 533 p.

Beyerle, G., and J. Williams. 1972. Survival, growth, and production by bluegills subjected to population reduction in ponds. Mich. Dep. Nat. Resour. Res. Dev. Rep. 273. 28 p.

Burd, A. C., and J. Bracken. 1965. Studies on the Dunmore herring stock 1 – a population assessment. J. Cons., Cons. Int. Explor. Mer. 29(3):277–301.

Butler, R. L. 1965. Freshwater drum *Aplodinotus grunniens* in the navigational impoundments of the upper Mississippi River. [See also Robert L. Butler. 1962. The status of the freshwater drum, *Aplodinotus grunniens*

Rafinesque, in the commercial fishery of the upper Mississippi River. Ph.D Thesis, Univ. Minnesota. 178 pp.]

Carson, R. 1955. The edge of the sea. Houghton Mifflin Co., Boston. 276 p.

Chapman, D. G. 1964. Reports of the committee of three scientists on the special scientific investigation of the Antarctic whale stocks. Rep. Int. Comm. Whales 14:32–106.

Christensen, K. E., and J. E. Williams. 1959. Status of the northern pike population in Fletcher Floodwater, Alpena and Montmorency Counties, 1948 and 1955–56. Mich. Dep. Conserv. Inst. Fish. Res. Rep. 1576. 13 p.

Christensen, S. W., W. Van Winkle, and P. C. Cota. 1975. Effects of Summit Power Station on striped bass populations. Testimony before the Atomic Safety and Licensing Board in the Matter of Summit Power Station, Units 1 and 2, USAEC Docket Nos. 50–450 and 50–451, March 1975.

Cooper, G. P., and W. C. Latta. 1954. Further studies on the fish population and exploitation by angling in Sugarloaf Lake, Washtenaw County, Michigan. Mich. Acad. Sci. 39:209–223.

Cucin, D., and H. A. Regier. 1965. Dynamics and exploitation of lake whitefish in southern Georgian Bay. J. Fish Res. Board Can. 23(2):221–274.

Cushing, D. H., and J. G. K. Harris. 1973. Stock and recruitment and the problem of density-dependence. Rapp. P.-V. Reun., Cons. Int. Explor. Mer. 164:142–155. 278 p.

Cushing, D. H. 1974. The natural regulation of fish populations. In H. Jones [ed.] Sea fisheries research. Paul Elek Ltd., London. 510 p.

————. 1975. Marine ecology and fisheries. Cambridge Univ. Press, London.

Dryer, W. R. 1964. Movements, growth and rate of recapture of whitefish tagged in the Apostle Islands area of Lake Superior. Fish. Bull. U.S. 63(3):611–618.

Eipper, A. W. 1964. Growth, mortality rates, and standing crops of trout in New York farm ponds. Cornell Univ. Ag. Exp. Sta. Mem. 388. 67 p.

Fish, G. R. 1968. An examination of the trout population of five lakes near Rotorus, New Zealand. N.Z. J. Mar. Freshwater Res. 2(2):333–362.

Foerster, R. E. 1944. The relation of lake population density to size of young sockeye salmon (Oncophynchus nerka). J. Fish. Res. Board Can. 6:267–280.

Forney, J. L. 1961. Growth, movements, and survival of smallmouth bass (Micropterus dolomieui) in Oneida Lake, New York. N.Y. Fish Game J. 8(2):88–105.

Fry, F. E. J. 1949. Statistics of a lake trout fishery. Biometrics 5:27–67.

Fukuda, Y. 1962. On the stocks of halibut and their fisheries in the northeast Pacific. Int. North Pac. Fish. Comm. Bull. 7:39–50.

Garrod, D. J. 1963. An estimation of the mortality rates in a population of Tilapia esculenta Graham (Pisces, Cichlidae) in Lake Victoria, East Africa. J. Fish. Res. Board Can. 20(1):195–227.

Gerking, S. D. 1950. Populations and exploitation of fishes in a marl lake. Invest. Indiana Lakes Streams 3(11):389–434.

———. 1953. Vital statistics of the fish population of Gordy Lake, Indiana. Trans. Am. Fish. Soc. 82:48–67.

Graham, G. M. 1935. Modern theory of exploiting a fishery and application to North Sea trawling. J. Cons., Cons. Int. Explor. Mer. 10(2):264–74.

Groebner, J. F. 1964. Contributions to fishing harvest from known numbers of northern pike fingerlings. Minn. Dep. Conserv. Invest. Rep. 280. 16 p.

Gulland, J. A. 1961. Fishing and the stocks of fish at Iceland. Fish. Invest. London Ser. 2, 23(4). 52 p.

———. 1968. Recent changes in the North Sea plaice fishery. J. Cons., Cons. Int. Explor. Mer. 31(3):305–322.

———. 1974. The management of marine fisheries. Univ. Wash. Press, Seattle. 198 p.

Haldane, J. B. S. 1953. Animal populations and their regulation. Penguin Books, London. New Biol. 15:9–24.

Hartman, W. L. 1959. Biology and vital statistics of rainbow trout in the Finger Lakes region, New York. N.Y. Fish Game J. 6(2):121–178.

Hjort, J., G. Jahn, and P. Ottestad. 1933. The optimum catch. Essays on population. Hvalradets Skr. 7:92–127.

Iversen, E. S., and A. W. Moffett. 1962. Estimation of abundance and mortality of a spotted seatrout population. Trans. Am. Fish. Soc. 91(4):395–398.

Jensen, A. L. 1971. Response of brook trout (*Salvelinus fontinalis*) populations to a fishery. J. Fish. Res. Board Can. 28(3):458–460.

Johnson, F. H., and A. R. Peterson. 1955. Comparative harvest of northern pike by summer angling and winter darkhouse spearing from Ball Club Lake, Itasca County, Minnesota. Minn. Dep. Conserv. Invest. Rep. 164. 11 p.

Kempinger, J. J., W. S. Churchill, G. R. Priegel, and L. M. Christensen. 1975. Estimate of abundance, harvest, and exploitation of the fish population of Escanaba Lake, Wisconsin, 1946–49. Wis. Dep. Nat. Resour. Res. Bull. 84. 30 p.

Kimsey, J. B. 1957. Largemouth bass tagging at Clear Lake, Lakes county, California. Calif. Fish Game 43(2):111–118.

Kipling, C., and W. E. Frost. 1969. Variations in the fecundity of pike (*Esox lucius L.*) in Windermere. J. Fish. Biol. 1:221–237.

Krebs, C. J. 1972. Ecology: the experimental analysis of distribution and abundance. Harper & Row, New York. 694 p.

Krykhtin, M. L. 1972. Changes in the composition and abundance of stocks of the great Russian sturgeon or kaluga [*Huso dauricus* (Georgi)] and of the Amur sturgeon [*Acipenser schrenck* (Brandt)] during the period of a fishery ban in the Amur Basin. J. Ichthy. 12(1):1–9.

LaFaunce, D. A., J. B. Kimsey, and H. K. Chadwick. 1964. The fishery at Sutherland Reservoir, San Diego County, California. Calif. Fish Game 59(4):271–291.

Latta, W. C. 1963. The life history of the smallmouth bass, *Micropterus dolomieue dolomieui*, at Wagoshance Point, Lake Michigan. Mich. Dep. Conserv. Inst. Fish. Res. Bull. 5. 56 p.

LeCren, E. D. 1958. Observations on the growth of perch (*Perca fluviatalis L.*) over twenty-two years, with special reference to the effects of temperature and changes in populations density. J. Anim. Ecol. 27(2):287–334.

Lindstrom, T., A. Fagerstrom, and K. J. Gustafson. 1970. Fishing pressure, growth, and recruitment in a small high mountain lake. Inst. Freshwater Res. Drottningholm Rep. 50:100–115.

Longhurst, A. R. 1964. Bionomics of the Scienidae of tropical West Africa. J. Cons., Cons. Int. Explor. Mer. 29(1):93–114.

Maloney, J. E., D. H. Shupps, and W. S. Scidmore. 1962. Largemouth bass populations and harvest, Glandstone Lake, Crow County, Minnesota. Trans. Am. Fish. Soc. 91(1):42–52.

Malthus, T. R. 1798. An essay on the principle of population. London, Johnson. (Various ed.; e.g., 1914, repro. of 7th ed., Everymans Library, 2 vol., 1909, New York, Macmillan, in Economic Classics).

Marshall, T. L., and H. R. MacCrimmon. 1970. Exploitation of self-sustaining Ontario stream populations of brown trout (*Salmo trutta*) and brook trout (*Salvelinus fontinalis*). J. Fish. Res. Board Can. 27(6):1087–1102.

Mathisen, O. A. 1969. Growth of sockeye salmon in relation to abundance in the Kvichak District, Bristol Bay, Alaska. Fish. Der. Skr. Ser. Hav. Unders. 15(3):172–185.

McCammon, G. W., and D. A. LaFaunce. 1961. Mortality rates and movement in the channel catfish population of the Sacramento Valley. Calif. Fish Game 47(1):5–26.

McFadden, J. T. 1961. A population study of the brook trout, *Salvelinue fontinalis.* Wildl. Monogr. 7. 73 p.

Merriman, D. 1941. Studies on the striped bass, *Roccus saxatilis*, of the Atlantic Coast. Fish. Bull. U.S. Fish Wildlife Serv. 50:1–77.

Miller, R. B. 1949. Problems of the optimum catch in small white fish lakes. Biometrics 5:14–26.

————. 1950. Observations on mortality rates in fished and unfished cisco populations. Trans. Am. Fish. Soc. 79:180–186.

Mraz, D., and C. W. Threinen. 1957. Angler's harvest, growth rate and population estimate of the largemouth bass of Brown's Lake, Wisconsin. Trans. Am. Fish Soc. 85:241–256.

Muir, B. S. 1963. Vital statistics of *Esox masquinongy* in Nogies Creek, Ontario. I. Tag loss, mortality due to tagging, and the estimate of exploitation. J. Fish. Res. Board Can. 20(5):1213–1230.

Murray, A. R. 1968. Smolt survival and adult utilization of Little Codroy River, Newfoundland, Atlantic salmon. J. Fish. Res. Board Can. 25(10):2165–2218.

Nicholson, A. J. 1933. The balance of animal populations. J. Anim. Ecol. 2:132–178.

Niemuth, W., W. Churchill, and T. Wirth. 1959. The walleye, its life history, ecology, and management. Wis. Conserv. Dep. Pub. 227. 14 p.

Olson, D. E. 1957. Statistics of a walleye sport fishery in a Minnesota lake. Trans. Am. Fish. Soc. 87:52–72.

Paloheimo, J. E., and A. C. Kohler. 1968. Analysis of the southern Gulf of St. Lawrence cod population. J. Fish. Res. Board Can. 25(3):555–578.

Parrish, B. B. [ed.] 1973. Fish stocks and recruitment. Rapports et proces – verbaux des reunions. Conseil Int. pour L'Exploration de la Mer, Charlottenlund Slot – Danemark 164. 372 p.

Patterson, D. L. 1953. The walleye population in Escanaba Lake, Vilas County, Wisconsin. Trans. Am. Fish. Soc. 82:34–41.

Paulik, G. J. 1973. Studies of the possible form of the stock-recruitment curve. p. 302–315. In B. B. Parrish [ed.] Fish stocks and recruitment. Rapports et proces – verbaux des reunions. Conseil Int. pour L'Exploration de la Mer, Charlottenlund Slot – Danemark 164. 372 p.

Pearl, R., and L. J. Reed. 1920. On the rate of growth of the population of the United States since 1790 and its mathematical representation. Proc. Nat. Acad. Sci. 6:275–288.

Pearl, R. 1930. The biology of population growth. Knopf, New York. 330 p.

Percival, E., and A. M. R. Burnet. 1963. A study of the Lake Lyndon rainbow trout (*Salmo gairdneri*). N.Z. J. Sci. 6(2):273–303.

Poweles, P. M. 1969. Size changes, mortality, and equilibrium yields in an exploited stock of American plaice (*Hippoglossoides platessoides*). J. Fish Res. Board Can. 26(5):1205–1235.

Quetelet, A. 1835. Sur l'homme et le development de ses facultes ou essai de physique sociale. Paris, Bachelier. 2 vol. p. 25–26.

Ricker, W. E. 1945. Abundance, exploitation, and mortality of the fishes in two lakes. Invest. Indiana Lakes Streams 2(17):345–448.

————. 1954. Stock and recruitment. J. Fish. Res. Board Can. 11:559–623.

————. 1955: Fish and fishing in Spear Lake, Indiana. Invest. Indiana Lakes Streams 4:117–162.

————. 1958. Handbook of computations for biological statistics of fish populations. Fish. Res. Board Can., Bull. 119. 300 p.

Rose, E. T. 1947. The population of yellow pikeperch (*Stizostedion v. vitreum*) in Spirit Lake, Iowa. Trans. Am. Fish. Soc. 77:32–42.

————. 1955. The fluctuation in abundance of walleyes in Spirit Lake, Iowa. Proc. Iowa Acad. Sci. 62:567–575.

Ryder, R. A. 1968. Dynamics and exploitation of mature walleyes, *Stizostedion vitreum*, in the Nipegon Bay region of Lake Superior. J. Fish. Res. Board Can. 25(7):1347–1376.

Schaefer, M. B. 1954. Some aspects of the dynamics of populations important to the management of the commercial fish populations. Int. Am.
Trop. Tuna Comm. Bull. 1(2):27–56.

Schneider, J. C. 1969. Results of experimental stocking of walleye fingerlings.
1951–63. Mich. Dep. Conserv. Inst. Fish Res. Rep. 1753. 31 p.

Slobodkin, L. B. 1973. Summary and discussion of the symposium, p. 7–14.
In B. B. Parish [ed.] Fish stocks and recruitment. Rapports et proces –
verbaux des reunions. Conseil Int. pour L'Exploration de la Mer,
Charlottenlund Slot – Danemark 164. 372 p.

Smith, M. W. 1956. Further improvement in trout angling at Crecy Lake,
New Brunswick, with predator control extended to large trout. Can. Fish
Cult. 19:13–16.

Snow, H. 1958. Northern pike at Murphy flowage. Wis. Conserv. Bull.
23(2):15–18.

Tanner, J. T. 1966. Effects of population density on growth rates of animal
populations. Ecology 47:733–737.

Threinen, C. W., C. Wistrom, B. Apelgren, and H. Snow. 1966. The northern
pike, its life history, ecology, and management. Wis. Conserv. Dep. Pub.
235. 16 p.

USAEC (United States Atomic Energy Commission). 1972. Final environmental statement related to operation of Indian Point Nuclear Generating
Plant, Unit No. 2. USAEC Docket No. 50-247, Vols. I and II, September
1972.

USAEC (United States Atomic Energy Commission). 1974. Final environmental statement related to the proposed Summit Power Station Units 1
and 2, USAEC Docket Nos. 50-450 and 50-451, July 1974.

USNRC (United States Nuclear Regulatory Commission). 1975. Final
environmental statement related to operation of Indian Point Nuclear
Generating Plant, Unit No. 3. USNRC Docket No. 50-286, Vol. I, February 1975.

Walburg, C. H. 1960. Natural mortality of American shad. Trans. Am. Fish.
Soc. 90:228–230.

Wesloh, M. L., and D. E. Olson. 1962. The growth and harvest of stocked
yearling northern pike, *Esox lucius* linnaeus, in a Minnesota walleye lake.
Minn. Dep. Conserv. Invest. Rep. 242. 9 p.

Impacts of Recent Power Plants on the Hudson River Striped Bass (*Morone saxatilis*) Population

K. Perry Campbell, Irvin R. Savidge,
William P. Dey, and James B. McLaren*

Texas Instruments, Incorporated
Buchanan, New York

ABSTRACT

The impact of the Bowline, Roseton, and Indian Point power plants has been studied by estimating entrainment and impingement mortality rates on striped bass at these plants and relating these mortality rates to reduction in equilibrium stock size under various hypothetical stock-recruitment relationships. The conditional rates of mortality due to entrainment by these three plants were 0.0811 in 1974 and 0.1188 in 1975. The conditional rates of mortality due to impingement were calculated to be 0.0426 in 1974 and 0.0229 in 1975. Based on a range of potential Ricker stock-recruitment curves, reduction in equilibrium stock size due to the estimated levels of power-plant-induced mortality were discussed.

This paper was presented orally at the Conference but was not available for publication in the *Proceedings*.

---

*Present address: Beak Consultants, Inc., 317 S.W. Alder, Portland, Oregon 97204.

Modeling of Compensatory Response to Power Plant Impact

John P. Lawler, Thomas L. Englert,
Robert A. Norris, and C. Braxton Dew

Lawler, Matusky and Skelly Engineers
Tappan, New York

ABSTRACT

Mechanisms and kinetics of compensatory response in fish populations subject to power plant cropping are presented. Topics discussed include (a) the biological and mathematical underpinning leading to nonlinear prey-predator and competitor relations; (b) the mathematical relationship between compensating and noncompensating systems for single species cases; (c) the relationship between Ricker-type stock-recruitment, population oscillation, and lagged logistic growth; (d) the use of Ricker and Beverton-Holt stock-recruitment curves to simulate compensation in single-species life cycle models; and (e) a model of density-dependent growth. Application of the last two topics to the assessment of power plant impact in the Hudson River is discussed.

This paper was presented orally at the Conference but was not available for publication in the *Proceedings*.

Assessing the Impact of Power Plant Mortality on the Compensatory Reserve of Fish Populations

C. Phillip Goodyear

U.S. Fish and Wildlife Service
Office of Biological Services
National Power Plant Team
Ann Arbor, Michigan

ABSTRACT

A technique is presented to quantify the concepts of compensation and compensatory reserve in exploited fish populations. The technique was used to examine the impact of power plant mortality on a hypothetical striped bass population. Power plant mortality had a more severe impact on the compensation ratio and compensatory reserve for an exploited stock than for an unexploited stock. The technique can be applied to determine a critical compensation ratio which could serve as a standard against which additional sources of mortality, such as those caused by power plants, could be measured.

Key words: compensation, compensatory reserve, critical compensation ratio, density–dependent factors, exploitation, fish population analysis, fish population dynamics, modeling, potential fecundity per recruit, power plant impact, reproduction curve, striped bass, viability index

INTRODUCTION

One of the critical problems in the assessment of the impact of power plants on populations of commercially or recreationally important fishes is our limited understanding of the mechanisms that control population change. It is comparatively easy to estimate the proportion of eggs, larvae, or juveniles which may be killed by power plant operation. Unfortunately, this information is not always sufficient to make sound management decisions related to siting or continued operation of individual power plants. Such decisions also require knowledge of the impact of the identified losses on the long-term health of the population and on the associated yield to fishermen.

The problem of predicting population responses to changes in mortality is not restricted to analysis of the impact of power plants but rather is a key

186

problem in fishery research in general (Cushing 1968, Hunter 1976). The same mechanisms that will dictate the long-term consequences of power plant mortality are responsible for the relationship between spawning stock and recruitment. These mechanisms can be broadly categorized as compensatory and include such phenomena as cannibalism and density-dependent starvation, growth, predation, fecundity, etc. The operation of these types of mechanisms enables a population to be exploited by man without its dwindling to extinction.

The continued existence of fish populations undergoing exploitation is ample evidence for the operation of compensatory mechanisms within the life cycles of those populations. However, there is a limit to the degree that these mechanisms can offset the additional mortality caused by increasing exploitation, as evidenced by the catastrophic failure of some fisheries (e.g., Burd 1974).

The purpose of this paper is to present a technique for quantifying the degree to which compensatory phenomena are exerted in exploited populations and to examine the impact of power-plant-induced mortality on this measurement of compensation.

COMPENSATION

Compensation, as used in this paper, is the sum of all density-dependent phenomena that act to stabilize the population. Consider, for example, a hypothetical Ricker-type stock recruitment relationship with recruits measured in terms of sexually mature progeny of the form in Fig. 1. Curve I is the reproduction curve for an unexploited population where the maximum sustainable yield can be achieved by exploiting the population such that the spawning stock is reduced by 73% of maximum surplus production (the distance AB/AC). If the stock is continuously exploited by this amount, then the reproduction curve for the exploited stock will correspond to Curve II. In each case the spawning stock will replace itself at a density corresponding to the point where the curve crosses the 45° axis. The difference between the two curves results from the fact that fishermen are removing the excess production, represented by the distance AB, so that the equilibrium spawning stock size is given by the distance BC (BC = 0C) for Curve II.

For both situations the combination of survival and fecundity parameters at equilibrium are such that, on the average, each egg spawned by the adult population will produce one egg in the recruit population. As a population is exploited the probability that one egg will survive and produce one egg must increase in order to maintain population stability. This increase in probability of survival is required because the average total number of eggs produced by each recruit over its lifetime is lessened due to fishing mortality.

The relative impact of exploitation on the egg production potential per recruit is illustrated in Fig. 2 for a hypothetical striped bass population. The

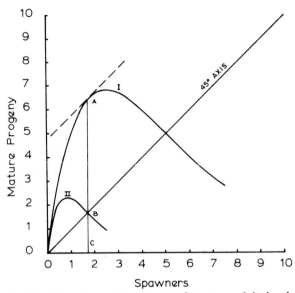

Fig. 1. Hypothetical stock recruitment curves for an unexploited and exploited fish population with recruits measured in terms of sexually mature progeny. Curve I represents the relationship between the abundance of spawners and mature progeny for the unexploited condition. Curve II represents the relationship between the abundance of spawners and mature progeny when the population of mature progeny is reduced by AB/AC. In both cases the population is in equilibrium where the curves cross the 45° axis.

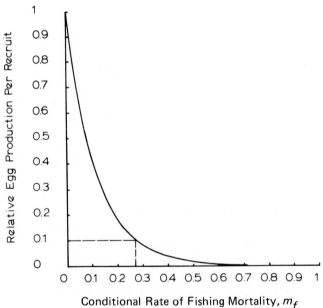

Fig. 2. Relationship between egg production per recruit and conditional rate of fishing mortality for a hypothetical striped bass population.

life history parameters (Table 1) used to generate this curve are those of a population in which 20% of the two-year-old fish and 100% of all subsequent age classes are vulnerable to the fishery. For a conditional rate* of fishing mortality, m_f, of 0.28, the egg production potential per recruit is reduced to only 10% of that for the unexploited population (Fig. 2). Thus, for the population to stabilize, the probability for a spawned egg to survive and reproduce itself must increase tenfold. This increase is a measure of the level of compensation required for the population to persist at that level of fishing mortality. This concept is developed further in the following discussion.

Table 1. Life history parameter values for the hypothetical striped bass population

Age class[a]	(F)[b]	(M)[c]	Survival rate[d]	(E)[e]
1	0.50	0.00	0.4	0
2	0.50	0.00	0.60	0
3	0.50	0.00	0.80	0
4	0.50	0.00	0.80	0
5	0.50	0.00	0.80	0
6	0.50	0.67	0.80	430347
7	0.50	1.00	0.80	559864
8	0.50	1.00	0.80	771958
9	0.50	1.00	0.80	956812
10	0.50	1.00	0.80	1179234
11	0.50	1.00	0.80	1414750
12	0.50	1.00	0.80	1682813
13	0.50	1.00	0.80	1899491
14	0.50	1.00	0.80	2135438
15	0.50	1.00	0.80	2458802

[a]Parameter values for age classes 15−25 are assumed to equal the values for age class 15.

[b]Fraction of fish that are female.

[c]Fraction of females that are sexually mature.

[d]The survival rate is the probability that an individual entering a given age class will survive the influence of natural mortality to enter the next age class.

[e]Fecundity as the mean number of eggs per mature female.

---

*A conditional mortality rate (Ricker 1975, p. 9) due to some single source of mortality (e.g., natural causes, entrainment, impingement, or fishing) is the probability of mortality due to that source in the absence of any other competing source of mortality.

COMPENSATION RATIO

The compensation ratio (CR) is an index of the degree of compensation being utilized by an exploited population at equilibrium and is defined as

$$CR = \frac{v_e}{v_0} ,$$ (1)

where v_e = viability index of the exploited stock and v_0 = viability index of the virgin stock.

The viability index (v) is a theoretical parameter which represents all of the density-dependent factors that operate throughout the life history of a population. It is defined as

$$v = \frac{1}{Ps_0} ,$$ (2)

where P = potential fecundity per recruit and s_0 = probability of survival from density-independent sources of mortality between the deposition of eggs and recruitment.

For purposes of this discussion we will now consider that an individual is recruited to the progeny population at the time it becomes one year of age. Thus, the power plant mortality resulting from entrainment and first-year impingement would be included in the parameter s_0 such that

$$s_0 = s_n s_p ,$$ (3)

where s_n = survival from natural causes of mortality for age class 0 and s_p = survival from power plant mortality for age class 0.

Potential fecundity per recruit (P) is defined as the average lifetime production of eggs per recruit at equilibrium population densities, plus those eggs that would have been produced under conditions of optimum growth and natural mortality. Thus, P is the maximum average lifetime production of eggs per recruit under optimum conditions. This definition is selected to incorporate into the parameter v the effect of density-dependent fecundity and survival that might exist in the adult population. These density-dependent phenomena would tend to decrease the actual average egg production per recruit at equilibrium below the levels that would exist at low population densities.

The potential fecundity per recruit (P) can be calculated from the following relationship:

$$P = \sum_{i=1}^{n} S_i E_i M_i F_i , \tag{4}$$

where

S_i = maximum probability of survival from recruitment to age class i,

E_i = maximum mean fecundity per mature female in age class i,

M_i = maximum fraction of age class i females that are mature,

F_i = fraction of age class i that is female,

n = number of age classes in the population.

The parameter values needed to solve for P can generally be assumed equivalent to their values in a heavily exploited stock. This procedure assumes that the density-dependent growth and survival that might occur within the adults has been fully utilized. Specifically, no further reduction in abundance would cause the recruits to be subject to less natural mortality, to be more fecund upon maturity, or to mature earlier.

The probability for a recruit to survive to age class i (S_i) is simply the product of the annual total survival probabilities to which the recruit is exposed prior to entering the ith age class. The total survival probabilities are higher in the case of an unexploited population than an exploited population. Thus, the value of P is higher for the unexploited population and the corresponding viability index, ν, is lower for the unexploited population (ν_0) than for the exploited population (ν_e). The relationship simply reflects that the probability of an egg surviving to produce a recruit egg has to be higher in the exploited population.

Substitution of Eqs. (2) and (3) into Eq. (1) yields the following:

$$CR = \frac{\nu_e}{\nu_0} = \frac{1/(P_e s_0)}{1/(P_0 s_0)} = \frac{1/(P_e s_n s_p)}{1/(P_0 s_n s_p)} = \frac{P_0 s_n s_p}{P_e s_n s_p} . \tag{5}$$

Since the value of s_p for a virgin stock is unity and the density-independent natural survival rates for both the exploited and unexploited populations are equal, Eq. (5) further reduces to

$$CR = \frac{P_0}{P_e s_p} . \tag{6}$$

Computed in this manner the compensation ratio is a measure of the overall change in the survival and fecundity parameters that must have occurred for a population undergoing exploitation to have stabilized at a new equilibrium. We can examine the impact of fishing mortality on the compensation ratio by plotting CR as a function of the conditional rate of fishing mortality, m_f, with no power plant mortality, that is, $s_p = 1$.

Results of the application of the technique to the striped bass population parameters previously used are presented in Fig. 3. For this set of population parameters the compensation ratio begins to increase sharply above a conditional rate of fishing mortality of about 0.35 (again, assuming that 20% of age class 2 and 100% of all older age classes are in the fishery). For example, $m_f = 0.5$ requires an increase in CR of 64-fold above the conditions existing in the virgin stock just for the population to persist (Fig. 3). At $m_f = 0.6$, this value rises to 160-fold.

Obviously there must be a finite limit to the compensatory ratio above which the population cannot maintain itself. The critical limit is defined as CR_{crit}. Beyond this critical limit the population tends toward extinction unless and until the rate of exploitation is reduced. The difference between the existing compensation ratio of a stock and the critical limit is a measure of the compensatory reserve available to allow the stock to sustain additional stresses such that

$$\text{compensatory reserve} = CR_{crit} - CR . \qquad (7)$$

The value of CR_{crit} may be impossible to ascertain accurately. However, it may be possible to employ estimates of the maximum sustainable rate of exploitation for a species. McFadden (1976) provided a summary of published exploitation rates* for a variety of fish stocks which range from about 0.05 to 0.7. Of the values presented only about 7% were above a value of 0.5. Using equations in Ricker (1975), it can be shown that the combination of a conditional rate of mortality from fishing, m_f, of 0.55 and a conditional rate of natural mortality, n, of 0.2 (see Table 1) results in an exploitation rate of 0.50. If we assume that $m_f = 0.55$ exceeds the compensatory capability of our hypothetical population, then we can compute a value for CR_{crit} that corresponds to that level of fishing. This calculation is done using Eq. (6) with $s_p = 1.0$. The variable P_e in the denominator of Eq. (6) is calculated using Eq. (4) with the S_i values reduced to reflect the additional mortality due to fishing. In our example this value would be about 100 (Fig. 3). In actual application the CR_{crit} value should be determined based on exploitation data specified for the species under investigation.

From a management standpoint it may not be necessary to estimate the CR_{crit} value above which the population can no longer maintain itself. Each CR value is associated with a specific equilibrium stock size. Thus, if we know

---

*Exploitation rate, denoted u by Ricker (1975, p. 5), is the fraction of the initial population killed by the fishery during a year.

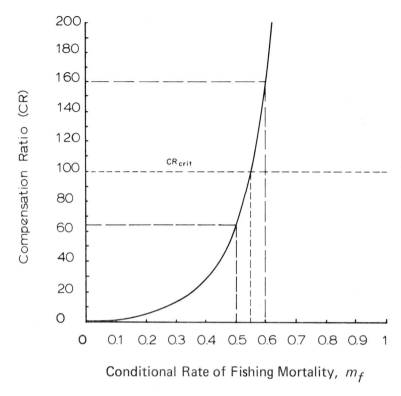

Fig. 3. Effect of fishing on the level of compensation that must occur to stabilize a striped bass population at the specified conditional rate of fishing mortality. A compensation ratio of 100 means that the combination of survival and fecundity parameters in the exploited population must be such that an egg is 100 times as likely to produce an egg in the succeeding generation as it would have been in the virgin population. The critical compensation ratio (CR_{crit}) is the level above which the population can no longer sustain itself.

that serious stock depletion occurs with fishing mortality rates in excess of a given level, we can solve for a CR_{crit} value which corresponds to that rate of exploitation. This procedure would provide a standard against which the impact of additional sources of mortality, such as those imposed by power plants, could be compared.

COMPENSATION RATIO AND POWER PLANT MORTALITY

The impact of a power plant on the compensation ratio of a fish stock is determined by including the probability of surviving power-plant-induced mortality (s_p) into Eq. (6) and plotting CR as a function of $1.0 - s_p^*$ for different levels of pre-existing fishing mortality. This procedure provides a technique for comparing the relative impact of the power plant both as a function of the level of mortality caused by the power plant and as a function of the pre-existing exploitation by fishermen.

This technique was applied with the striped bass population parameters previously used (Table 1) for various levels of fishing mortality (Fig. 4). The

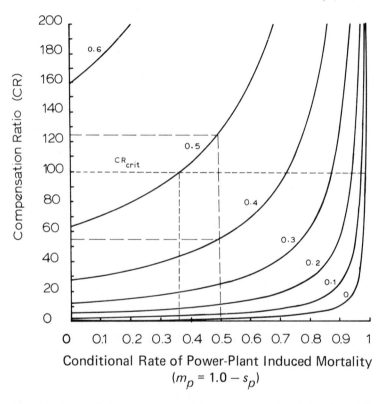

Fig. 4. Relationship between conditional rate of power-plant-induced mortality and the compensation ratio at various levels of the conditional rate of fishing mortality for a hypothetical striped bass population.

impact of power plant mortality on the compensation ratio increases as the fishing mortality rate increases. With no fishing mortality $(m_f = 0.0)$ the compensation ratio is about 2 for a conditional rate of power-plant-induced mortality, m_p, of 0.5. However, for $m_f = 0.5$ the value of CR increases from

---

*This is, $(1.0 - s_p) = m_p$, the conditional rate of power-plant-induced mortality.

62 at m_p = 0.0 to 124 at m_p = 0.5. As a consequence of this observation, it is clear that the probability for power plants to induce sufficient mortality for CR_{crit} to be exceeded increases with increasing fishing mortality.

Assuming that we have determined a CR_{crit} value of 100, as developed in the preceding discussion, we can examine the impact of various levels of power plant mortality on the compensatory reserve of the stock. For example, with m_f = 0.4 and no power plant mortality (m_p = 0.0), the compensatory reserve in the population would be approximately 72 compensatory units (i.e., 100 − 28; Fig. 4). The introduction of a power plant mortality of m_p = 0.5 in the first year of life would decrease the reserve to about 45 units (i.e., 100 − 55). However, if m_f = 0.5, the compensatory reserve of 37 units that exists with m_p = 0.0 (i.e., 100 − 63; Fig. 4) would be entirely consumed by a power plant mortality of m_p = 0.36.

Any analysis of the effects of power-plant-caused mortality on a fish population must consider all of the other sources of mortality that are being imposed by man's activities. Each additional source of mortality will remove some of the compensatory reserve of the stock until ultimately it can sustain no more. When this point is surpassed the stock will dramatically decline and may never be able to recover.

From the standpoint of managing our fish resources we should seek to maintain sufficient compensatory reserves to allow the populations to sustain additional unpredicted mortality. For exploited populations, the mortality caused by power plant operations will, at a minimum, reduce this margin of safety and make the affected resource more vulnerable to destruction.

LITERATURE CITED

Burd, A. C. 1974. The north-east Atlantic herring and the failure of an industry, p. 167–191. *In* F. R. Harden Jones [ed.] Sea fisheries research. John Wiley, New York.

Cushing, D. H. 1968. Fisheries biology: a study in population dynamics. University of Wisconsin Press, Madison, Wisconsin. 200 p.

Hunter, J. R. [ed.] 1976. Report of a colloquium on larval fish mortality studies and their relation to fishery research, January 1975. NOAA Technical Report NMFS CIRC-395. National Marine Fisheries Service, Department of Commerce, Seattle, Washington. 5 p.

McFadden, J. T. 1976. Environmental impact assessment for fish populations, p. 89–137. *In* R. K. Sharma, J. D. Buffington, and J. T. McFadden [eds.] Proc. Workshop Biol. Significance Environ. Impacts. NR-CONF-002. U.S. Nuclear Regulatory Commission, Washington, D.C.

Ricker, W. E. 1975. Computation and interpretation of biological statistics of fish populations. Fish. Res. Board Can., Bull. 191. 382 p.

Development of a Stock-Progeny Model for Assessing Power Plant Effects on Fish Populations*

S. W. Christensen, D. L. DeAngelis, and A. G. Clark†

Environmental Sciences Division
Oak Ridge National Laboratory
Oak Ridge, Tennessee

ABSTRACT

A multi-age-class model, based on simple but general biological principles, is developed to assess the impact of power plants on fish populations. The model is then parameterized in order to produce a variety of stock-progeny relationships, assuming that the stock is always at stable age distribution. The predicted response of the fish stock to power plant cropping of young-of-the-year fish is investigated for each of these stock-progeny relationships. In general, the sensitivity of the equilibrium stock size to cropping is positively related to the slope of the stock-progeny curve at the equilibrium point and, to a lesser extent, negatively related to the slope of the curve at the origin. In addition, the timing of power-plant-induced mortality in relation to the timing of compensation is important. The maximum amount of power-plant-induced mortality that can be tolerated by the stock can be calculated from the slope of the curve at the origin. Application of the model to specific cases will likely need to utilize time-series simulations in addition to the steady-state approach investigated here.

Key words: compensation, fish population dynamics, Leslie matrix, mathematical model, *Morone saxatilis,* power plant impacts, stock-recruitment model, striped bass

INTRODUCTION

In recent years, substantial scientific effort has been expended on methodologies for the biological aspects of environmental impact assessment. In a

---

*Research sponsored in part by the Energy Research and Development Administration under contract with Union Carbide Corporation and in part by the Nuclear Regulatory Commission. ESD Publication No. 1017, Environmental Sciences Division, Oak Ridge National Laboratory.

†Present address: Department of Biology, Stanford University, Stanford, California.

typical situation, ecologists are charged with predicting the ecological effects of constructing a proposed source of impact (the "plant") and operating it for a specified period of time. The kind, quantity, and quality of data which are provided or available for this task vary greatly from case to case, as does the time and level of effort available for analysis.

Van Winkle et al. (1976a) have proposed that the primary role of the ecologist is to identify the components of the ecosystem which are most likely to be affected and to predict, as quantitatively as possible, the magnitude of the effects. As Christensen et al. (1976) have noted, the present state-of-the-art is such that single populations of selected species are most amenable to this kind of analysis, and simulation modeling has been a useful analytical tool. For example, field sampling of the temporal and spatial distributions of young-of-the-year (y-o-y) life stages of fish has been used as input to models that predict an average annual percentage reduction in the number of one-year-old fish of a given species, due to entrainment and impingement at power plants. Life-cycle models can then be linked to the y-o-y models in an attempt to predict the effects of the annual "cropping" of the y-o-y fish on adult population levels over many years (Christensen et al. 1975, Eraslan et al. 1976, Goodyear 1976, Hess et al. 1975, Lawler 1972, 1974, and 1976, USAEC 1972 and 1974, USNRC 1975, Van Winkle et al. 1974).

These "impact models" nearly always contain compensatory (density-dependent) mechanisms in some form as a part of the population dynamics. In general, mortality and/or growth rates somewhere in the life cycle are made to increase as density increases. This causes the impact models to exhibit a degree of stability, corresponding to observations about stability in nature. These impact models have usually been implemented by constructing compensatory functions, and parameterizing these functions with varying degrees of "strength." The long-term changes in adult population size which result from cropping of y-o-y fish by sources of impact acting in a density-independent manner to increase mortality rates early in the life cycle are then examined. The functions necessarily incorporate judgments concerning the nature and degree of compensation, based on a consideration of the dynamics of particular life stages. These judgments result in predicted decreases in the adult population size, which in turn, can vary by well over one order of magnitude, depending on the assumed strength of compensation (Van Winkle et al. 1976b). The primary difficulty with this approach is the general dearth of appropriate data to guide the formulation, and particularly the parameterization, of the compensatory functions. Different magnitudes of predicted impact from the same source thus can result in a confrontation of judgments, often with little apparent hope of using science to settle the issue.

One way in which more information can be incorporated into such impact models is suggested by the classical stock-recruitment approach as developed for numerous fish species (Beverton and Holt 1957, Paulik 1973,

Ricker 1954, 1958, and 1975). In general, stock-recruitment theory is particularly applicable to the simulation modeling of fish populations if recruitment (the probability of survival of an egg to a yearling or to an adult stage) is a decreasing function of stock size, that is, if compensation is operating in the population. The implementation of a specific compensatory function in an impact model implies a certain type of stock-recruitment relationship in the modeled population. Conversely, a particular causal stock-recruitment relationship restricts plausible compensatory mechanisms in an impact model, if the model is to be capable of reproducing that relationship.

Figure 1 shows two forms of stock-recruitment curves which have been widely used in analyzing fish populations. In general, the stock-recruitment theory and models which underlie such curves have been used to analyze the effects of fishing. Since fishing mortality acts on fish older than those that are generally killed by power plants, we felt a need to extend stock-recruitment theory via the development of a new model. The purposes of this paper are (1) to develop a simple model, based on broad biological principles and capable of producing stock-recruitment relationships similar to those in Fig. 1; and (2) to examine the potential impact of power plant cropping of y-o-y fish for

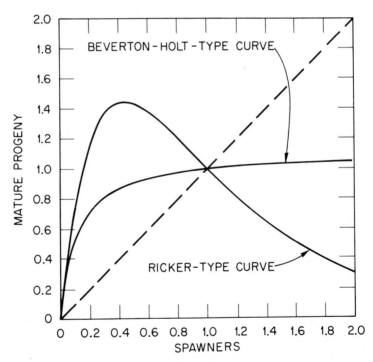

Fig. 1. Two forms of classical stock-recruitment (reproduction) curves: the Beverton-Holt-type curve and the Ricker-type curve. The intersection of such a curve with the (dashed) 45° line is the equilibrium point.

a range of stock-progeny relationships. We use the phrase "stock-progeny" rather than "stock-recruitment" because we have elected to measure reproduction in units of stock value (defined in Eq. 8), rather than as recruitment in the sense of Ricker (1975) (i.e., recruitment to a fishery). If different indices were used in the same model, it would be fully appropriate to refer to it as a "stock-recruitment" model.

DEVELOPMENT OF THE MATHEMATICAL MODEL

The model incorporates age classes (year classes), each representing a cohort of fish which were all born at the same time. Age class 0 (the y-o-y) begins with eggs spawned at the beginning of a year. During their development and metamorphosis through their first year, the y-o-y are subjected to both density-independent and density-dependent mortality. The equation chosen to describe the y-o-y population dynamics is

$$\frac{dX_0(t)}{dt} = -[d_0 + g_0 X_0(t)] X_0(t) - \sum_{i=1}^{N} c_{0,i} X_0(t) X_i(t_s) , \qquad (1)$$

with the terms defined as follows:

N = number of age classes, excluding age class 0 (the y-o-y),

$X_i(t_s)$ = number of fish in the ith age class at the time of spawning, t_s; $i = 1$, N,

$X_0(t)$ = instantaneous number of y-o-y fish at any time t during the year; $t_s \leq t \leq t_s + 1$,

d_0 = rate of density-independent mortality (units of t^{-1}). It incorporates mortality from physical stresses, power plant cropping, the density-independent component of predation, and the density-independent component of mortality ascribable to food limitations,

g_0 = per-individual rate of density-dependent mortality [units: (number $\cdot$ time)$^{-1}$] due to intra-year-class competition and cannibalism, as well as the density-dependent component of predation by other species (e.g., if some predators exhibit "switching" predatory behavior; see Murdoch et al. 1975),

$c_{0,i}$ = density-dependent rate of mortality due to cannibalism by, or competition with, the ith age-class fish of the same species; $i = 1$, N [units: (number $\cdot$ time)$^{-1}$].

The initial size of the y-o-y population is the number of eggs spawned, given by

$$F \equiv X_0(t_s) = \sum_{i=1}^{N} f_i X_i(t_s) , \qquad (2)$$

where the f_i's are age-specific fecundity values with units of eggs per fish. In a population with a 1:1 sex ratio, these values would be one-half the number of eggs per female.

At the end of the year $(t_s + 1)$, mortality is applied to the older age classes according to Eq. (3):

$$X_i(t_s + 1) = [\exp(-d_i)] X_i(t_s) = s_i X_i(t_s) \quad (i = 1, N) , \qquad (3)$$

where d_i is the age-specific instantaneous mortality rate and s_i is the annual probability of survival from age i to age $i + 1$. In the present application of the model, survival rates of older fish are held constant (i.e., are density-independent), and include the effects of both natural and fishing mortality. The assumption that $X_i(t) = X_i(t_s)$, where $t = t_s, ..., t_s + 1$, allows Eq. (1) to be solved analytically. The approximation is acceptable, since annual survival of the y-o-y fish is typically several orders of magnitude lower than that of any older age classes.

The older fish are then advanced into the next age classes:

$$X_i(t_s + 1) \rightarrow X_{i+1}(t_s + 1) \quad (i = 1, N - 1) , \qquad (4)$$

where $t_s + 1$ denotes the beginning of the next spawning year. The oldest age class thus disappears, and the first age class is left vacant. The number of progeny (PR) recruited into the first age class at $t = t_s + 1$ is obtained, in terms of variables at $t = t_s$, by integrating Eq. (1) to yield:

$$PR \equiv X_1(t_s + 1) = \frac{(d_0 + C_0) X_0(t_s) \exp[(-d_0 - C_0)\Delta t]}{d_0 + C_0 + g_0 X_0(t_s) \{1 - \exp[(-d_0 - C_0)\Delta t]\}} , \qquad (5)$$

where Δt is one year and C_0, the total mortality rate due to competition with and cannibalism by older fish, is defined as

$$C_0 \equiv \sum_{i=1}^{N} c_{0,i} X_i(t_s) . \qquad (6)$$

The cycle is completed with the spawning of this next year's egg complement, following Eq. (2). Although we use a differential equation to describe survival through the first year of life, the model is otherwise conceptually a Leslie matrix model.

APPLICATION OF THE MODEL

In the present application of the model we base some parameter estimates on data and vary the remaining parameters to enable the model to take on a variety of stock-progeny relationships. We selected the Hudson River striped bass (*Morone saxatilis*) population as our test case because of our familiarity with this population, the availability of data to estimate a number of the life-cycle parameters, and the present high level of interest in this particular fish population.

A life table (Table 1) was constructed using a reasonable set of adult survival rates (which encompass both natural and fishing survival) and an initial number of one-year-old fish sufficient to sustain plausible fishing yields (USNRC 1975, pp. V-166 to V-178 and XI-84 to XI-86). Fecundity estimates (f_i's, Eq. 2) were based on Texas Instruments (1975, p. VIII-6), but were halved for consistency with an assumed 1:1 sex ratio. In addition, all of the five-year-olds were assumed to be mature. The life table assumes a stable age distribution (SAD). We wish to stress that Table 1 is only a grossly approximate treatment of the Hudson River striped bass stock, and furthermore, that we do not necessarily endorse the validity of these data. As will be demonstrated later, the conclusions reached in the present application of the model are not substantially dependent on the precision or accuracy of these values.

Table 1. Hypothetical life table for Hudson River striped bass stock[a]

Age	Fraction surviving to the next age class[b]	Number[c]	Fecundity rate (eggs per fish)[d]	Number of eggs produced (billions)[e]
1	0.400	7,000,000[f]	0	0
2	0.600	2,800,000	0	0
3	0.689	1,680,000	0	0
4	0.572	1,157,520	0	0
5	0.572	662,101	390,000	258.2
6	0.572	378,722	325,000	123.1
7	0.640	216,629	415,000	89.90
8	0.640	138,643	685,000	94.97
9	0.685	88,731	770,000	68.32
10	0.685	60,781	905,000	55.01
11	0.727	41,635	885,000	36.85
12	0.727	30,269	885,000	26.79
13	0.757	22,005	885,000	19.47
14	0.757	16,658	885,000	14.74
15	0.0	12,610	885,000	11.16
Total		14,306,304[g]		798.52[h]

Table 1 (continued)

Calculated quantities

$$\text{Fecundity per unit stock} \equiv F/S = \frac{7.9852 \times 10^{11}}{1.4306304 \times 10^7} = 55{,}816$$

$$\text{Stock value per yearling} \equiv SV = \frac{1.4306304 \times 10^7}{7 \times 10^6} = 2.044$$

[a]As noted in the text, specified values in this table are based on a combination of data and assumptions. Neither the data nor the assumptions have been rigorously analyzed, since this was not necessary for the purposes of this paper.

[b]Specified values (see text).

[c]For this table, the number of one-year-old fish was specified as seven million. Numbers at other ages are calculated using the values for the fraction surviving to the next age class. For example, the number of two-year-old fish (2.8 million) is obtained by multiplying the number of one-year-old fish (7.0 million) by the fraction of one-year-olds surviving to two years (0.4). Tabulated numbers include both sexes.

[d]It is assumed that 100% of fish age five and older are mature, and that 50% of the fish are male and 50% female. These values are one-half of the values previously tabulated for the number of eggs per female [Texas Instruments 1975 (p. VIII-6)]. No data were given for fecundities of fish age 12 and older; these fecundities were assumed to remain constant at the value tabulated for age-11 fish.

[e]Obtained by multiplying the number of age-i fish by the fecundity rate of age-i fish.

[f]Progeny $\equiv PR = X_1(t_s)$ [see Eq. (5)].

[g]Stock $\equiv S$. In applying the model, we actually chose 14.5 million for the reference equilibrium stock size (EQ). At stable age distribution, this corresponds to a population with slightly more than seven million one-year-old fish, and therefore with slightly changed values for the numbers of older fish and the numbers of eggs produced. At stable age distribution, however, F/S and SV (the only values from the life table used by the model) are dependent only on the age-specific fecundities and survival rates, and not on the numbers of fish.

[h]Fecundity $\equiv F$.

"Stock" is defined as the total of all fish that are at least one year old (see Table 1):

$$S(t_s) = \sum_{i=1}^{N} X_i(t_s) . \tag{7}$$

Progeny (PR from Eq. 5) are evaluated in units of stock value. The stock value (SV) of a single yearling is its total expected contribution to subsequent stocks. This is calculated as

$$SV = 1 + \sum_{i=2}^{N} \prod_{j=1}^{i-1} s_j . \tag{8}$$

Equation (8) simply takes the yearling through all the older age classes (applying mortality at each transition point) and produces the sum of these contributions. The stock value of a group of yearlings can be obtained by multiplying the number of yearlings by the right-hand side of Eq. (8).

For the population depicted in Table 1, S is the total number of Hudson River—spawned fish one year old and older, or 14.306×10^6. The stock value of the seven million age-1 fish at the top of the table is likewise 14.306×10^6. The stock value of one yearling (SV) can be found from Eq. (8), or alternatively from the population sizes in Table 1, by dividing 14.306×10^6 by 7×10^6 to yield 2.044. This value (SV = 2.044) is the stock value per yearling. Stock-progeny relationships in this paper are obtained by arbitrarily varying stock size, solving Eq. (5) for the number of one-year-old progeny (PR), and multiplying this number by SV. The "stock value of progeny" is then plotted on the ordinate, as a function of stock (plotted on the abscissa). This procedure imposes the condition of stable age distribution (SAD) on the population.

At SAD, fecundity per unit stock (F/S) is also a constant. It is calculated as

$$F/S = \frac{f_1 + \sum_{i=2}^{N} \left(f_i \prod_{j=1}^{i-1} s_j \right)}{1 + \sum_{i=2}^{N} \prod_{j=1}^{i-1} s_j} . \tag{9}$$

F/S can also be calculated from Table 1 by dividing 7.9852×10^{11} by 14.306×10^6 to yield 5.582×10^4. If values on the axes of stock-progeny curves in this paper were multiplied by F/S, the curves would be expressed in terms of stock egg production vs eventual progeny egg production (still assuming stable age distribution).

Figure 2 illustrates a parameterization of the model which yields a Ricker-type stock-progeny relationship. The point where the 45° line cuts the curve is the *equilibrium point*, where the stock replaces itself. The *equilibrium stock size* (EQ) can thus be found by moving vertically to the x-axis. The *slope at equilibrium* (ES) is depicted by the dashed line tangent to the

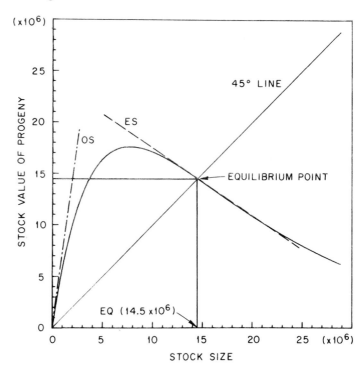

Fig. 2. **A Ricker-type stock-progeny curve generated by the model.** Lines have been added to illustrate the initial slope, OS, the slope at equilibrium, ES, and the technique for finding the equilibrium stock size, EQ.

curve at the equilibrium point. The *slope at the origin* (OS) is depicted by the dot-and-dashed line tangent to the curve at the origin. In applications of the model, we arbitrarily specified values for these three attributes of the curve (OS, EQ, and ES). With the stock survival and fecundity rates specified, the remaining variables which could be adjusted in the model to produce the desired curve were d_0, g_0, and the N values for $c_{0,i}$. Since stable age distribution is assumed and we are not concerned here with the dynamic behavior of the model, it is not necessary to specify the distribution of cannibalistic behavior among adult age classes. Rather, a single variable ($c_{0,s}$) corresponding to the concept of "cannibalism rate per unit stock" was defined. Equation (6) was then rewritten as

$$C_0 \equiv c_{0,s} \sum_{i=1}^{N} X_i . \tag{10}$$

We have found empirically, and confirmed analytically, that if OS, EQ, and ES are specified, values of d_0, g_0, and $c_{0,s}$ are uniquely determined for the range of specified values used in this paper, provided that biologically unreasonable values (e.g., negative g_0's and $c_{0,s}$'s) are excluded. Specifying the slope at the origin uniquely determines d. Increasing the ratio of $c_{0,s}/g_0$ makes the slope at equilibrium steeper (i.e., more negative). Changing both $c_{0,s}$ and g_0 by the same factor changes the equilibrium stock size by that factor. These principles were used in a series of iterative convergence schemes to fit the model to a variety of stock-progeny curves.

To examine the effect of power-plant cropping of y-o-y fish, it was assumed that the potential amount of "density-independent depletion" (DID) was known. The DID percentage is defined as the potential percentage of reduction, after one year, in the number of one-year-old fish due to the impact, assuming compensatory (density-dependent) processes are not operating. If stock S would produce PR yearlings with no impact and no compensation and PR' yearlings with the impact and no compensation, the DID percentage is defined as:

$$\text{DID percentage} = 100 \left(\frac{\text{PR} - \text{PR}'}{\text{PR}} \right). \tag{11}$$

In application to an impact assessment, the DID percentage will usually be available from some sort of model [a "young-of-the-year" model (USNRC 1975), an "entrainment model" (USAEC 1974), etc.]. The present stock-recruitment model can then be used to predict the resulting changed equilibrium stock size. To do this, the DID percentage is incorporated into the stock-recruitment model by increasing the d_0 term by Δd_0, calculated as

$$\Delta d_0 = -\ln \left(1 - \frac{\text{DID percentage}}{100} \right). \tag{12}$$

Increasing d_0 by Δd_0 will result in a reduction of PR in a one-year run of the stock-progeny model without compensation (i.e., with g_0 and C_0 equal to zero) by an amount equal to the DID percentage. Equation (12) is derived in the Appendix.

The eventual percentage reduction in stock size resulting from the impact is calculated as

$$\text{stock depletion percentage} = 100 \left(\frac{\text{EQ} - \text{EQ}'}{\text{EQ}} \right), \tag{13}$$

where EQ is the former equilibrium stock size and EQ' is the new equilibrium stock size when Δd_0 has been added to d_0. EQ and EQ' are calculated from the stock-progeny model, which includes compensation, since in Eq. (5) g_0 and (for Ricker-type cases) $c_{0,s}$ now have positive values (i.e., they are no longer zero). The analytic condition for equilibrium in the model is $SV[\text{PR}(t_s + 1)] = S(t_s)$.

One final quantity of interest merits definition. An "impact factor" can be defined as

$$\text{impact factor} = \frac{\text{stock depletion percentage}}{\text{DID percentage}}. \qquad (14)$$

An impact factor larger than 1.0 indicates that the eventual predicted effect of an impact on the stock size, with compensation operating, is greater than the annual predicted effect of the impact on the yearlings without compensation. The impact factor is a direct measure of the sensitivity of the stock, in terms of relative change, to a given level of impact.

The primary focus of our application of the model was to examine the range of impact factors which are predicted by the model, given a range of stock-progeny relationships at constant stable age distribution. The next section describes the results of this analysis and of an analysis of the sensitivity of the impact factor to the specified variables (e.g., EQ, ES, OS, fecundity, and survival rates of older fish). For the baseline investigations, the equilibrium stock size (i.e., one-year-old and older fish) without impact was arbitrarily specified as 14.5 million fish, and the fecundity and survival rates shown in Table 1 were used. A 10% DID percentage was used as the reference impact case.

RESULTS

Beverton-Holt-type stock-progeny curves are obtained from the model by setting $c_{0,s}$ equal to zero. The term $c_{0,s}$ represents cannibalism by older age classes, and since the size of these older age classes is dependent on prior events, $c_{0,s}$ introduces a "lag time" or "memory" into the system. When it is set equal to zero, larger stock sizes never produce smaller absolute numbers of one-year-old offspring. In this case, our model (Eq. 1 with $c_{0,s} = 0$) is identical to the starting point for the derivation of the Beverton-Holt formulation (see Beverton and Holt 1957, p. 48).

Table 2 and Fig. 3 present the three Beverton-Holt-type curves we investigated. In our model, when $c_{0,s}$ is set to zero and the slope at the origin (OS) and the equilibrium stock size (EQ) are specified, d_0, g_0, and the slope at equilibrium (ES) are uniquely determined. Since the slope is everywhere positive in a Beverton-Holt-type curve, the slope at equilibrium is also positive. It can be shown analytically that the slope at equilibrium is the inverse of the slope at the origin (see Ricker 1975, p. 346).

Values of the impact factor for a 10% DID percentage range from about 2 for the "mild" curve in Case 1 to about 0.9 for the "severe" curve in Case 2a. Figure 4 shows the baseline curve and the impacted curve for Case 2. The impacted equilibrium stock size is reduced about 10% by the 10% DID percentage, as implied by the impact factor value of 1.01 (Table 2).

Table 2. Beverton-Holt-type cases[a]

Case	OS[b]	ES	EQ	d_0	g_0	$c_{o,s}$	For 10% density-independent depletion				
							New d_0 value	OS'	ES'	EQ'	Impact factor
1	2.00	+0.50	14.5E6	10.95	1.36E-11	0	11.06	1.80	+0.55	11.7E6	1.92
2	10.0	+0.10	14.5E6	9.34	1.04E-10	0	9.45	9.02	+0.11	13.0E6	1.01
2a	100.0	+0.01	14.5E6	7.04	8.65E-10	0	7.14	90.26	+0.011	13.2E6	0.876

[a]Cases 1, 2, and 2a are Beverton-Holt-type cases which differ in OS and ES.

[b]Explanation of symbols: OS – slope at the origin; ES – slope at equilibrium; EQ – equilibrium stock size; the prime symbol (') indicates a value which, because of the impact, has changed; d_0 – density-independent mortality rate for fish during their first year of life; g_0 – density-dependent mortality rate (per individual) for y-o-y fish, due to intra-year-class competition and cannibalism; $c_{o,s}$ – density-dependent mortality rate (per unit stock) for y-o-y fish, due to competition with, or cannibalism by, fish one-year-old or older.

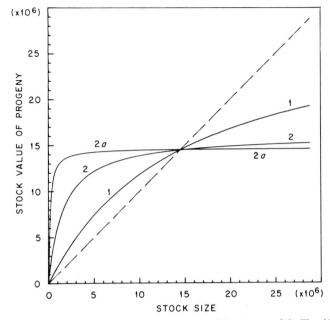

Fig. 3. Three Beverton-Holt-type cases generated by the model. The identifying numbers correspond to the case numbers in Table 2.

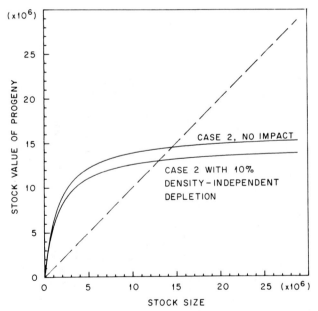

Fig. 4. Stock-progeny curves for Case 2 with and without a power plant impact. Parameters are given in Table 2.

Table 3 and Fig. 5 show the fit of the model to a variety of Ricker-type stock-progeny curves. In these cases, OS, EQ, and ES were specified, thus determining the appropriate values of d_0, g_0, and $c_{0,s}$. Complete freedom in specifying OS was not available, since specifying ES (and to a lesser extent, EQ) restricts OS to a somewhat limited range. EQ was held constant at 14.5 million fish, and ES values were chosen in decrements of $-\frac{1}{3}$ between the range of $+\frac{2}{3}$ and -2. OS values were chosen to approximately correspond to slopes at the origin of stock-recruitment curves with similar slopes at equilibrium presented by Ricker (1975, p. 284).

Examination of the impact factor values for these cases shows that the impact factor increases as the slope at equilibrium increases (i.e., becomes more positive). Furthermore, an interesting approximate generalization can be drawn from these cases, namely:

$$\text{impact factor} \approx \frac{1}{1 - ES}. \qquad (15)$$

(Note that the limiting value of ES is 1.0, which is the slope of the 45° line in Fig. 2.) This relationship (which we will refer to as the empirical approximation for the impact factor) holds equally well (i.e., approximately) for the

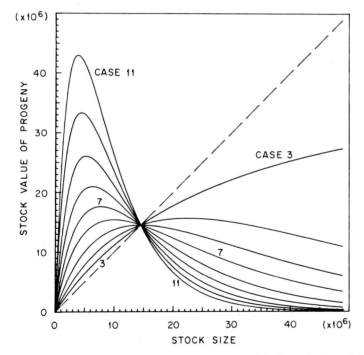

Fig. 5. **Nine Ricker-type cases generated by the model.** Cases 3, 7, and 11 are labeled. Parameters for all of these cases (3 through 11) are given in Table 3.

Table 3. Ricker-type cases[a]

Case	OS[b]	ES	EQ	d_o	g_o	$c_{o,s}$	For 10% density-independent depletion				
							New d_o value	OS'	ES'	EQ'	Impact factor
3	1.50	+0.67	14.5E6	11.24	6.93E-12	5.15E-11	11.34	1.35	+0.74	10.2E6	2.93
4	2.00	+0.33	14.5E6	10.95	2.93E-12	3.48E-8	11.06	1.80	+0.43	12.3E6	1.83
5	3.02	0.0	14.5E6	10.53	8.24E-12	4.46E-8	10.64	2.74	+0.088	13.0E6	1.01
6	5.00	-0.33	14.5E6	10.03	1.57E-11	5.77E-8	10.14	4.50	-0.255	13.4E6	0.745
7	7.00	-0.67	14.5E6	9.70	1.50E-11	8.27E-8	9.80	6.30	-0.584	13.6E6	0.599
8	10.0	-1.0	14.5E6	9.34	1.53E-11	1.06E-7	9.45	9.00	-0.915	13.8E6	0.499
9	15.0	-1.33	14.5E6	8.94	1.78E-11	1.28E-7	9.04	13.51	-1.25	13.9E6	0.426
10	23.0	-1.67	14.5E6	8.51	2.12E-11	1.50E-7	8.61	20.70	-1.58	14.0E6	0.371
11	34.0	-2.00	14.5E6	8.12	2.30E-11	1.73E-7	8.22	30.61	-1.92	14.0E6	0.329

[a] Cases 3-11 are Ricker-type cases which differ in OS and ES.

[b] For an explanation of the symbols, see Footnote b to Table 2.

Ricker-type and the Beverton-Holt-type cases examined in this paper. We put forward the hypothesis that it also would hold for other models, including Ricker's (1975) and Beverton-Holt's (1957), if the terms in those models corresponding to d_0 are similarly increased to simulate power plant impacts. Figure 6a is a graph of the empirical approximation for the impact factor.

Examination of Table 3 reveals a number of other interesting relationships. For example, the slope at the origin is decreased in direct proportion to the level of impact, according to Eq. (16):

$$OS' = OS \left(\frac{DID \text{ percentage}}{100} \right), \tag{16}$$

where OS' is the eventual new slope at the origin for the impacted populations. Since the equilibrium point is at the origin when the slope at the origin is unity and disappears when the slope at the origin is less than unity, a "critical impact" (CI) can be defined as that level of DID which is just large enough to cause eventual extinction of the stock (i.e., to cause the slope at the origin to decrease to unity). The critical impact (CI) expressed as a percentage can be calculated as

$$CI = 100 \left(\frac{OS - 1}{OS} \right). \tag{17}$$

This same concept, in the context of harvests of pre-reproductive fish, was expressed by Ricker (1954, p. 582). Figure 6b is a graph of the critical impact as a function of the slope at the origin. Another interesting observation from Table 3 is that the slope at equilibrium of an impacted population (ES') is always greater (i.e., more positive) than that of the nonimpacted population (ES) (Table 3). This implies that the impacted population has less "compensatory reserve," and will be more vulnerable to additional impacts (i.e., will decrease by a proportionally greater amount in response to an additional impact of the same magnitude as the original impact).

Case 7, with an equilibrium stock size of 14.5 million fish, a slope of 7 at the origin, and a slope of −0.67 at equilibrium was chosen as the baseline case for a series of analyses of sensitivity of the impact factor to various parameters in the model. These are presented below.

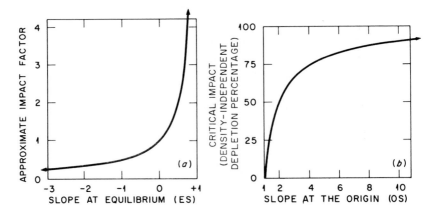

Fig. 6. **Relationships between stock-progeny model parameters and impact parameters.** (*a*): Empirical approximation for the impact factor as a function of ES, the slope at equilibrium. This is a graph of Eq. (15). The maximum possible value for ES is 1. (*b*): Critical impact (in terms of the density-independent depletion percentage) as a function of OS, the slope at the origin. This is a graph of Eq. (17). The minimum possible value for OS is 1.

Sensitivity to the DID Percentage

Table 4 shows the relationship between the DID percentage and the impact factor for Case 7. For each DID percentage, the d_0 value of 9.699 in the nonimpacted baseline case was increased according to Eq. (12). The values for the stock depletion percentage (Eq. 13) and the impact factor (Eq. 14) in Table 4 confirm the previous generalization that the vulnerability of a stock increases as the level of impact increases. Figure 7 shows the stock-progeny curves for most of these levels of impact.

Sensitivity to the Slope at the Origin

Table 5 shows the effect of varying d_0 in order to change the slope at the origin, while specifying unchanged values for EQ and ES. The impact factor is seen to be very slightly sensitive to the choice of the slope at the origin.

Sensitivity to the Slope at Equilibrium

Table 6 shows the effect of varying g_0 and c_0 to attain different slopes at equilibrium (ES), while specifying unchanged values for OS and EQ. The fractional changes in ES here are similar to the fractional changes in OS in

Table 4. Sensitivity analysis for density-independent depletion level

DID[a] (%)	OS	ES	EQ	$d_o{}^a$	$g_o{}^a$	$c_{o,s}{}^a$	Stock depletion percentage of Eq. (13) (%)	Impact factor
None (baseline Case 7)	7.00	−0.67	14.5E6	9.699	1.50E-11	8.27E-8		
1	6.93	−0.66	14.4E6	9.709	1.50E-11	8.27E-8	0.57	0.573
10	6.30	−0.58	13.6E6	9.804	1.50E-11	8.27E-8	6.0	0.599
25	5.25	−0.44	12.1E6	9.987	1.50E-11	8.27E-8	16	0.649
50	3.50	−0.11	8.94E5	10.392	1.50E-11	8.27E-8	38	0.766
75	1.75	0.47	3.80E6	11.085	1.50E-11	8.27E-8	74	0.984
90	0.700		None	12.002	1.50E-11	8.27E-8	100 (extinction)	

[a]DID − density-independent depletion level; for an explanation of the other symbols, see Footnote b to Table 2.

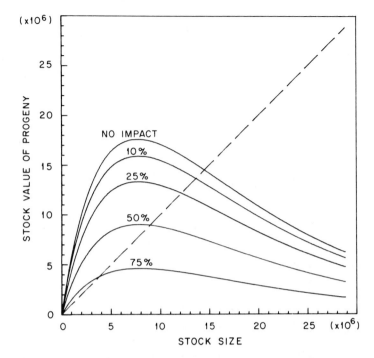

Fig. 7. Stock-progeny curves for Case 7 with no impact (top curve) and with 10, 25, 50, and 75% levels of density-independent depletion. Parameters are given in Table 4.

Table 5. The impact factor is roughly two to three times more sensitive to changes in the slope at equilibrium (Table 6) than to changes in the slope at the origin (Table 5).

Sensitivity to the Equilibrium Stock Size

Table 7 shows the effect of varying the specified value of EQ while holding OS and ES constant. The impact factor is not measurably affected as the assumed equilibrium stock size is varied by a factor of 4. Despite this insensitivity, it would not be correct to conclude that EQ is a totally unimportant quantity in the model, since to at least a small extent it determines the range of permissible OS values for a specified value of ES.

Sensitivity to Stock Fecundity Values

In all cases until now, the fecundity per unit stock has been held constant at the values shown in Table 1. Table 8 shows the effect of doubling or

Table 5. Sensitivity analysis for the slope at the origin[a]

Case[b]	OS	Base conditions			1% DID (d_0 increased by 0.01005)			10% DID (d_0 increased by 0.10536)		
		F/S ES EQ	d_0 g_0 $c_{0,s}$		New d_0 value	OS' ES' EQ'	Impact factor	New d_0 value	OS' ES' EQ'	Impact factor
7	7.00	55,816 -0.67 14.5E6	9.699 1.50E-11 8.27E-8		9.709	6.93 -0.66 14.4E6	0.573	9.804	6.30 -0.58 13.6E6	0.599
7h	6.00	55,816 -0.67 14.5E6	9.853 7.55E-12 9.36E-8		9.863	5.94 -0.66 14.4E6	0.584	9.958	5.40 -0.57 13.6E6	0.611
7i	8.00	55,816 -0.67 14.5E6	9.565 2.16E-11 7.66E-8		9.575	7.92 -0.66 14.4E6	0.568	9.670	7.20 -0.59 13.6E6	0.592

[a]Groups of three numbers appearing under columns identified by three labels correspond in sequence to the sequence of the labels. For example, for Case 7h under base (no impact) conditions, $d_0 = 9.853$, $g_0 = 7.55E-12$, and $c_{0,s} = 9.36E-8$. Explanation of symbols: DID – density-independent depletion; OS – slope at the origin; ES – slope at equilibrium; EQ – equilibrium stock size; F/S – fecundity per unit stock, the prime symbol (') indicates a value which, because of the impact, has changed; d_0 – density-independent mortality rate for y-o-y fish; g_0 – density-dependent mortality rate (per individual) for y-o-y fish, due to intra-year-class competition and cannibalism; $c_{0,s}$ – density-dependent mortality rate (per unit stock) for y-o-y fish, due to competition with, or cannibalism by, fish one-year-old or older.

[b]Cases 7h and 7i are the same as Case 7 except for different slopes at the origin (OS).

Table 6. Sensitivity analysis for the slope at equilibrium[a]

Case[b]	ES	Base conditions		1% DID (d_0 increased by 0.01005)			10% DID (d_0 increased by 0.10536)		
		F/S OS EQ	d_0 g_0 $c_{o,s}$	New d_0 value	OS' ES' EQ'	Impact factor	New d_0 value	OS' ES' EQ'	Impact factor
7	−0.67	55,816 7.00 14.5E6	9.699 1.50E-11 8.27E-8	9.709	6.93 −0.66 14.4E6	0.573	9.804	6.30 −0.58 13.6E6	0.599
7m	−0.57	55,816 7.00 14.5E6	9.699 2.02E-11 7.03E-8	9.709	6.93 −0.56 14.4E6	0.605	9.804	6.30 −0.49 13.6E6	0.631
7n	−0.77	55,816 7.00 14.5E6	9.699 9.97E-12 9.65E-8	9.709	6.93 −0.76 14.4E6	0.547	9.804	6.30 −0.68 13.6E6	0.572

[a]For an explanation of the table layout and the meaning of the symbols, see Footnote a to Table 5.
[b]Cases 7m and 7n are the same as Case 7 except for different slopes at equilibrium (ES).

Table 7. Sensitivity analysis for the equilibrium stock size[a]

Case[b]	Base conditions			1% DID (d₀ increased by 0.01005)			10% DID (d₀ increased by 0.10536)		
	EQ	F/S OS EQ	d_0 g_0 $c_{o,s}$	New d_0 value	OS' ES' EQ'	Impact factor	New d_0 value	OS' ES' EQ'	Impact factor
7	14.5E6	55,816 7.00 −0.67	9.699 1.50E-11 8.27E-8	9.709	6.93 −0.66 14.4E6	0.573	9.804	6.30 −0.58 13.6E6	0.599
7f	7.25E6	55,816 7.00 −0.67	9.699 2.99E-11 1.65E-7	9.709	6.93 −0.66 7.21E6	0.573	9.804	6.30 −0.58 6.82E6	0.599
7g	29.0E6	55,816 7.00 −0.67	9.699 7.48E-12 4.13E-8	9.709	6.93 −0.66 28.8E6	0.573	9.804	6.30 −0.58 27.3E6	0.599

[a] For an explanation of the table layout and the meaning of the symbols, see Footnote a to Table 5.

[b] Cases 7f and 7g are the same as Case 7 except for different equilibrium stock sizes (EQ).

Table 8. Sensitivity analysis for the fecundity per unit stock[a]

Case[b]	F/S	Base conditions OS ES EQ	Base conditions d_0 g_0 $c_{0,s}$	1% DID (d_0 increased by 0.01005) New d_0 value	1% DID OS' ES' EQ'	1% DID Impact factor	10% DID (d_0 increased by 0.10536) New d_0 value	10% DID OS' ES' EQ'	10% DID Impact factor
7	55,816	7.00 −0.67 14.5E6	9.699 1.50E-11 8.27E-8	9.709	6.93 −0.66 14.4E6	0.573	9.804	6.30 −0.58 13.6E6	0.599
7a	27,908	7.00 −0.67 14.5E6	9.005 2.78E-11 8.31E-8	9.015	6.93 −0.66 14.4E6	0.572	9.110	6.30 −0.58 13.6E6	0.597
7b	111,632	7.00 −0.67 14.5E6	10.392 8.05E-12 8.22E-8	10.402	6.93 −0.66 14.4E6	0.575	10.497	6.30 −0.58 13.6E6	0.600

[a] For an explanation of the table layout and the meaning of the symbols, see Footnote a to Table 5.
[b] Cases 7a and 7b are the same as Case 7 except for different values of F/S (eggs per unit stock).

halving the assumed fecundity values in Table 1. It can be seen that the impact factor is only very slightly sensitive to variations in assumed fecundity when OS, EQ, and ES are held constant.

Sensitivity to Survival Rates of Older Age Classes

Table 9 presents the reference case (Case 7), together with a case having high survival rates of older age classes and consequently higher fecundity (Case 7c) and a case with high survival rates but the reference level of total fecundity (Case 7cl). The "high survival" cases are altered in that the fraction surviving to the next age class is held constant at 0.8 for age classes 3–14 (see Table 1). Examination of Table 9 shows that the impact factor value is very slightly sensitive to the values used for adult survival rates. Case 7cl shows that this sensitivity is actually a result of the fact that fecundity changes as the survival rates change.

Sensitivity to Timing of DID in Relation to Compensation

For this analysis, the model was modified slightly to provide for three separate periods within age class 0. During the first (f) and last (l) periods, only density-independent mortality was included (i.e., g_0 and $c_{0,s}$ were set to zero). During the middle (m) period, both density-independent and density-dependent mortality were included (i.e., d_0, g_0, and $c_{0,s}$ terms all positive). In this version of the model, Eq. (5) is rewritten as:

$$X_1(t_s + 1) = \frac{(d_m + C_0)X_0(t_s)\exp(-d_f\Delta t_f)}{d_m + C_0 + g_0 X_0(t_s)\exp(-d_f\Delta t_f)}$$

$$\times \frac{\exp[(-d_m - C_0)\Delta t_m]\exp(-d_l\Delta t_l)}{1 - \exp[(-d_m - C_0)\Delta t_m]}, \tag{18}$$

where the f, m, and l subscripts denote the first, middle, and last periods, respectively, and C_0 is calculated from Eq. (10). The first period was defined as 10 days ($\Delta t_f = 0.0274$ year), the middle period as 91 days ($\Delta t_m = 0.249$ year), and the last period as 264 days ($\Delta t_l = 0.723$ year). The three d values for the nonimpacted case were chosen so that the overall probability of surviving each period would be the same if compensatory mechanisms were not operating (i.e., g_0 and $c_{0,s}$ equal zero, which of course they are already in Eq. 18 except for the middle period). Values for g_0 and $c_{0,s}$ (Case 7jo, Table 10) were then calculated in order to produce the "reference" Case 7 values of

Table 9. Sensitivity analysis for survival rates of older age classes[a]

Case[b]	Adult survival rates	Base conditions			1% DID (d_o increased by 0.01005)			10% DID (d_o increased by 0.10536)		
		F/S	OS ES EQ	d_o g_o $c_{o,s}$	New d_o value	OS' ES' EQ'	Impact factor	New d_o value	OS' ES' EQ'	Impact factor
7	Standard	55,816	7.00 −0.67 14.5E6	9.699 1.50E-11 8.27E-8	9.709	6.93 −0.66 14.4E6	0.573	9.804	6.30 −0.58 13.6E6	0.599
7c	High	161,123	7.00 −0.67 14.5E6	10.974 5.91E-12 8.18E-08	10.984	6.93 −0.66 14.4E6	0.576	11.079	6.30 −0.58 13.6E6	0.602
7cl	High	55,816	7.00 −0.67 14.5E6	9.913 1.53E-11 8.27E-8	9.923	6.93 −0.66 14.4E6	0.574	10.018	6.30 −0.59 13.6E6	0.599

[a]For an explanation of the table layout and the meaning of the symbols, see Footnote a to Table 5.

[b]See text for an explanation of the differences among these cases.

Table 10. Sensitivity analysis for timing of density-independent depletion in relation to compensation[a]

Case[b]	Depletion conditions	OS ES EQ	t_f t_m t_i (year)	d_f d_m d_i (year^{-1})	g_0[c] $c_{o,s}$	Impact factor
7jo	No DID	7.00 −0.67 14.5E6	0.0274 0.249 0.723	118.00 12.968 4.4699	4.86E-10 3.70E-7	
7j	10% DID before compensation	6.30 −0.60 13.9E6	0.0274 0.249 0.723	121.85 12.968 4.4699	4.86E-10 3.70E-7	0.355
7k	10% DID concurrent with compensation	6.30 −0.59 13.7E6	0.0274 0.249 0.723	118.00 13.390 4.4699	4.86E-10 3.70E-7	0.571
7l	10% DID after compensation	6.30 −0.58 13.6E6	0.0274 0.249 0.732	118.00 12.968 4.6155	4.86E-10 3.70E-7	0.629
7jl	10% total depletion, distributed equally	6.30 −0.58 13.7E6	0.0274 0.249 0.723	119.29 13.018 4.518	4.86E-10 3.70E-7	0.517

[a]See Footnote a to Table 5 for explanation of the table layout and the meaning of DID, OS, ES, EQ, g_0, and $c_{o,s}$. The other symbols are explained in the text. In this table, F/S (fecundity per unit stock) is constant at 55,816.

[b]See text for an explanation of the differences among these cases.

[c]These values (g_0 and $c_{o,s}$) are applied only during the middle period, and are zero for the first and last periods.

OS, ES, and EQ (Table 3); note that since these terms now operate during only part of the y-o-y peroid, they have different values from Case 7, where these terms (and hence compensation) operate during the entire y-o-y period. When only one period was to be impacted in the model (Cases 7*j*, 7*k*, and 7*l* in Table 10), the impact was then introduced into this period via the following modified form of Eq. (12):

$$\Delta d_x = \frac{-\ln\left(1 - \dfrac{\text{DID percentage}}{100}\right)}{\Delta t_x} \qquad (x = f, m, \text{ or } l) . \qquad (19)$$

For Case 7*jl* in Table 10, the same total degree of DID (10%) was distributed equally among the three periods by solving Eq. (19) for each period with a DID percentage of 3.451%, and for each period adding this increment to the base (Case 7*jo*) value for d_x. Thus, density-independent survival for each period was 96.55% as high as in Case 7*jo*, and since the periods are sequential, density-independent survival for the total y-o-y period was $(0.9655)^3 = 0.900$ as high, corresponding to the desired 10% DID percentage.

Table 10 shows the results of this analysis for a 10% level of DID, introduced during various periods of the y-o-y age class. It is clear that the impact factor is sensitive to the timing of imposition of the impact in relation to the timing of compensation. Vulnerability of the stock to an impact is greatest when the impact is imposed on the population after the compensatory period, and is least when the impact precedes the compensatory period. This result is intuitively sensible. The lower value of the impact factor when the impact is imposed concurrently with compensation (Case 7*k* in Table 10) compared with a 10% impact for the standard case (e.g., Case 7 in Table 8) is likely a reflection of the fact that the density-independent mortality which precedes compensation in Case 7*k* is functionally equivalent to a decrease in fecundity, which reduces the impact factor. (See Case 7b in Table 8).

DISCUSSION

The application of our model is intentionally "results oriented" in the sense that the model was "forced" to assume a variety of arbitrary stock-progeny relationships. For each set of parameters which produced a desired form of relationship, the impact factor was evaluated at one or more levels of impact. While the reasons for many of the results are not intuitively obvious, some interesting and relevant generalizations emerge:

1. The impact factor is a measure of the sensitivity (or vulnerability) of the stock to power plant cropping. In general, the impact factor is positively related to the slope at the equilibrium point (i.e., increases as the slope becomes more positive; see Fig. 6a) and to a lesser extent is negatively related to the slope at the origin. Thus, the greater the deviation of the Ricker-type or Beverton-Holt-type stock-progeny relationship from the 45° reference line, the less sensitive the stock is to power plant cropping. The timing of cropping in relationship to compensation is, however, quite important in determining the impact factor.

2. For a wide range of stock-progeny relationships, power plant cropping in our model always acts to decrease, and never to increase, the equilibrium stock size from its unimpacted value. This finding is investigated analytically in DeAngelis, Christensen, and Clark (*in preparation*).

3. The numbers, fecundities, and death rates of the older age classes are not particularly important in determining the sensitivity to power plant cropping, if the stock-progeny relationship is known.

Taken together, these results indicate that if a casual relationship can be established between stock size and production of progeny in a population of interest, some inferences may be possible about the eventual effect of increased mortality during the y-o-y period. The empirical approximation for the impact factor (Eq. 15) will indicate the magnitude of the eventual effect of modest impacts, if the slope at equilibrium is known and the model describes reasonably well the population and the forces acting on it. In addition, the critical impact (Eq. 17) indicates a level of impact which, if approached, would entail a substantial risk to the population in the long term.

The difficulty in applying the results should, however, not be underestimated. These results assume stable age distribution, whereas the information needed to infer the SAD stock-progeny relationship will necessarily come from populations which are varying in size. Such populations will almost surely not have a stable age distribution. It would be necessary to have much more detailed information than is generally available, and over a long period of time, to establish the actual relationship between "stock size" and the "stock value of progeny."

Practical application of this model to specific cases will require an approach based on dynamic simulations. Allowing parameters in the model to vary randomly can result in different kinds of periodic behavior, depending on the mean values of the parameters (DeAngelis, Christensen, and Clark, *in preparation*). If the available data consist of a time-series of catch data which represents sub-samples of only certain ages of the stock and from which specific age-structure information is not available, a catch index can be constructed in the model. Determination of possible values for the unknown parameters (d_0, g_0, and the N values of $c_{0,i}$) can be based on the ability of the simulated catch index to reproduce the observed catch data. It is important to note that age-specific fecundity values and death rates, which are relatively unimportant in determining the sensitivity to power plant cropping when the stock-progeny relationship is known, may be a quite important component in the dynamic simulations needed to establish the stock-progeny relationship. In addition, more than one stock-progeny relationship (or none at all) may be consistent with the available information, and the prediction of impact may be indeterminate. Nonetheless, given the lack of direct data to serve as a guide to parameterizing detailed y-o-y compensatory functions, we believe that further work with this and similar models will provide additional useful insights into the question of the potential effects of power plants on fish stocks.

ACKNOWLEDGMENTS

The authors wish to express their sincere appreciation to L. W. Barnthouse, R. V. O'Neill, and particularly to W. Van Winkle, for their thoughtful and helpful comments on the manuscript.

LITERATURE CITED

Beverton, R. J. H., and S. J. Holt. 1957. On the dynamics of exploited fish populations. Fish. Invest., Ser. 2, 19. 533 p.

Christensen, S. W., W. Van Winkle, and P. C. Cota. 1975. Effects of Summit Power Station on striped bass populations. Testimony before the Atomic Safety and Licensing Board in the Matter of Summit Power Station, Units 1 and 2, USAEC Docket Nos. 50-450 and 50-451, March 1975.

Christensen, S. W., W. Van Winkle, and J. S. Mattice. 1976. Defining and determining the significance of impacts: concepts and methods, p. 191–219. In R. K. Sharma, J. D. Buffington, and J. T. McFadden [eds.] Proc. Workshop Biol. Significance Environ. Impacts, NR-CONF-002. U.S. Nuclear Regulatory Commission, Washington, D.C.

Eraslan, A. H., W. Van Winkle, R. D. Sharp, S. W. Christensen, C. P. Goodyear, R. M. Rush, and W. Fulkerson. 1976. A computer simulation model for the striped bass young-of-the-year population in the Hudson River. ORNL/NUREG-8. Oak Ridge National Laboratory, Oak Ridge, Tennessee.

Goodyear, C. P. 1976. Testimony of C. Phillip Goodyear. Testimony in United States Environmental Protection Agency (Region IV) Proceedings in the Matter of Brunswick Steam Electric Plant, NPDES Permit No. NC0007064, Docket No. AHNC512NR, June 2–16, 1976.

Hess, K. W., M. P. Sissenwine, and S. B. Saila. 1975. Simulating the impact of the entrainment of winter flounder larvae, p. 1–29. In S. B. Saila [ed.] Fisheries and energy production: a symposium. Lexington Books, D. C. Heath & Co., Lexington, Massachusetts.

Lawler, J. P. 1972. Effect of entrainment and impingement at Indian Point on the population of the Hudson River striped bass. Testimony before the Atomic Safety Licensing Board, Indian Point Unit 2, USAEC Docket No. 50-247, October 30, 1972.

―――. 1974. Effect of entrainment and impingement at Cornwall on the Hudson River striped bass population. Testimony before the Federal Power Commission in the matter of Cornwall, USFPC Project No. 2338, October 1974.

―――. 1976. Testimony of Dr. John P. Lawler for the Carolina Power and Light Company. Testimony in United States Environmental Protection Agency (Region IV) Proceedings in the Matter of Brunswick Steam Electric Plant, NPDES Permit No. NC0007064, Docket No. AHNC512NR, June 2–16, 1976.

Murdoch, W. W., S. Avery, and Michael E. B. Smyth. 1975. Switching in predatory fish. Ecology **56**:1094–1105.

Paulik, G. J. 1973. Studies of the possible form of the stock-recruitment curve. Rapp. P.-V. Reun. Cons. Perm. Int. Explor. Mer. **164**:302–315.

Ricker, W. E. 1954. Stock and recruitment. J. Fish. Res. Board Can. **11**:559–623.

―――. 1958. Handbook of computations for biological statistics of fish populations. Fish. Res. Board Can., Bull. 119. 300 p.

―――. 1975. Computation and interpretation of biological statistics of fish populations. Fish. Res. Board Can., Bull. 191. 382 p.

Texas Instruments, Inc. 1975. First annual report for the multiplant impact study of the Hudson River Estuary, Vol. I, July 1975.

U.S. Atomic Energy Commission (USAEC). 1972. Final environmental statement related to operation of Indian Point Nuclear Generating Plant, Unit No. 2, USAEC Docket No. 50-247, Vols. I and II, September 1972.

―――. 1974. Final environmental statement related to the proposed Summit Power Station, Units 1 and 2, USAEC Docket Nos. 50-450 and 50-451, July 1974 (Appendix G).

U.S. Nuclear Regulatory Commission (USNRC). 1975. Final environmental statement related to operation of Indian Point Nuclear Generating Station, Unit No. 3, USNRC Docket No. 50-286, Vols. I and II, February 1975.

Van Winkle, W., B. W. Rust, C. P. Goodyear, S. R. Blum and P. Thall. 1974. A striped bass population model and computer programs. ORNL/TM-4578. Oak Ridge National Laboratory, Oak Ridge, Tennessee.

Van Winkle, W., S. W. Christensen, and J. S. Mattice. 1976a. Two roles of ecologists in defining and determining the acceptability of environmental impacts. Int. J. Environ. Stud. **9**:247–254.

Van Winkle, W., S. W. Christensen, and G. Kauffman. 1976b. Critique of the compensation function used in the LMS Hudson River striped bass models. ORNL/TM-5437. Oak Ridge National Laboratory, Oak Ridge, Tennessee.

APPENDIX

This Appendix presents a derivation for Eq. (12) in the text. We begin with Eq. (5) from the text, but with the terms C_0 and g_0 set to zero and a Δt of one year:

$$X_1(t_s + 1) = PR = \frac{d_0 X_0(t_s)\exp(-d_0)}{d_0} . \tag{A.1}$$

Simplifying, taking the natural logarithm of both sides, and rearranging, yields

$$d_0 = -\ln\left(\frac{PR}{X_0(t_s)}\right) . \tag{A.2}$$

For the impacted population, but still with C_0 and g_0 set equal to zero, we obtain from Eq. (5):

$$X_1'(t_s + 1) = PR' = \frac{(d_0 + \Delta d_0)X_0(t_s)\exp(-d_0 - \Delta d_0)}{d_0 + \Delta d_0} , \tag{A.3}$$

which, after manipulation, yields

$$d_0 + \Delta d_0 = -\ln\left(\frac{PR'}{X_0(t_s)}\right) . \tag{A.4}$$

Subtracting Eq. (A.2) from Eq. (A.4), we obtain

$$\Delta d_0 = \ln\left(\frac{PR}{X_0(t_s)}\right) - \ln\left(\frac{PR'}{X_0(t_s)}\right)$$

$$= -\ln\left(\frac{PR'}{PR}\right)$$

$$= -\ln\left(1 - \frac{PR - PR'}{PR}\right)$$

$$= -\ln\left[1 - \frac{100\left(\frac{PR - PR'}{PR}\right)}{100}\right] . \tag{A.5}$$

After substituting Eq. (11) into Eq. (12), Eq. (A.5) is the same as Eq. (12).

Part IV

Monitoring Programs
and Data Analysis

The Quality of Inferences Concerning the Effects of Nuclear Power Plants on the Environment*

Donald A. McCaughran

Center for Quantitative Science
in Forestry, Fisheries and Wildlife
University of Washington
Seattle, Washington

ABSTRACT

The analysis of variance is the most common statistical procedure applied to nuclear power plant monitoring data at the present time. Under the assumption that it is appropriate, minimum detectable difference is defined and its relationship with the power of the statistical test is computed. The relationship between station number, replicate number, variance, and minimum detectable difference with power = 0.90 and α = 0.10 for phytoplankton, zooplankton, smelt, and alewives at the Zion Nuclear Power Plant are given.

Key words: analysis of variance, experimental designs, minimum detectable differences, number of replicates, number of stations, power of a statistical test, probability of detection

INTRODUCTION

Over the past ten years large quantities of data have been collected by various groups for the purpose of detecting environmental changes induced by nuclear power plants. The utilities have spent millions of dollars for these data, which have subsequently been used in the process of obtaining construction and operating licenses and 316a and 316b variances.

Two considerations are of paramount importance in this process: the level of change in the environment which is considered unacceptable and the ability to detect that change with the data collected. The second of these considerations depends upon the first. Both are the business of the regulatory agencies.

---

*Work supported under Contract AT(49-24)-0222 with the Nuclear Regulatory Commission, Washington, D.C.

229

Without regulation regarding intensity of studies, the logical approach for a utility is to spend as little money as possible on data collection to ensure insufficient data to detect differences and then to conclude there are no differences. On the other hand, the strategy of opponents of nuclear plants should be to insist that great effort is expended in sampling – ensuring that small differences are found statistically significant, to disregard whether the differences are real or biologically important, and then to claim unacceptable damage to the environment. Both strategies are obviously undesirable. However, until unacceptable change is clearly set forth, no scientifically acceptable procedure is obvious.

The present work deals with minimum detectable differences and the probability of detection in several experimental designs. The relationship of cost and detectable change is discussed.

MINIMUM DETECTABLE DIFFERENCES

The typical data collection and analysis scheme for an environmental monitoring program at a nuclear power plant is represented in Fig. 1. The concepts to be considered are shown as inputs to the various phases of the monitoring study. The data are most often collected at fixed points in time; consequently, the application of intervention analysis in time series models (Box and Tiao 1975) may seem appropriate. However, given the small amount of preoperational data and such large seasonal effects, the utility of intervention analysis is not clear. The present "state of the art" relies on statistical tests such as the analysis of variance. The present work deals with the use of the analysis of variance in determining change.

At the beginning of a monitoring study an attempt is made to select stations which will be in control areas and affected areas once the plant is in operation. Stations are sampled periodically, and the resulting data are analyzed to see if there are differences in density or catch per effort between stations. Such analyses prior to operation of the plant are useful for examining the adequacy of the stations. The same analyses applied to operational data often give some measure of plant effect, but the sensitivity of the analyses depends upon the unperturbed similarity of the stations. It is worthwhile examining the ability of these tests to detect change.

The analysis performed on the data from any one sampling time uses a completely random experimental design with t stations and n replicates at each station. The ability to detect differences in the station means (power of the statistical test) is a function of the magnitude of the differences, of n, of t, and of the significance level of the test. The relationship between these variables can be seen from examination of the noncentrality parameter under an alternative hypothesis.

A reasonable alternative hypothesis to the null hypothesis of no differences is: The largest difference between station means equals Δ – that is,

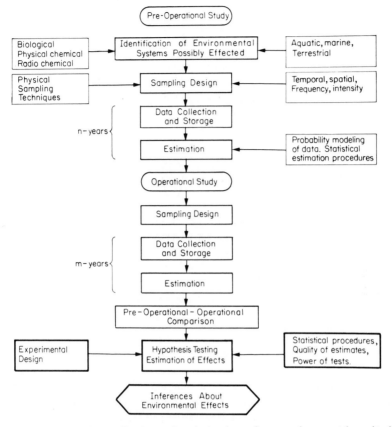

Fig. 1. Typical data collection and analysis scheme for an environmental monitoring program at a nuclear power plant.

$$\mu_{MAX} - \mu_{MIN} = \Delta , \tag{1}$$

and all other station means are equal to the average $(\mu_{MAX} + \mu_{MIN}/2)$, where μ_{MAX} is the largest station mean and μ_{MIN} is the smallest station mean. This alternative hypothesis will produce minimum power for a maximum difference, Δ.

The noncentrality parameter, δ, for this simple analysis of variance is

$$\delta = \left[\frac{n}{\sigma^2} \sum_{i=1}^{t} (\mu_i - \bar{\mu})^2 \right]^{1/2} , \tag{2}$$

where $\bar{\mu}$ is the mean of the station means. Imposing the alternative hypothesis,

$$\delta = \left(\frac{n}{2}\right)^{\frac{1}{2}} \left(\frac{\Delta}{\sigma}\right), \qquad (3)$$

where σ is the common within-station variance.

As the number of stations, t, changes, the treatment and residual degrees of freedom change. The Pearson-Hartley (1954) charts for power incorporate this change by requiring the computation of a parameter ϕ, where $\phi = \delta/\sqrt{t}$. Hence,

$$\phi = \left(\frac{n}{2t}\right)^{\frac{1}{2}} \left(\frac{\Delta}{\sigma}\right), \qquad (4)$$

and as ϕ gets larger the power increases. For fixed minimum detectable difference Δ, the probability of detection (power) increases as n increases, as t decreases, and as σ decreases.

To illustrate the relationship between n, t, Δ, and power, a variance of 0.25 will be used. This is the approximate variance if a square root transform is applied to observations modeled with the Poisson distribution.

Number of Replicates (n)

Power (as a statistical test) as a function of the minimum detectable difference Δ for a different number of samples per station is illustrated for $t = 2$ (Fig. 2) and for $t = 4$ (Fig. 3). For a given power, considerable reduction is made in the difference detectable (Δ) with an increase from two to eight replicates. For example (Fig. 3), with power of 0.90, $\Delta = 3.15$ with two replicates and $\Delta = 1.00$ with eight replicates.

The relationship between minimum detectable difference and number of replicates per station for different levels of the probability of type I error, α, is illustrated for power of 0.85 (Fig. 4) and power of 0.95 (Fig. 5). As the number of replicates increases, Δ decreases. The decrease in Δ is large in going from $n = 2$ to $n = 8$, but after $n = 8$ the change is much less. It is also observed that for a fixed number of replicates the minimum detectable difference decreases as α increases.

Number of Stations (t)

Figure 6 shows the relationship between power, n, and t. The total number of samples (nt) is held constant at 24. For a fixed minimum detectable difference, power is seen to increase as t decreases and as n increases.

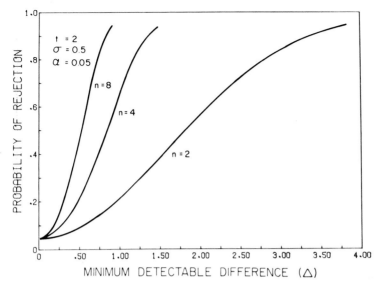

Fig. 2. Probability of rejection of the null hypothesis of no statistically significant difference between two station means (i.e., power of the statistical test) as a function of the minimum detectable difference between the two station means at any one sampling time for 2, 4, or 8 replicates at each station; $\sigma = 0.5$ and $\alpha = 0.05$.

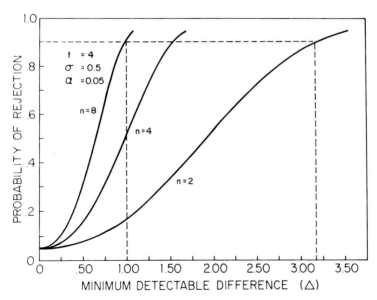

Fig. 3. Probability of rejection of the null hypothesis of no statistically significant difference between station means as a function of the minimum detectable difference between the largest and smallest station means (i.e., $\mu_{MAX} - \mu_{MIN} = \Delta$) at any one sampling time for 2, 4, or 8 replicates at each of four stations; $\sigma = 0.5$ and $\alpha = 0.05$.

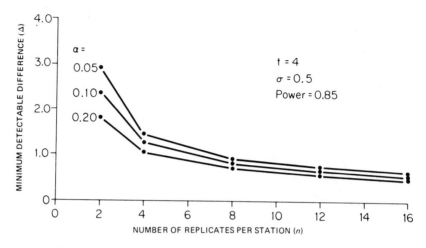

Fig. 4. Minimum detectable difference between the largest and smallest station means at any one sampling time as a function of the number of replicates per station at each of four stations; $\sigma = 0.5$, power = 0.85, and $\alpha = 0.05, 0.10,$ or 0.20.

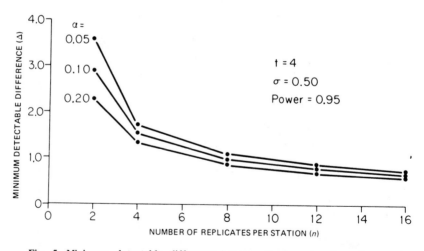

Fig. 5. Minimum detectable difference between the largest and smallest station means of any one sampling time as a function of the number of replicates per station at each of four stations; $\sigma = 0.5$, power = 0.95, and $\alpha = 0.05, 0.10,$ or 0.20.

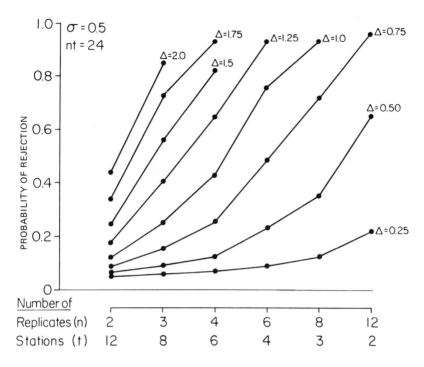

Fig. 6. Probability of rejecting the null hypothesis of no statistically significant difference between station means (i.e., power) at any one sampling time as a function of the number of replicates per station (n) and the number of stations (t) for different levels of minimum detectable difference (Δ). Total number of samples (nt) is constant at 24, and $\sigma = 0.5$.

This relationship is relevant to monitoring designs, whether two or twelve stations are used, if the stations with μ_{MAX} and μ_{MIN} are always included in the design.

Variance (σ^2)

Figure 7 shows the relationship between the detectable difference and the square root of the variance (σ) for $n = 2, 4$, and 8 and for $t = 2$ and 4 with power of 0.95. The increase in the minimum detectable difference is almost a linear function of increasing standard deviation. It is also observed that the smaller the replicate number the greater the effect of increasing standard deviation.

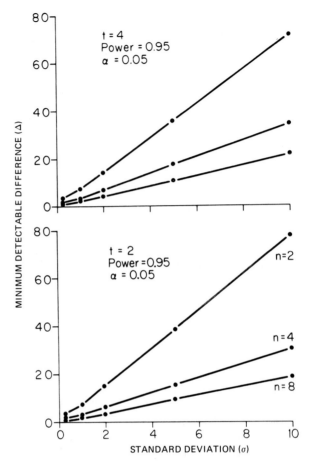

Fig. 7. Minimum detectable difference between stations as a function of the common within-station standard deviation for 2, 4, or 8 replicates (*n*) at each of 2 or 4 stations (*t*).

Percent Minimum Difference

The minimum detectable difference is in absolute units of measurement, for example, numbers of organisms per milliliter, number per Ponar dredge, milligrams per liter, or catch per minute. The interpretation of a detectable difference of Δ units will vary with the magnitude of the means. If $\Delta = 2$, with power of 0.95 and $\mu_{MAX} = 20$, a reduction of 10% could be detected with probability 0.95. If, however, $\mu_{MAX} = 200$ and all other parameters were still the same, then a 1.0% change could be detected with 0.95 probability. Consequently, minimum detectable difference should always be related to the magnitude of the station means. This is a problem in monitoring designs, since the means fluctuate seasonally. A certain sample size might ensure the detection of a 10% change in summer but only a 50% change in winter. Regulatory agencies will have to address this problem.

Preoperational-Operational Comparisons

The comparison of preoperational and operational data can be accomplished in several ways. Preoperational and operational data from stations in influenced areas can be compared, but since the data are collected in different years environmental effects are often confounded with plant effects. If environmental differences between years affect both control and influenced stations in a similar manner, plant effect can be determined by comparing preoperational-operational differences between control and influenced stations. A factorial treatment design can be used for this comparison by testing the interaction hypothesis in the following 2 X 2 design.

	Control	Influenced
Preoperational	μ_{11}	M_{12}
Operational	μ_{21}	M_{22}

$$\text{Interaction hypothesis: } \mu_{11} - \mu_{21} = \mu_{12} - \mu_{22}. \tag{5}$$

To determine the probability of detecting an effect Δ — that is,

$$\mu_{11} - \mu_{21} - (\mu_{12} - \mu_{22}) = \Delta, \tag{6}$$

the noncentrality parameter becomes

$$\delta = \left(\frac{n}{4}\right)^{1/2}\left(\frac{\Delta}{\sigma}\right) = \left(\frac{n}{2}\right)^{1/2}\left(\frac{\Delta}{\sqrt{2}\sigma}\right). \tag{7}$$

Consequently, the previous power curves can be used with variance $2\sigma^2$ instead of σ^2, $t = 2$, and number of replicates $= 2n - 1$.

APPLICATION TO IMPACT STUDIES

The major problem in using the analysis of variance with monitoring data is that the normal distribution is not often a good model. There are exceptions, however, since the central limit theorem applies if means are used as observations, and many discrete distributions converge to the normal when the values of parameters get large. Often the sampling can be arranged to increase parameter values so that the normal distribution can be used. This can be accomplished with the Poisson or negative binomial by increasing the sample size, that is, more water, more substrate, or more trawling time.

When the normal distribution can be used, we are often faced with unequal variances. With equal sample size this is less of a problem since the robustness is greater with equal sample sizes. Transformations are often applied to transform the observations into random variables with equal variance. Transformations, however, will interfere with the test of impact in some situations. For example, if a $\log_e$ transform is applied to catch-per-effort data modeled with the log-normal distribution, the test of the interaction hypothesis may be invalid, since

$$H_0: \mu_{11} - \mu_{21} = \mu_{12} - \mu_{22} \tag{8}$$

in the $\log_e$ scale does not imply

$$H_0: e^{\mu_{11}} - e^{\mu_{21}} = e^{\mu_{12}} - e^{\mu_{22}} \tag{9}$$

in the untransformed scale. Hence transformations may remove or create statistical significance of impact in factorial treatment designs. The problem is not encountered in the one-way designs. In most situations, particularly with equal sample sizes, it is better to rely on the robustness of the analysis of variance and not transform the data.

The other problem with using power functions to determine minimum detectable differences and sample size is the unknown variance. Figures 8 and 9 present estimated monthly variances computed from the 1973 monitoring data from the Zion Nuclear Power Plant. The data are total zooplankton and total phytoplankton densities and catch per minute for alewives and smelt. It is observed that considerable variability exists in the estimates. No particular seasonal pattern is obvious, although a pattern might be expected for the seasonal change in density. To be able to approximate the power function, the variances can be pooled to arrive at a reasonable estimate. The power functions are conditional on the choices of σ^2.

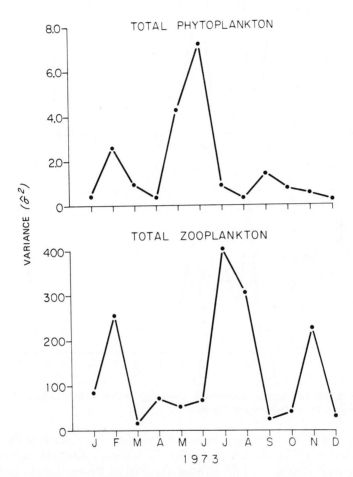

Fig. 8. Monthly sample variances for phytoplankton and zooplankton estimated from the 1973 Zion Nuclear Power Plant monitoring data.

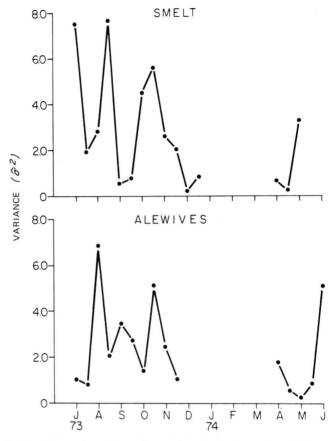

Fig. 9. Monthly sample variances for smelt and alewives estimated from the 1973 Zion Nuclear Power Plant monitoring data.

Review of the 1973 Zion Nuclear Power Plant data (Industrial Bio-Test Laboratories 1975) allows a comparison of minimum detectable difference with average density of all the stations. These values are presented in Table 1. The data for phytoplankton and zooplankton were square-root transformed and the catch-per-effort data was $\log_e$ transformed prior to analysis.

A computation of percent detectable difference can be obtained by considering

$$\left(\frac{\Delta}{\overline{X}}\right) 100 = \text{percent change detectable,} \tag{10}$$

where $\overline{X}$ is the annual average for the variable of interest.

This quantity is useful in relating the detectable difference to the average density found in the environment. These quantities are given in Table 1.

Table 1. Annual average, $\overline{X}$, sample variance, $\hat{\sigma}^2$, sample standard deviation, $\hat{\sigma}$, minimum detectable difference in the transformed scale, Δ, minimum detectable difference in the untransformed scale, e^{Δ}, and percent change detectable

Computed from the 1973 monitoring data from the Zion Nuclear Power Plant[a]

	Zooplankton	Phytoplankton	Alewives	Smelt
$\overline{X}$	179.5	38.23	3.33	1.30
$\hat{\sigma}^2$	133.10	16.73	2.40	2.95
$\hat{\sigma}$	11.54	4.09	1.55	1.72
Δ[b]	28.0	10.9	4.1	4.7
e^{Δ}[c]			60.3	110.0
Percent change detectable	15.6	28.5		

[a]Power = 0.90; α = 0.10. Zooplankton and phytoplankton data (number per cubic meter) are square-root transformed and CPUE (catch per minute) are $\log_e$ transformed prior to analysis.

[b]$\Delta = \mu_{MAX} - \mu_{MIN}$.

[c]$e^{\Delta} = e^{\mu MAX}/e^{\mu MIN}$.

With catch per effort the log normal distribution is often used as a model. The $\log_e$ transform is applied and the analysis of variance is used to make inferences about the parameter μ. Since the median of the log normal distribution is e^{μ}, making inferences about μ in the transformed scale allows us to make inferences about the median in the untransformed scale:

$$\Delta = \mu_{MAX} - \mu_{MIN}, \text{ transformed scale;}$$

$$e^{\Delta} = e^{\mu MAX}/e^{\mu MIN}, \text{ untransformed scale.} \tag{11}$$

Consequently, the measure of change is the ratio of the largest to smallest median, which is clearly a reasonable measure of change.

The usual number of replicates varies between one and four for most monitoring studies reviewed. With one replicate very little can be said, since no change can be detected. With four replicates considerable sensitivity results. A review of Figs. 3 and 4 indicates that above eight replicates small decreases in the minimum detectable difference are accomplished by further increasing the number of replicates. However, as the number of replicates increases, so does the cost of the study. Since the cost of the study is passed on to the public, this cost must be compared with the costs associated with changes in the environment due to plant operation.

It is recommended that the regulatory agencies should make a concerted effort in defining unacceptable environmental change and then insist upon monitoring studies capable of detecting such changes. Examination of power functions will be helpful in this regard.

LITERATURE CITED

Box, G. E. P., and G. C. Tiao. 1975. Intervention analysis with applications to economic and environmental problems. J. Am. Stat. Assoc. **70**:70–79.

Pearson, E. S., and H. O. Hartley. 1954. Biometrika tables for statisticians, Vol. 1, Cambridge University Press, London.

Industrial Bio-Test Laboratories. 1975. Evaluation of thermal effects in southwestern Lake Michigan, special studies 1972–1973. Report to Commonwealth Edison Company, Chicago, Illinois.

Factors to Consider in Monitoring Programs Suggested by Statistical Analysis of Available Data*

John M. Thomas

Battelle, Pacific Northwest Laboratories
Richland, Washington

ABSTRACT

Based on experience gained in the statistical analysis of data collected during monitoring programs at three nuclear power plants, as well as on other studies in the area of impact assessment, I have attempted to outline what has been done and what I believe can be done in assessing environmental changes.

Procedural changes that I suggest include the implementation of a stopping rule so field studies are terminated after a negotiated period of time and the commitment of all resources to studies of one or two species. Simulation models are suggested as a useful tool in an iterative process where results of field studies are routinely incorporated until a negotiated stopping time is reached or until acceptable results are attained. Finally, I describe the statistical analyses we have used and their limitations, and I give some sample-size estimates needed to detect changes of specified sizes in population numbers. To detect changes in population numbers of the size we have encountered, calculated sample sizes are found to be much larger than in current use.

Key words: environmental impact, power, sample size, simulation, statistical analyses, stopping rule

INTRODUCTION

A considerable amount of money has been spent on extensive ecological studies carried out to determine the environmental impact of a myriad of construction projects. We have previously commented (Eberhardt 1976a, 1976b; Thomas and Eberhardt 1976) on the limitations inherent in such field studies, given some preliminary suggestions for field designs, and offered

---

*This paper is based on work conducted for both the Nuclear Regulatory Commission and the Energy Research and Development Administration (under contract EY-76-C-06-1830).

some methods for statistical analysis of the ensuing data. In a separate effort, Eberhardt and Gilbert (1975) gave sample sizes necessary to detect changes in population indices of various kinds.

During the past year I have been involved in the quantitative evaluation of monitoring data required by Environmental Technical Specifications for three nuclear power plants (Gore et al. 1976, 1977a, 1977b). Workers at other laboratories have used statistical methods to evaluate data obtained from other nuclear power stations (Murarka et al. 1976, Adams et al. 1977). In contrast, others (e.g., Rosen 1976, Skutsch and Flowerdew 1976) have suggested rating schemes as a basis for evaluation of impacts. However, since my definition of impact for this paper is the detection (in a statistical sense) of a change in some ecological community, species, or trophic level, I will not consider such methods further here. In addition, simulation models have often been used to predict effects of impacts, but Eberhardt (1976b) perceives that such efforts should be viewed with great care.

The purpose of this paper is to outline past experience and to suggest what might reasonably be done to assess environmental changes induced by nuclear power plants. To this end I discuss (1) desirable changes in emphasis on what to measure and on how to conduct the field work, including a proposed "stopping rule," (2) a few quantitative issues which impinge on the choice of field design(s), (3) suggestions for methods of statistical analysis, and (4) estimation of sample sizes necessary to detect changes in population numbers based on data we have evaluated.

SOME CHANGES IN CURRENT PRACTICE

We have previously suggested (Thomas and Eberhardt 1976) that, prior to starting preoperational surveys, a competent field ecologist should identify species where effects may be expected. Based on the evaluation of data from three nuclear power plants, I believe that after the initial selection of possible candidate species by such an expert, a rigorous review of any evidence about expected effect for each species should be conducted. Even though the statistical power (i.e., ability to detect differences when they exist) of the tests used to assess data from the three power plants was probably low in our studies, no potentially disastrous short-term effects were detected (Gore et al. 1976, 1977a, 1977b). Therefore, for every candidate field study proposed, I suggest that a list of reasons why the study is important and of the concomitant environmental effects expected be prepared. If the reasons for each proposed study are comparably vague and general, then the one or two economically important species, usually those near the top of the food web, should be selected for the most intensive study. If existing regulations require surveys of other trophic levels, then these surveys must also be conducted, but at the lowest acceptable level of effort.

The size of the change, or the precision of estimates that the study will be designed to detect, should be established (negotiated), even though the estimated magnitude of change may be based only on an educated guess. Major questions will probably concern harvestable adult fish or shellfish, and intensive population studies may be required. However, because effects on earlier life stages (e.g., eggs, larvae) may influence population size at later times, other studies may be required to estimate the magnitude of the loss of these earlier forms. In this case, the precision of the estimates can be negotiated (i.e., we will determine within some band of confidence the fraction of "available" larvae entrained). However, several estimates should be obtained, each with the agreed upon precision. The number and quality of these and future estimates should be evaluated at periodic intervals. In contrast, comparison of preoperational and operational shellfish numbers would necessitate an agreement about the size of the changes the study should detect, whether or not studies on other life stages were concurrently conducted. In both cases, the studies should at least be conducted for a negotiated period of time. The resulting data should be examined periodically to ensure that expensive monitoring does not continue without producing useful biological information and that feedback into new or improved field studies occurs at least yearly.

One way to enforce at least yearly examination of data is to modify carefully, as required, and to use, prior to preoperational surveys, existing simulation models to *aid* in establishing the size of changes to be detected or the precision of estimates. Models, with more site- and species-specificity, can be devised during the actual preoperational period. These models should be used to suggest additional needed data and/or field studies identified during the construction and running of the model. Evaluation of model output could result in renegotiating the precision of estimates and/or size of change to be detected. For example, where projected (predicted) population effects were still well below the assumed ecologically detrimental level, studies could be terminated. One negotiated level of effect might be an order of magnitude below the presumed detrimental level. When model projections are made based on other field estimates, then a range of these projections must also be within agreed error bounds (i.e., fraction entrained ± an error).

In the case where a change must be detected in the number of shellfish between preoperational and operational years, the study could be terminated after failing to detect the negotiated changes within an agreed number of years. If a change as large as or larger than the agreed upon was detected, then ameliorative schemes could be instituted (i.e., artificial rearing or hatcheries). Species from other trophic levels, if selected as being important, could be studied using the same format of negotiated change and yearly data examination. The logic of the feedback system outlined above and negotiation of the number of years of study are suggested as the basis for a stopping rule.

Another area in need of a priori negotiation is to establish that some areas, other than intake screens, within close proximity to the intakes and outfalls will be affected by actual construction and possibly by changed currents and discharges after plant operation begins. These areas should not be required to have sampling stations for all biota (plankton, benthic organisms, etc.), unless a rare species is thought to be present or unless some other unique site-specific factor, such as concentration of chemicals in sediments, must be evaluated. The size of such areas is also site specific, so it would be an extra expense to accurately design another field sampling survey to define it. In effect, these outfall areas are agreed on in advance as being "impacted." The resources normally devoted to sampling such areas could be used elsewhere.

DESIGNING AND CONDUCTING THE FIELD PROGRAM

In order to decide on the number of samples necessary to detect changes in population numbers, (hopefully, established as suggested above), a statistical model or symbolic depiction of how the analysis will be performed, accompanied by the necessary assumptions, is needed. Both the statistical methods we have used, as well as most of the assumptions implicit in each method, are discussed below. In addition, I note that different designs might be used for different components of the biota. For example, a regression approach where numbers of benthic organisms are modeled with distance from the discharge structure as the independent variable, or a ratio or difference analysis of zooplankton numbers from stations likely to be somehow affected vs control stations might be used.

If sampling is carried out without a statistical model and concomitant design considerations, our ability to provide sample sizes a priori is severely impaired. On the other hand, by using a statistical model, sample sizes can be chosen such that changes of a stated size are not labeled as statistically significant when they are not. Most biologists are usually willing to accept this kind of error (i.e., type I error) 5% of the time. Also, we can be protected from missing differences of the same size when they are present (called the power of the test). Unfortunately, protection against both kinds of errors is expensive in terms of increased sample size.

In addition to considering the statistical model and sample sizes necessary to minimize the errors discussed above, several other practical matters should be considered. First, data should be collected from more than one control station (considerations such as those just above might indicate how many more) to obtain a measure of change due to the stochastic vagrancies in the environment and to be sure the direction of any change is the same in two unaffected locations. Aside from statistical considerations, biotic populations can be very different over an area defined as a control area, so measurements at a single station may be a poor descriptor of the entire area. A single area used for trawling may sometimes be an exception, depending on the size of the area, but usually more than one control area should be sampled. Secondly, all stations should be monitored for at least two preoperational years, excepting operationally dependent studies like entrainment or impingement. One can then assess the similarity of changes for two seasons (reproductive cycles for some species) at the control site(s) (i.e., detect interactions). Two years of observation would also provide information about longer-term changes in species numbers at both control and affected stations. For these affected stations, population reductions observed during preoperational years could explain continued, further reductions observed during operation, if population numbers continue to decline at the same rate. However, similar measurements from more than one control station are necessary, so that the concomitant decline in numbers of a population at both control and affected stations cannot be attributed to the power plant.

The temporal frequency of measurements, providing time series methods are not used, needs to be considered from two points of view. First, since most species numbers appear in a cyclic manner over the year, measurements should be taken during periods when the most reliable information about change can be obtained. For some fish species, zooplankton, etc., this period occurs during the growing season. However, care should be taken so that very frequent sampling is not the real impact. In addition, frequent sampling in time is not equivalent to additional replication. The problem is that observations through time are probably correlated and not independent as required by most statistical methods. Samples which contain few organisms or which frequently contain none are almost impossible to use in a statistical analysis. Therefore, the counting may be expedited by grouping various species in a sample or sometimes by actually weighing groups and calibrating weight against numbers. In the latter case, an analysis of covariance (see below) might be used.

Finally, where possible, control stations should be selected to be as ecologically similar as a counterpart affected station, even if some sort of paired method of statistical analysis is not used. Selection of stations by random procedures, insofar as possible, is encouraged, but this matching requirement may make randomization impossible. In addition, sampling methods should remain consistent over the entire study period or extensive parallel intercalibration of methods will be necessary.

SOME ASPECTS OF THE STATISTICAL ANALYSIS OF DATA

Most of the data collected will be in the form of indices of population size (i.e., numbers per area, volume, or effort). Many of the factors which influence these indices appear to operate as multipliers. Therefore, a log transform makes the statistical errors additive* and also serves to "normalize" the skewed underlying population distribution as well as to stabilize variances (Gore et al. 1976, 1977a). The matter of the probable lack of independence of observations taken serially in time can be mitigated somewhat (but not really resolved) by the nature of the experimental design and by using logarithms of ratios [i.e., log(control station/affected station)] or arithmetic differences (i.e., control station − affected station). Most statistical models on which analyses are based require that the data be obtained so that the observations are independent and normally distributed with constant variances and with an additive error structure. However, environmental studies at nuclear power plants have often been conducted without considering the foregoing statistical assumptions or most of the suggestions listed in the preceding sections. Consequently, subsequent statistical analyses (e.g., Gore et al. 1976, 1977a, 1977b; Murarka et al. 1976; Adams et al. 1977) are inevitably handicapped, so that the results of these analyses may be criticized.

In the following discussion, some of these statistical procedures and their limitations are reviewed. The analysis of variance using logarithmic transformed data was used to supply an overview to investigate most interactions in two separate field studies. Each study design included both upstream and downstream control as well as treated sites. We found that when more than one control and one treated site were sampled frequently (at least once per nonwinter month) during operational and preoperational periods, that the statistical model for such analyses was unbalanced and was very difficult conceptually and computationally to analyze. Even when the analyses were done on a large computer, difficulties arose because of computer core-size requirements and questions about the validity of F-tests. Decisions as to whether an effect was random or fixed were sometimes arbitrarily made. Other problems arose because of the many interactions in the model and the fact that some main effects were "nested" or observed within another effect, making testing and interpretation the province of a professional statistician.

The source of variation in the analysis of variance of primary interest to us as a measure of change attributable to power plant operation was the interaction between plant operational status (i.e., operating or not operating) and location (i.e., control or treated sites). Almost all the other statistically significant interactions or main effects (i.e., location, time, sites, subsites, operational status, and most of their interactions), while interesting, were not directly germane to the question of assessing change due to a power plant. However, as a research tool to explore all main effects and interactions, this analysis was most advantageous (Gore et al. 1977a), because it was necessary to examine seemingly irrelevant (to the question of change) nested terms to "explain" some statistically significant change(s) in number of biota during

---

*L. L. Eberhardt. Appraising variability in population studies (*draft manuscript*). Battelle, Pacific Northwest Laboratories, Richland, Washington.

preoperational and operational periods for both control and treated stations. Unfortunately, changes in physical factors at certain stations, not directly associated with reactor operations, sometimes explained the observed changes.

Thus, we conclude that (1) an analysis-of-variance overview of environmental-impact data sets may be expensive, (2) the statistical model will be difficult to construct and implement on a computer, (3) the computer output must be professionally interpreted, and (4) care must be exercised not to lose sight of the datum of real interest — that is, is there a change in biota attributable to plant operation? Since we have not assessed data from a well-designed and balanced field study, we are not able to make a recommendation for or against the use of the analysis of variance.

We have used the ratio of measurements at the "impacted" site to those at a control site as an index change (Eberhardt 1976b). The preoperational period is used to establish a baseline ratio which is then compared to the ratio during the operational period. The scheme has the advantage of "smoothing out" the wide fluctuations in abundance of species, like plankton, because the population cycles and/or natural variations are usually similar in comparable sites. In addition, the ratio variables may be more nearly statistically independent than the original variables.

The statistical analysis involved comparing log-transformed, pre-stress ratios (treated/control) to similarly transformed post-stress ratios to assess such changes. The major problem was that no attempt to match the physical or ecological characteristics of treated and control sites was made before the study. Thus, we had to construct unplanned "after the fact" ratios. Further, in some cases, only one control and two treated stations (or the converse) were sampled. In such a situation, an average of two site or two control stations was used in the numerator or denominator of the ratio, or two separate ratios were calculated in which either the treated site(s) or control site(s) appeared twice.

A second problem occurred when no data were reported for either the control or treated station, which resulted in either an undefined ratio or a value of zero. As a practical matter these ratios were not used. Examination of counts from other stations on the same sampling date usually indicated few organisms had been present, so it is possible that many blank or missing sample values were either lost or mishandled in the laboratory. Thus, the resulting statistical analysis of the remaining ratios was probably biased. We assessed data on zooplankton, benthos, chlorophyll a, and fish seining from one of the three nuclear power plants using the ratio procedure, but no statistically significant changes attributable to power plant operation were found. In general, sample sizes ranged from 4 to 10 ratios per year, with data from either one or two preoperational years and two or more operational years. Thus, the statistical power of the test was probably not adequate to detect moderate changes (see the next section, on sample sizes). However, in data sets we examined, changes in the mean preoperational to operational ratio were usually small, suggesting that future population changes would probably be negligible.

Our efforts have also included several attempts to use nonparametric tests (to avoid assumptions about the underlying probability distributions) to statistically analyze data collected as a function of time from matched control and treated stations. These tests were used because the data sets usually were incomplete or contained only a few seasonal cycles, so we could not use the methods of time series analysis. Time series methods are particularly attractive because the assumption of no correlation between successive observations is unnecessary. The two principal drawbacks of nonparametric tests (we have used Wilcoxon's Signed Ranks Test) seems to be their lack of sensitivity relative to parametric counterparts and the known intercorrelations in the data which can invalidate the test results.

We have also used regression analysis where station temperatures progressively decreased with distance from a discharge area. Distance was used as the independent variable (to mimic temperature change), and the logarithm of the measure of abundance was the dependent variable. The principal disadvantage of this procedure was that data from only a few stations (distances) were available to decide if changes in the logarithm of the dependent variable were linear or nonlinear. In addition, the seasonal nature of abundance for most organisms resulted in different parameter values for similar models fit to data obtained during different seasons. Finally, the level of statistical significance of the parameter estimates seemed to depend on data from stations closest to the discharge.

For certain estimates, such as fraction of available organisms entrained, a confidence interval may suffice, so that usually only the type I error alluded to in the preceding section need be considered. In most cases, a confidence interval of plus or minus some percentage of the mean is acceptable. However, since fraction entrained can vary seasonally and is probably different for different species, several estimates may be needed. Each estimate should be based on data collected over as short a time period (e.g., days or less) as possible. Not enough estimates were available to give us a feel for such problems, so we can offer little practical guidance on the number of such estimates needed either to define a range or the likely underlying probability distribution. However, it may be necessary, in constructing error bounds for the ratio, to consider the distribution of data used for the denominator.

In some specific instances we have tried to use time series analysis, multivariate techniques, and analysis of covariance (in the sense of removing by regression certain effects not controlled by the experimental design). These techniques were not applicable because of the lack of suitable data sets or because some of the assumptions discussed in the preceding sections were violated. However, each of these techniques may have a use if designed surveys, with larger sample sizes than those currently employed, are used. In addition, other statistical techniques, such as intervention analysis (Box and Tiao 1975), have been suggested and may offer some additional ways to detect changes.

SAMPLE SIZES

The determination of sample size depends on the objectives of the study, the statistical model under consideration, and the sensitivity (power) of a test based on some alternative hypothesis derived from the statistical model. Some preliminary methods for deriving sample sizes are given in Eberhardt and Gilbert (1975), and rules of thumb with examples are given in Eberhardt.* We have used the method outlined in this second paper [based on the Tables of Kastenbaum et al. (1970)], in addition to an approximate procedure for estimating sample size, in situations involving an unbalanced analysis of variance. We derived this method from a combination of the methodologies in Guenther (1964) and Scheffé (1959). Most of the problems had to do with the fact that the effect for which estimates of sample size were desired was an interaction. In addition, the statistical model contained both nested and crossed terms in an unusual way.

The approximate number of sampling stations per area required for analysis-of-variance models of the type that we have used is shown in Table 1. The additional assumptions necessary to construct such a table are listed at the bottom of the table. Based on these assumptions, and in light of the previous discussion, the sample-size estimates in Table 1 are considered provisional. Clearly, estimates of an expected logarithmic mean square for the interaction between operational status of the power plant (i.e., operating or not operating) and area (i.e., control or treated) are not readily available; thus tables like Table 1 cannot easily be used prior to designing and carrying out a monitoring program. However, we found that the error mean square obtained in the statistical analyses of two benthic studies was relatively stable (Gore et al. 1976, 1977a), so that the mean squares observed for the interaction may be representative. Mean square values usually lay between 0.50 and 0.85, but in a few instances were as high as 1.50. Therefore, it appears that to judge such an interaction statistically significant, a sample size of between 7 and 11 stations per area would be needed (Table 1).

Figure 1 was prepared (based on the lognormal distribution) so that knowledge of the coefficient of variation and the ratio of means could be used to obtain estimates of the number of sampling stations required in each area (i.e., control and treated areas) to detect statistically significant differences between the two areas. The coefficient of variation is the standard

---

*L. L. Eberhardt. Appraising variability in population studies (*draft manuscript*). Battelle, Pacific Northwest Laboratories, Richland, Washington.

Table 1. Approximate number of sampling stations per area for either two or three areas, needed to obtain an 80 or 90% chance of detecting a statistically significant power plant effect ($\alpha = 0.05$) as a function of the magnitude of the mean square for the interaction between operational status of the power plant (i.e., operating or not operating) and area (i.e., control or treated)[a]

Interaction mean square (logarithmic units)	For two areas: number of sampling stations per area	
	Power[b] = 0.80	Power = 0.90
0.25	14	18
0.50	7	9
0.75	5	6
1.00	5	5
1.50	3	3
2.00	2	3

Interaction mean square (logarithmic units)	For three areas[c]: number of sampling stations per area	
	Power = 0.80	Power = 0.90
0.125	21	27
0.250	11	14
0.375	7	9
0.500	6	7
0.750	4	5
1.00	3	4

[a]Assumptions: (1) additive model, homogeneous variances, independence of observations, and normality of data; (2) balanced design — no confounding; (3) at least two years each of preoperational and operational data; (4) at least six observations per year for each station; and (5) random effects model.

[b]The power of a statistical test is its ability to detect the alternative hypothesis when the alternative hypothesis is true; that is, it is the probability of rejecting the null hypothesis when the null hypothesis is false, given a particular alternative hypothesis.

[c]Upstream and downstream control areas and one treated area.

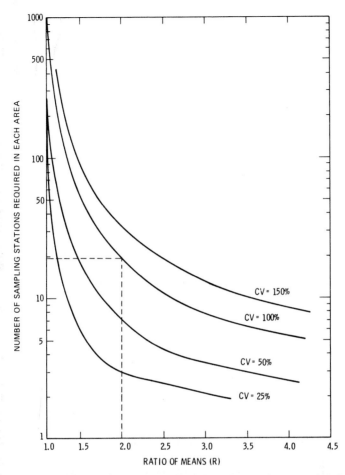

Fig. 1. Number of sampling stations required in each area (i.e., control and treated areas) to detect statistically significant differences between the two areas at the $\alpha = 0.05$ level with a power of 0.70 as a function of the ratio of means for each of four levels of coefficient of variation (CV).

deviation divided by the mean and expressed as a percentage. The ratio of means is either the mean number of organisms in a control area divided by the mean number in a treated area or the reciprocal of this ratio, since use of Fig. 1 requires that the ratio of means be greater than or equal to 1.0. Again, the reader is cautioned that many assumptions are used in constructing a diagram like Fig. 1 and that the field design and statistical analysis should be based on these assumptions. In our analyses, coefficients of variation ranged from 56 to 143% for benthic samples and ratios of means usually ranged from 1.0 to about 4.0 depending on the time of the year, station, and species. However, except for areas very close to outfalls, the largest ratio of means has been about 2. If we assume a coefficient of variation of 100% and a ratio of means of 2.0, reference to Fig. 1 shows that about 20 samples for stations in the control and affected areas would be necessary to declare this difference statistically significant. This number of samples is nearly double the number obtained using Table 1 and the analysis of variance model, although the statistical powers used in Fig. 1 (0.70) and in Table 1 (0.80 and 0.90) are slightly different. Reference to Fig. 1 indicates that sample sizes much larger than those currently used in most monitoring programs are needed to detect even modest changes (say a 50% change), given the variability (usually greater than 100% coefficient of variation) commonly found in environmental samples.

Much of the effort currently spread over many reactor monitoring programs should be directed to one or two programs and adequate numbers of samples should be collected. Efforts should also concentrate on field studies designed to estimate parameters (with confidence intervals of an agreed-upon size) for simulation models. If stopping rule(s) such as we have discussed gain acceptance, then environmental studies at power plants can be terminated after a negotiated time or when projected effects are consistently much lower than ecologically detrimental levels.

ACKNOWLEDGMENTS

The sample sizes in Table 1 were obtained by Judy Mahaffey of the Energy Systems Department, Pacific Northwest Laboratories. I thank Marji Cochran for programming aid in preparing Fig. 1 and L. L. Eberhardt, Judy Helbling, and D. Watson for reading a draft manuscript.

LITERATURE CITED

Adams, M. A., P. A. Cunningham, D. D. Gray, and K. D. Kumar. 1977. A critical evaluation of nonradiological environmental technical specifications. Vols. 1–4, *draft reports*, Oak Ridge National Laboratory, Oak Ridge, Tennessee.

Box, G. E. P., and G. C. Tiao. 1975. Intervention analysis with applications to economic and environmental problems, J. Am. Stat. Assoc. **70**:70–79.

Eberhardt, L. L. 1976a. Some quantitative issues in ecological impact evaluation, p. 307–315. *In* R. K. Sharma, J. D. Buffington, and J. T. McFadden [eds.] Proc. Workshop Biol. Significance Environ. Impacts, NR-CONF-002, U.S. Nuclear Regulatory Commission, Washington, D.C.

————. 1976b. Quantitative ecology and impact assessment. J. Environ. Manage. 4:27–70.

Eberhardt, L. L., and R. O. Gilbert. 1975. Biostatistical aspects. Part 8 (p. 783–918, vol. 2) of Environmental impact monitoring of nuclear power plants, source book of monitoring methods. Atomic Industrial Forum, Inc., New York.

Gore, K. L., J. M. Thomas, L. D. Kannberg, and D. G. Watson. 1976. Evaluation of Monticello Nuclear Power Plant, environmental impact prediction, based on monitoring programs. BNWL-2150, NRC-1. Battelle, Pacific Northwest Laboratories, Richland, Washington. 127 p.

Gore, K. L., J. M. Thomas, L. D. Kannberg, J. A. Mahaffey, and D. G. Watson. 1977a. Evaluation of Haddum Neck (Connecticut Yankee) Nuclear Power Plant, environmental impact prediction, based on monitoring programs. BNWL-2151, NRC-1. Battelle, Pacific Northwest Laboratories, Richland, Washington. 181 p.

Gore, K. L., J. M. Thomas, L. D. Kannberg, and D. G. Watson. 1977b. Evaluation of Millstone Nuclear Power Plant, environmental impact prediction, based on monitoring programs. BNWL-2152, NRC-1, Battelle, Pacific Northwest Laboratories, Richland, Washington. 120 p.

Guenther, W. C. 1964. Analysis of variance. Prentice-Hall, Englewood Cliffs, New Jersey.

Kastenbaum, M. A., D. G. Hoel, and K. O. Bowman. 1970. Sample size requirements: one-way analysis of variance. Biometrika 57(2):421–430.

Murarka, I. P., A. J. Policastro, J. G. Ferrante, E. W. Daniels, and G. P. Marmer. 1976. An evaluation of environmental data relating to selected nuclear power plant sites. ANL/EIS-1 (Kewaunee), ANL/EIS-2 (Quad-Cities), ANL/EIS-3 (Duane Arnold), ANL/EIS-4 (Three Mile Island), ANL/EIS-5 (Zion), ANL/EIS-6 (Prairie Island), ANL/EIS-7 (Nine Mile Point). Argonne National Laboratory, Argonne, Illinois.

Rosen, S. J. 1976. Manual for environmental impact evaluation. Prentice-Hall, Englewood Cliffs, New Jersey.

Scheffé, H. 1959. The analysis of variance. John Wiley and Sons, Inc., New York. 477 p.

Skutsch, M. M., and R. T. N. Flowerdew. 1976. Measurement techniques in environmental impact assessment. Environ. Conserv. 3:209–217.

Thomas, J. M., and L. L. Eberhardt. 1976. Ecological impact assessment, p. 181–197. In S. Fernbach and H. M. Schwartz [eds.] Proc. Conf. Comput. Support Environ. Sci. Anal., Albuquerque, New Mexico, July 9–11, 1975. CONF-750706, U.S. Energy Research and Development Administration, Washington, D.C.

Estimation of Age Structure of Fish Populations from Length-Frequency Data*

K. Deva Kumar and S. Marshall Adams

Environmental Sciences Division
Oak Ridge National Laboratory
Oak Ridge, Tennessee

ABSTRACT

A probability model is presented to determine the age structure of a fish population from length-frequency data. It is shown that when the age-length key is available, maximum-likelihood estimates of the age structure can be obtained. When the key is not available, approximate estimates of the age structure can be obtained. The model is used for determination of the age structure of populations of channel catfish and white crappie. Practical applications of the model to impact assessment are discussed.

Key words: age-length key, age structure, channel catfish, length-frequency distribution, maximum likelihood estimates, mixture of normals, probability model, simulation study, white crappie

INTRODUCTION

Fish monitoring programs at nuclear power plants are designed to detect impacts and to ensure that these plants operate without causing unacceptable environmental damage. However, these monitoring programs, in general, are not comprehensive enough to detect impacts on fish populations due to power plant operations. Problems in the design of meaningful monitoring programs occur primarily because of the difficulties in sampling and interpreting monitoring data collected in open aquatic systems under highly variable environmental conditions. The inherent problems in sampling fish populations, such as size selectivity of sampling gear, mobility of fishes, and some of the behavioral characteristics of fish such as schooling, all tend to increase the effort and cost of an effective monitoring program.

Several approaches have been tried to evaluate the extent of changes in fish populations due to perturbations. One approach attempts to detect changes in density but is very difficult to implement due to inherent problems in quantitative sampling. Tag and release methods seem best for

---

*Research sponsored in part by the Energy Research and Development Administration under contract with Union Carbide Corporation and in part by the Nuclear Regulatory Commission Office of Nuclear Reactor Regulation under Interagency Agreement ERDA 40-554-75. ESD Publication No. 1040, Environmental Sciences Division, Oak Ridge National Laboratory.

estimating the size of a fish population, but these programs are expensive and time consuming. Methods currently employed to monitor potential impacts of power plant operations on the density of fish populations include comparing the catch per unit effort between control and discharge areas and between preoperational and operational periods. However, interpretation of catch-per-unit-effort data and extrapolation to effects on a population are difficult. For example, higher catches of fish in a thermal area compared to catches in a control area or variations in catches from year to year may mean either that some redistribution of the population has occurred, or that a change in population structure has occurred, or both. For meaningful interpretation, this method requires obtaining estimates of the fish density of the entire aquatic system, which is difficult.

Another approach to monitoring fish population dynamics involves the species structure. The best example of this is the comparison of fish diversity indices and similarity coefficients between discharge and control areas or between preoperational and operational periods. It is difficult, however, to relate changes in these indices to changes in the functional aspects of populations such as population growth, production, mortality, and age structure. Functional parameters, particularly production, can be used to quantify the ecological success of a species and to provide a measure of stress in aquatic ecosystems (Chadwick 1976). Functional parameters for evaluation of potential perturbations are also desirable because they reflect the dynamics of the entire system rather than of a single component. A change in density may be reflected in changes in mortality and growth rates, unless mortality and growth are density independent. If operation of a power plant is having a major impact on a fish population, then significant changes could possibly be expected in the mortality rate, growth rate, and production rate which in turn will be reflected in the age structure of the population.

In this paper we are interested in estimating the age structure of a fish population. Knowledge of the age structure is useful for the estimation of natural mortality, fishing mortality, and production (Ricker 1975, Seber 1973). The age-length key* is usually based on a subsample of the catch (Allen 1966), and the age distribution in the subsample is assumed to be an estimate of the age structure in the catch and, therefore, of the age structure in the population. The length-frequency distribution of a fish population is usually derived from either the total catch or a very large subsample. However, due to the cost and effort involved, the age-length key is based on a much smaller sample. If the subsample chosen for aging is "optimal" (based on some criteria), then the age-length key method will yield reliable estimates of the age structure.

In the following sections we discuss a probability model for the length of a fish which takes into consideration the age-structure of the population and

---

*A double frequency table, usually with age in the columns and lengths in the rows (Ricker 1975, p. 206).

is estimated from length-frequency data. If an estimate of the age-length key is available, reliable estimates of the age structure can be obtained, and even when this key is not available, the model gives a first approximation for the age distribution. While the Allen estimator (see above) for the age structure is a straight extrapolation from a subsample, this model attempts to use the larger sample associated with the length-frequency data to estimate the age distribution.

MODELS FOR FISH LENGTH

In this section we discuss a probability model for the length of fish. For a historical review of the model, see Ricker (1975).

Assume that there are k age classes in a fish population, and that ρ_i is the proportion of the population of age i. Let the lengths x of fish of age i be distributed normally with mean μ_i and variance σ_i^2. The probability density function $f(x: \Theta)$ of x for the population can be written as

$$f(x: \Theta) = \sum_{i=1}^{k} \rho_i N(x; \theta_i) , \qquad (1)$$

where

$$\Theta = \langle \theta_1, \theta_2, ..., \theta_k, \rho_1, \rho_2, ..., \rho_k \rangle ,$$
$$\theta_i = \langle \mu_i, \sigma_i^2 \rangle ,$$

$N(x; \theta_i) =$ density function of a normal distribution with mean μ_i and variance σ_i^2 ,

$\rho_i =$ proportion of the population of age i,

$k =$ number of age classes in the population which is generally assumed to be known.

The parameter vector Θ is related to some well-known fishery statistics. For example, the catch curve* can be represented by the plot of $\log_e \rho_i$ against age i, and the plot of μ_i against age i is the growth curve. If the variances of the individual normal distributions are small, then the length-frequency plot will have k well-defined modes at $\mu_1, \mu_2, ..., \mu_k$. If there are differential growth rates for various year classes, the modes may not be well defined. MacDonald (1971) attempted to estimate modes from the length-frequency data for fast-growing pike (*Esocidae*), and he was able to identify

---

*A graph of the logarithm of the number of fish taken at successive ages or sizes (Ricker 1975, p. 2).

five separate age classes. For these data, the average distance between the mean lengths for successive age classes was 100 mm. Separations this distinct would not be expected for most smaller or slower-growing fish. In many cases small separations occur between age classes, especially between the middle and upper age class.

Age-length data for white crappie (*Pomoxis annularis*) presented in Table 1 show that distinct and large differences between age classes after age class 2 do not occur. For example, the mean lengths for age classes 2 and 3 differ by only 21 mm and the mean lengths for age classes 3 and 4 differ by only 28 mm. If the growth curve is modeled by the von Bertalanffy equation (Ricker 1975, p. 22), the average distance between the mean lengths of successive age classes decreases as age increases. As a result, the estimation of all the parameters in the vector Θ becomes very difficult.

Table 1. Age-length relationships for white crappie collected by trap nets in Conowingo Pond from 1966–1973[a]

Year of capture	Average length (mm) at ages 1–5				
	1	2	3	4	5
1966	117	188	220	247	278
1967	101	171	209	236	306
1968	114	183	209	230	240
1969	127	191	204	240	273
1970	116	173	196	221	229
1971	121	184	193	215	257
1972	112	168	173	202	252
1973	114	171	197	215	221
Mean	115	179	200	228	257

[a]Data from Mathur et al. (1975).

In Table 2, the average yearly increment in length for one-year-olds (two years old at the end of the given growing season) is presented for white crappie. Identification of the good and poor recruitment years is taken from Mathur et al. (1975). The average increase in length when recruitment was poor was about 75 mm and the average increase when recruitment was good was only 48 mm. This inverse relationship between growth rate and recruitment success is probably due to inter- and intra-age-class competition for the same food source in the pond. Young white crappie (age classes 0–3) in the pond consume mainly zooplankton (Euston 1976). When there is a large recruitment of young into the pond, individuals in age class 0 are competing for zooplankton among themselves and also with age classes 1 and 2. Consequently, during a good recruitment year, growth of these three age classes is slow, whereas during a poor recruitment year, growth of these age classes is greater due to reduced competition for food. Also, during high recruitment

Table 2. The effect of recruitment success on the growth rate
of age class 1 white crappie in Conowingo Pond[a]

Year	Recruitment success	Length of age class 1, mm	Length of age class 2, mm	Growth increment,[b] mm
1966	Good	117	188	
1967	Poor	101	171	54
1968	Poor	114	183	82
1969	Good	127	191	77
1970	Poor	116	173	46
1971	Good	121	184	68
1972	Poor	112	168	47
1973	Poor	114	171	59
1974	Average	129	202	88
1975	Poor	86	175	46

[a]Data from Mathur et al. (1975).

[b]Calculated as length of age class 2 in year t minus length of age class 1 in year t minus length of age class 1 in year $t - 1$.

years for white crappie, the predator-prey ratio in this system probably decreases. Therefore, predation on young-of-the-year white crappie would be reduced and probably would not be intensive enough to ultimately reduce the competition for food. Some large predators such as walleye, largemouth bass, and smallmouth bass, are present in the system, but seemingly in small numbers (Mathur et al. 1975).

The previous discussion has indicated that there is not always a direct correspondence between the modes in length-frequency data and the ages present in the catch. Because the age-length key is based on a subsample, it would be, in general, more efficient to use the larger sample associated with the length-frequency data to estimate the age-structure. This can be accomplished by obtaining the mean lengths of the various age classes (μ_i's). Once the μ_i's are known, the reduced parameter vector $\Omega = \langle \sigma_1^2, \sigma_2^2, ..., \sigma_k^2, \rho_1, \rho_2, ..., \rho_k, \rangle$ can be estimated; that is, instead of estimating k means, k variances, and $k - 1$ proportions, we estimate k variances and $k - 1$ proportions.

Additionally, if the variances σ_i^2 can be estimated from the subsample employed for obtaining the age-length key, then the only unknowns are the $k - 1$ proportions. The estimate of Ω will be referred to as the conditional maximum likelihood (CML) estimate, and the estimate of Θ defined in Eq. (1) will be referred to as the maximum likelihood (ML) estimate. This method will be applied to some fish catch data.

ESTIMATION OF THE PARAMETERS OF THE MODEL

The estimation of the parameters of a mixture of two normal distributions (also referred to as a compound normal distribution) has been investigated by Cohen (1967), Pearson (1894), and Rao (1952). Pearson (1894) derived the moment estimator for $k = 2$ and Cohen (1966) found that for sample sizes less than 400 the moment estimators were not very reliable.

Hasselblad (1966) has investigated the properties of the maximum likelihood estimators for a mixture of k normal distributions. He reported that (1) as $d_{ij} = |\mu_i - \mu_j|$ decreases, the variances of the proportions ρ increase rapidly; (2) as d_{ij} decreases, the variances of the location (i.e., μ) and scale (i.e., σ) estimators increase; and (3) as ρ_i approaches 0, the variances of the location and scale estimates increase. Hasselblad suggested an iterative method for the maximum likelihood estimator. Dick and Bowden (1973) used the Newton iterative scheme for a mixture of two normal distributions.

MacDonald (1971) obtained estimates that minimized the Cramer–von Mises statistic* (Cramer 1974, p. 451), and Kumar and Nicklin (1976) obtained estimates that minimized the mean squared distance between the empirical and theoretical characteristic function. In this paper, we restrict ourselves to the maximum likelihood estimator. It is beyond the scope of this paper to compare the various estimators.

The likelihood function is nonlinear in the parameters, Θ. It was felt that when k is greater than 2, methods based on derivatives become inefficient and the amount of computer time required tends to render this method impractical. Described below is an alternative method for estimation of the parameters (E. H. Nicklin, General Foods Corporation, White Plains, New York, *personal communication*). The negative of the natural logarithm of the likelihood function L given by

$$L = -\sum_{i=1}^{n} \ln f(x_i; \Theta) \qquad (2)$$

was minimized with respect to Θ (or Ω) using the derivative-free minimization technique developed by Nelder and Mead (1965). This method is called the simplicial triangulation method. The method involves the evaluation of the function at the vertices of an $m + 1$ dimensional moving simplex (m is the number of parameters to be estimated). Based on the functional values at the vertices, the simplex moves away from the maximum calculated value. This technique has been found to be efficient for most well-behaved surfaces.

In order to keep the search for optimal values in the feasible region, transformed parameters were estimated. The location parameters (i.e., μ_i)

---

*The Cramer–von Mises statistic is a goodness of fit criterion for testing distributional assumptions.

were restricted to the interval $[x_{(1)}, x_{(n)}]$, where $x_{(1)}$ and $x_{(n)}$ are the observed minimum and maximum values. The transformed means are given by

$$\alpha_i = \ln[\mu_i - x_{(1)}] - \ln[x_{(n)} - \mu_i] .$$ (3)

This transformation is called the logistic transformation. The scale parameters have to be nonnegative, and this was achieved by using the transformation

$$\lambda_i = \ln(\sigma_i^2/2) .$$ (4)

Finally, since the proportions ρ_i must sum to unity, we need to estimate only $k - 1$ proportions. However, these proportions must also satisfy the condition

$$\sum_{i=1}^{k-1} \rho_i \leq 1 - b ,$$ (5)

where b is some small positive constant less than 1. The constant b is introduced to prevent the pathological case of one of the proportions going to 0 or 1, and it is usually chosen to be some function of the sample size n. Inequality (5) can be rewritten as

$$b \leq \rho_1 \leq (1 - b) = l_1 ,$$

$$b \leq \rho_i \leq \left(1 - b - \sum_{j=1}^{i-1} \rho_i\right) = l_i \quad 2 \leq i \leq k - 1 .$$ (6)

The transformed proportion is

$$\delta_i = \ln(\rho_i - b) - \ln(l_i - \rho_i) \quad 1 \leq i \leq k - 1 .$$ (7)

Note that the ordering of i is not important.

EMPIRICAL EVALUATION OF THE ESTIMATION TECHNIQUE

In the preceding section the problems involved in the estimation of the parameters were discussed. An empirical study of a hypothetical population was conducted to determine the advantages of knowing, a priori, the mean length for each age class. The growth of a fish was assumed to follow the Bertalanffy growth curve. The average lengths for ages 1–5 were assumed to be 117, 156, 190, 220, and 245 mm, respectively, and the proportions of each of the five age classes were set to 0.13, 0.40, 0.24, 0.144, and 0.086, respectively. These proportions were assigned by assuming that the annual survival rate between ages 2 and 5 was 0.60 per year and that age 2 was the first fully recruited age class and constituted 40% of the total fish population.

The estimates of the proportions in each age class were obtained when (a) none of the parameters were known (Θ), and (b) when mean lengths were assumed known (Ω). The common variance $\sigma^2 = \sigma_i^2; i = 1, 2, ..., 5$ was varied between 100 and 400.

A sample of 1000 fish was generated from the population described above. Five sets of expected order statistics* from a normal distribution (Harter, 1952) with sample sizes 130, 400, 240, 144, and 86 were obtained, and the parameter vectors Θ and Ω were estimated with the appropriate variance. These estimates represent the "expected" behavior of the estimates, in the sense that if the methodology does not work for this sample, it is highly doubtful that it will work with a random sample. The results are summarized in Table 3 and Fig. 1 for the parameter vector Θ [case (a)] and in Table 4 and Fig. 2 for the parameter vector Ω [case (b)].

For case (a), the estimated values of the elements of the parameter vector Θ show considerable variation from the "true" values. When the variance is 400, the proportion ρ_2 is estimated to be 14%, whereas in fact, the true value is 40%, and ρ_3 is estimated to be 40%, whereas the true value is 24% (Table 3). The estimated means and variances also show considerable departure from the true values. The estimated catch curves (Fig. 1) are also significantly different from the "true" catch curve. When the variance is 300, the estimated ρ_5 is 0.20 and ρ_4 is 0.15, giving a false impression that there is differential recruitment and mortality for the two years; also, when the variance is 400, the estimated model indicates that age classes 1 and 2 are partially recruited and age 3 is the first fully recruited age class. In the assumed population, only age 1 is partially recruited (the initial ascending limb of the curves represent partially recruited age classes, Ricker 1975, p. 33).

The results for case (b) are summarized in Table 4. Considerable improvement is observed in the estimates of the proportions. For example, when the variance is 400, the estimated proportions are reasonably close to the true values, as opposed to case (a). The estimated catch curves are plotted in Fig. 2. For smaller variances the estimated catch curves almost overlap the true catch curve. As the variance increases, the estimated catch curves tend to depart from the true curve, though this departure is not significant. As the variance increases, the estimated variances also exhibit greater departure from the true values, as one would expect.

---

*An expected order statistic is defined as follows. If $x_1, x_2, ..., x_n$ is a sample of size n such that $x_1 < x_2 < ... < x_n$, then $E[x_i]$ = expected order statistic, where $E[\cdot]$ is the usual expectation operator.

Table 3. Case (a): Estimates of the parameter vector Θ from a sample of expected order statistics ($n = 1000$)[a]

True variance (σ^2)	Estimator[b]	Age class				
		1	2	3	4	5
100	$\hat{\mu}$	117.1	156.2	188.11	207.5	233.7
	$\hat{\sigma}^2$	98.1	103.9	70.8	145.9	225.8
	$\hat{\rho}$	0.126	0.407	0.178	0.114	0.175
200	$\hat{\mu}$	115.5	152.0	175.9	201.8	234.5
	$\hat{\sigma}^2$	174.2	170.4	219.2	218.8	302.4
	$\hat{\rho}$	0.11	0.31	0.22	0.18	0.18
300	$\hat{\mu}$	116.1	150.8	174.9	201.7	232.1
	$\hat{\sigma}^2$	293.4	240.0	229.0	188.8	412.4
	$\hat{\rho}$	0.125	0.296	0.231	0.146	0.202
400	$\hat{\mu}$	114.8	142.2	169.6	207.8	242.4
	$\hat{\sigma}^2$	337.2	191.9	368.8	323.6	368.0
	$\hat{\rho}$	0.126	0.144	0.399	0.207	0.124

[a]The true parameter vector $\Theta = \langle 117, 156, 190, 220, 245, \sigma_1{}^2, \sigma_2{}^2, ..., \sigma_5{}^2, 0.13, 0.40, 0.24, 0.144, 0.086 \rangle$ and $\sigma^2 = \sigma_i{}^2, i = 1, 2, ..., 5$.

[b]$\hat{\mu}$ = estimated mean length; $\hat{\sigma}^2$ = estimated variance; and $\hat{\rho}$ = estimated proportion.

Table 4. Case (b): Estimates of the parameter vector Ω from a sample of expected order statistics ($n = 1000$)[a]

True variance (σ^2)	Estimator[b]	Age class				
		1	2	3	4	5
100	$\hat{\mu}$	117	156	190	220	245
	$\hat{\sigma}^2$	98.2	102.7	92.3	115.4	99.3
	$\hat{\rho}$	0.128	0.406	0.232	0.152	0.082
200	$\hat{\mu}$	117	156	190	220	245
	$\hat{\sigma}^2$	196.1	198.6	173.8	173.9	189.0
	$\hat{\rho}$	0.13	0.405	0.235	0.144	0.087
300	$\hat{\mu}$	117	156	190	220	245
	$\hat{\sigma}^2$	301.9	273.4	285.9	212.6	266.6
	$\hat{\rho}$	0.137	0.383	0.260	0.123	0.097
400	$\hat{\mu}$	117	156	190	220	245
	$\hat{\sigma}^2$	348.6	207.0	287.0	241.8	351.0
	$\hat{\rho}$	0.134	0.403	0.236	0.121	0.107

[a]The true parameter vector $\Theta = \langle 117, 156, 190, 220, 245, \sigma_1{}^2, \sigma_2{}^2, ..., \sigma_5{}^2, 0.13, 0.40, 0.24, 0.144, 0.086 \rangle$ and $\sigma^2 = \sigma_i{}^2, i = 1, 2, ..., 5$.

[b]$\hat{\mu}$ = estimated mean length; $\hat{\sigma}^2$ = estimated variance; and $\hat{\rho}$ = estimated proportion.

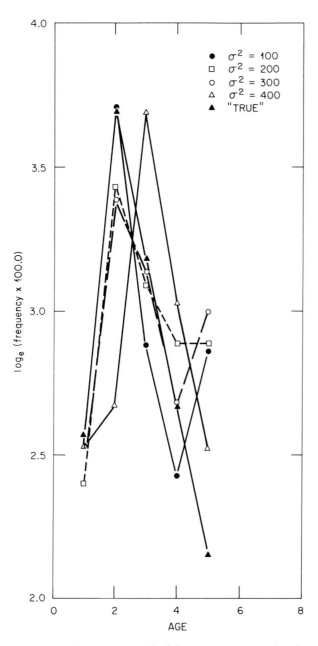

Fig. 1. Estimated catch curves when the full parameter vector Θ is estimated for the expected order statistic and for $\sigma^2 = 100, 200, 300,$ and 400.

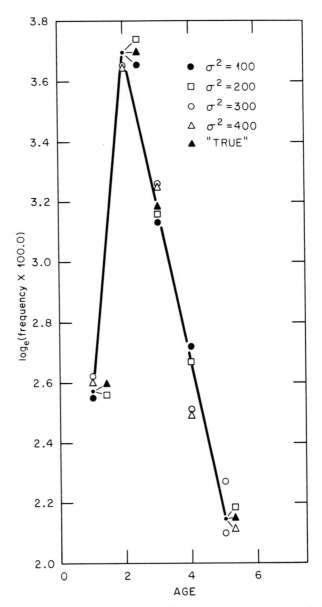

Fig. 2. Estimated catch curve when the location is assumed known and the reduced vector Ω is estimated for the expected order statistic and for σ^2 = 100, 200, 300, and 400.

These results show that if one has an age-length key available (case (b)], the length-frequency data can be used to obtain reliable estimates of the age structure of the population. When the key is not available, the estimates must be viewed as first approximations. In the next section, the results of a small scale simulation study for case (b) are presented.

SIMULATION RESULTS

The fish population was assumed to be the same as the above and the simulations were run for two cases:

$$\text{(Case 1) } \sigma^2 = \sigma_i^2 = 100 \quad i = 1, 2, ..., 5 .$$

$$\text{(Case 2) } \sigma^2 = \sigma_i^2 = 200 \quad i = 1, 2, ..., 5 .$$

The random numbers were generated as follows. The sample size was fixed at 1000. The sample sizes for the five normal distributions were 130, 400, 240, 144, and 86, respectively, based on the proportions defined above. Twenty replicate samples of 1000 fish each were obtained for each of the two cases described above, and the estimates of the elements of the parameter vector, Ω, were obtained using the method discussed in the preceding sections. In order to reduce the number of replicates required, the replicates were rejected if the Cramer—von Mises statistic based on the true values exceeded 0.275. Hence, the replicates used in the analysis represent a stratified sample from the population.

The results of the simulations for cases (1) and (2) are summarized in Tables 5 and 6 respectively. In both cases, the estimated proportions are in

Table 5. Estimates of the parameter vector Ω based on 20 replicates and a sample size of 1000 when the mean lengths are set equal to 117, 156, 190, 220, and 245 mm and $\sigma^2 = \sigma_i^2 = 100, i = 1, 2, ..., 5$

Parameter	True value	Mean estimate	Standard deviation
σ_1^2	100	97.41	16.50
σ_2^2	100	100.28	10.08
σ_3^2	100	95.76	18.70
σ_4^2	100	95.03	25.38
σ_5^2	100	99.93	16.50
p_1	0.13	0.130	$0.509(-2)^a$
p_2	0.40	0.399	$0.111(-1)$
p_3	0.24	0.241	$0.157(-1)$
p_4	0.144	0.145	$0.143(-1)$
p_5	0.085	0.085	$0.760(-2)$

[a] $0.509(-2) = 0.00509$.

Table 6. Estimates of the parameter vector Ω based on 20 replicates and a sample size of 1000 when the mean lengths are equal to 117, 156, 190, 220, and 245 mm and $\sigma^2 = \sigma_i^2 = 200$, $i = 1, 2, ..., 5$

Parameter	True value	Mean estimate	Standard deviation
σ_1^2	200	190.75	35.62
σ_2^2	200	190.25	30.06
σ_3^2	200	187.55	39.14
σ_4^2	200	188.01	58.85
σ_5^2	200	185.95	34.82
ρ_1	0.13	0.130	$0.896(-2)^a$
ρ_2	0.40	0.397	$0.167(-1)$
ρ_3	0.24	0.241	$0.169(-1)$
ρ_4	0.144	0.146	$0.192(-1)$
ρ_5	0.086	0.085	$0.125(-1)$

[a] $0.896(-2) = 0.00896$.

very good agreement with the "true" values. This result implies that if the means of the populations are known a priori, then the proportions (catch curve) can be estimated with some reliability. The variances, which are nuisance parameters, show greater variability than the proportions and are generally biased downward. The standard deviations of the estimates of σ_i^2 increase with increasing σ^2, as would be expected.

These results support the conclusion we reached with the expected order statistics. They demonstrate the feasibility of obtaining the CML estimates from the length-frequency data when the age-length key is available.

APPLICATIONS OF THE MODEL

We now apply the techniques discussed above to two species of fish, channel catfish (*Ictalurus punctatus*) and white crappie (*Pomoxis annularis*). Length-frequency data for these two species were collected as part of the ecological monitoring program at the Peach Bottom Atomic Power Station. The station, consisting of two units of 1065 MW(e) each, is located in southeastern Pennsylvania on the west bank of Conowingo Pond, an impounded reach of the Susquehanna River. Fish surveys with trap nets and trawls have been conducted at a series of stations located upstream of the power plant (controls) and in areas expected to be within the heated plume. Monitoring stations were located on the basis of available habitats and physical constraints imposed by topography. Length-frequency data and life-history information for the major species of fish in the pond are given in Robbins and Mathur (1974, 1976). In our study, length-frequency data of fish captured in trap nets were utilized. Trawl samples generally were dominated by age classes 0–2, whereas fish caught by trap nets were generally older and larger.

Figure 3 shows the length-frequency plot for the trap net catch of channel catfish in August 1971. Since the age-length key is not available for this species, we estimated all the parameters, Θ, defined by Eq. (1). It was emphasized earlier that these estimates are only a first approximation. When none of the parameters are known, the estimation problem is more complicated since in addition to the unknown parameter vector Θ, and additional parameter k, the number of age classes in the probability model, is also unknown. Several values of k, therefore, will be used in attempting to fit the model, and the one that is most "satisfactory" will be used. The estimates of the elements of the parameter vector Θ based on the data shown in Fig. 3 are summarized for $k = 5$ and $k = 6$ in Table 7.

Table 7. Estimates of parameter vector Θ for trap net data for channel catfish collected in August 1971 in Conowingo Pond

k^a	Sample size	Estimator[b]	Modes					
			1	2	3	4	5	6
5	345	$\hat{\mu}$	117	153	230	261	346	
		$\hat{\sigma}$	9.4	9.0	28.1	51.1	37.4	
		$\hat{\rho}$	0.31	0.18	0.36	0.11	0.04	
6	345	$\hat{\mu}$	117	153	203	236	254	320
		$\hat{\sigma}$	9.4	8.4	33.3	21.3	35.0	42.4
		$\hat{\rho}$	0.31	0.16	0.15	0.22	0.09	0.07

[a]k = number of age classes in the probability model.

[b]$\hat{\mu}$ = estimated mean length; $\hat{\sigma}$ = estimated standard deviation; and $\hat{\rho}$ = estimated proportion.

For both $k = 5$ and $k = 6$, the first two modes are the same and the proportions are comparable. The mode at 230 mm for $k = 5$ is separated into two modes at 203 and 236 mm for $k = 6$. Clearly, the mode at 346 mm for $k = 5$ represents all the remaining older age classes. The estimated proportion for this mode is only about 4% of the sample of 345 channel catfish (i.e., about 14 fish). Either the species is short-lived or the trap nets are selective in that they do not capture the older age classes. The modes at 261 and 346 mm for $k = 5$ are represented by modes at 254 and 320 mm for $k = 6$. Also, some of the observations classified into modes 5 and 6 for $k = 5$ are classified into mode 4 with a mean length of 236 mm for $k = 6$. Even though we have associated a mode with a specific age class (e.g., mode 1 corresponds to age class 1), this is only a convenient way of referring to the modes.

Mathur et al. (1975) reported that the average lengths for channel catfish in Conowingo Pond between the years 1966–1969 for age classes 1–6 were 88, 149, 198, 226, 254, and 288 mm, respectively. For age classes 2–5 the agreement between the estimated mean lengths and these reported values is

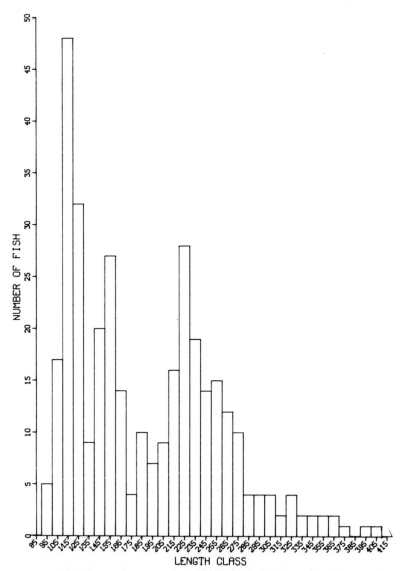

Fig. 3. Length-frequency data for channel catfish collected by trap nets in Cono-wingo Pond during August 1971.

reasonable (Table 7). The large discrepancy for age class 1 may be due to the selectivity of trap nets for the larger individuals in age class 1 or to a higher than normal growth rate for the 1970 year class.

The estimated standard deviations for age classes 3–6 are relatively large (Table 7). Since few channel catfish in the older age classes were caught, the estimated variances are only approximate. If the species is short-lived, the sample could be treated as representative of the true population. Otherwise, collection methods to obtain a more representative sample of the older age classes may be needed. The majority of specimens captured in Conowingo Pond are less than eight years old, but a few fish in age classes 8–16 have been captured (Robbins and Mathur 1974). Davis (1959) and Finnell and Jenkins (1954) found that channel catfish in Kansas and Oklahoma, respectively, seldom live longer than seven years. In contrast, 60% of the channel catfish collected by Stevens (1959) from two aquatic systems in South Carolina were eight years old or older. Age class structure and longevity of channel catfish may therefore be a function of the particular aquatic system concerned. Under the assumption that the catch was representative of the channel catfish population in Conowingo Pond in August 1971, the estimated catch curve is given by curve (*a*) in Fig. 4 for $k = 6$. The effect of variable recruitment among year classes is pronounced, as indicated by the departure of the curve from a straight line.

Cases could be encountered where the modes are too close to each other to represent separate age classes. Such would be the case when the spawning period of a species extends over a long period. Fish spawned in the early part of the period might have a higher mean length than those spawned towards the end of the period. The availability of food in different locations and inter- and intra-species competition for food may also play a strong role in creating multiple modes for the same age class. This situation probably exists in Conowingo Pond for white crappie and particularly for channel catfish. The growth of older channel catfish is relatively slow, and the growth of younger catfish is relatively rapid in Conowingo Pond. The average lengths of catfish in age classes 1 and 2 are greater than the lengths of the corresponding age classes in other aquatic systems. Conversely, the average lengths for age classes 3 and older are less than in other systems. Young catfish feed on zooplankton for about two years; thereafter, they assume a more benthic existence and consume greater quantities of benthos. Since the standing crop biomass of benthos is very sparse in the system, competition for this food source is high, and consequently, growth is slow. If two modes are close to each other due to differential growth within a single age class, it may become necessary to subdivide the age classification further. Alternatively, one could take the mean of the two modal lengths.

Another example for channel catfish is given in Fig. 5 and Table 8 for the June 1975 trap net data for $k = 5$ and $k = 6$. The modal lengths are the same for mode 1 and for mode 2. The large standard deviation associated with the

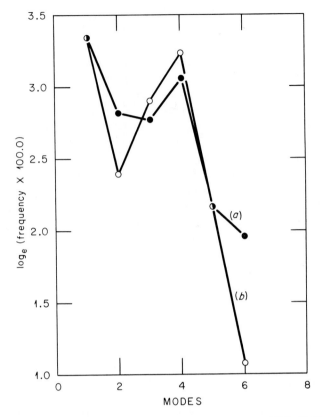

Fig. 4. Estimated catch curves for channel catfish: (*a*) August 1971 (Table 7), (*b*) June 1975 (Table 8).

Table 8. Estimates of parameter vector Θ for trap net data for channel catfish collected in June 1975 in Conowingo Pond

k^a	Sample size	Estimator[b]	Modes					
			1	2	3	4	5	6
5	314	$\hat{\mu}$	103	152	195	275	369	
		$\hat{\sigma}$	16.3	7.9	26.0	28.7	30.0	
		$\hat{\rho}$	0.32	0.14	0.46	0.06	0.02	
6	314	$\hat{\mu}$	103	153	177	201	255	349
		$\hat{\sigma}$	16.2	7.2	22.5	22.7	30.9	42.5
		$\hat{\rho}$	0.31	0.12	0.18	0.27	0.09	0.03

$^a k$ = number of age classes in probability model.

$^b \hat{\mu}$ = estimated mean length; $\hat{\sigma}$ = estimated standard deviation; and $\hat{\rho}$ = estimated proportion.

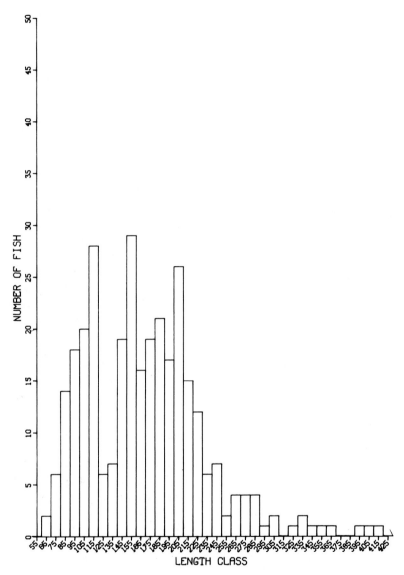

Fig. 5. Length-frequency data for channel catfish collected by trap nets in Conowingo Pond, June 1975.

mode at 103 mm is due to the presence of age class 0, as indicated by a minimum length of 65 mm for fish in the sample. When $k = 5$ there appears to be a large separation between modes 3 and 4 and between modes 4 and 5. Also, for mode 3, $p_3 = 0.46$, a proportion too high to be true. When $k = 6$, the modes are realigned, and there is a mode at 177 mm, which is 24 mm from the previous mode at 153 mm. Based on the estimated standard deviation at mode 2, a length difference of 24 mm represents approximately 3σ. The standard deviation at mode 3 is much larger (22.5 mm) than the standard deviation at mode 2 (7.2 mm), which indicates that mode 2 is about 1σ away from mode 3. It is evident that due to differential recruitment and growth patterns, the age classes 3–5 have overlapping length distributions. The estimated catch curve for $k = 6$ is given by curve (b) in Fig. 4. The peak at mode 4 indicates differential growth rate.

A third example is the trap net data for white crappie collected in Conowingo Pond during January, February, and March of 1974. The length-frequency data for each of the three months are shown in Figs. 6–8. The age-length key (Table 1) is based on samples caught during the nongrowing season (i.e., the winter months). Since January–March 1974 would be part of the nongrowing season of 1973, the age-length key should apply to the January–March 1974 data. The reported difference in mean lengths between age classes 4 and 5 is only 6 mm for 1973 (Table 1). However, since this value is based on a small sample, it is not very reliable. The length-frequency curves for these three months (Figs. 6–8) show that there are few fish between 160 and 180 mm (less than 2% of the catch in January). Since 1972 was classified as a year of poor recruitment (see Table 2), the proportion of age class 2 in the population in 1974 would be expected to be small.

In Table 9 the estimates of the full parameter vector Θ for each of the three months are summarized. In January, modes 1 and 2 occur at 125 and 136 mm, and these two modes together (year classes 1972 and 1973) constitute 59% of the catch, which is in direct conflict with the recruitment classification of Table 2. At the present time we are unable to resolve this conflict. The same pattern for modes 1 and 2 is also observed in February (68%) and March (80%). In February and March of 1974 mode 1 is at 116 mm, which is reasonably close to the reported mean length for 1973 of 114 mm in Table 1.

There exists, however, a second mode at approximately 135 mm for all three months. According to the age-length key (Table 1), this mode must represent age class 1 fish, since age class 2 had an average length of 171 mm in 1973. This result indicates that the mode at 135.0 mm represents the upper range for age class 1 in 1973. The mode at 116 mm that is present in February and March does not appear in January (Table 9). There are two possible explanations for the apparent discrepancy; either the average length for age class 1 was 130 mm in January 1974 or the catchability of fish less than 100 mm in length was low. Since the smallest fish caught was about 75 mm, the average length for age class 1 is probably 130 mm.

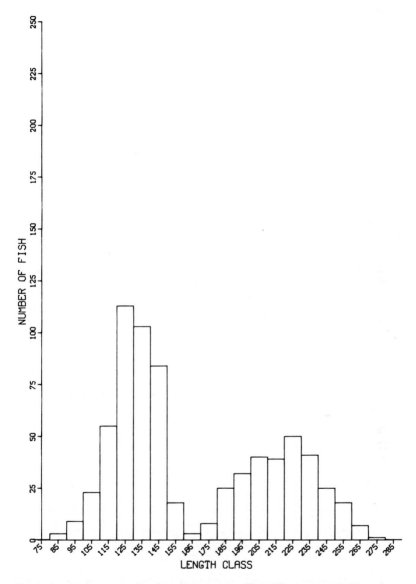

Fig. 6. Length-frequency data for white crappie collected by trap nets in Conowingo Pond, January 1974.

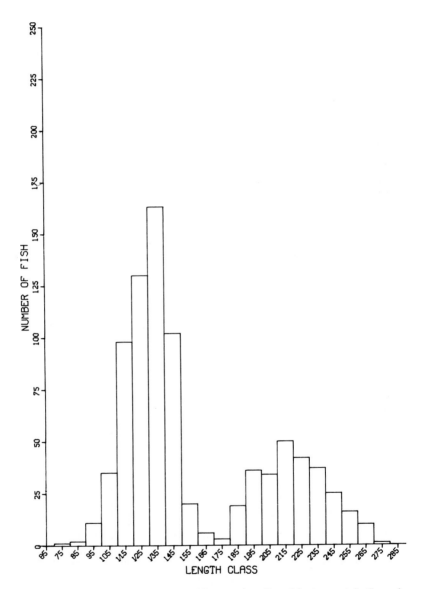

Fig. 7. Length-frequency data for white crappie collected by trap nets in Conowingo Pond, February 1974.

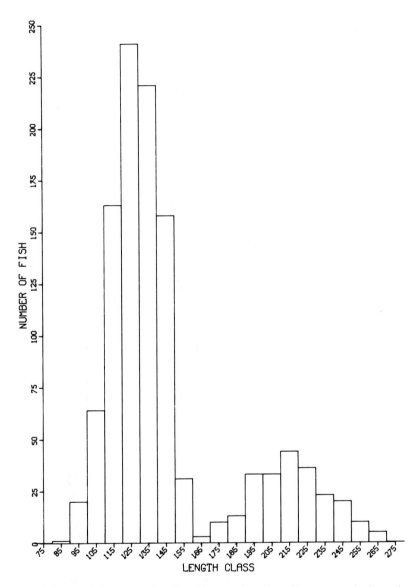

Fig. 8. Length-frequency data for white crappie collected by trap nets in Conowingo Pond, March 1974.

Table 9. Estimates of parameter vector Θ for trap net data for white crappie collected in January, February, and March 1974 in Conowingo Pond

Month	Sample size	Estimator[a]	Modes				
			1	2	3	4	5
January	697	$\hat{\mu}$	125	136	198	223	237
		$\hat{\sigma}$	15.1	10.2	13.5	16.1	15.4
		$\hat{\rho}$	0.33	0.26	0.15	0.13	0.13
February	841	$\hat{\mu}$	116	133	199	214	238
		$\hat{\sigma}$	12.9	11.7	13.2	15.2	15.8
		$\hat{\rho}$	0.16	0.52	0.09	0.10	0.13
March	1129	$\hat{\mu}$	116	135	194	208	232
		$\hat{\sigma}$	10.5	10.5	16.4	13.8	16.9
		$\hat{\rho}$	0.28	0.52	0.04	0.07	0.09

[a] $\hat{\mu}$ = estimated mean length; $\hat{\sigma}$ = estimated standard deviation; $\hat{\rho}$ = estimated proportion.

The estimated modes for the three months are 198, 199, and 195 mm for age class 3, which is in good agreement with the value of 197 mm from Table 1. For age class 4, the estimated modes for the three months are 223, 214, and 208 mm. The weighted average length in mm for age class i, $\bar{x}_i$, is computed by

$$\bar{x}_i = \sum_j \frac{(\rho_{ij}C_j)\mu_{ij}}{\sum_j \rho_{ij}C_j}, \tag{8}$$

where

ρ_{ij} = proportion of age class i in the catch at time j,

C_j = total number of fish in the catch at time j,

μ_{ij} = the ith estimated modal length (in mm) at time j.

The number of white crappie of all age classes caught in January, February, and March was 697, 841, and 1129, respectively. The estimated weighted mean lengths for ages 3, 4, and 5 are 198, 215, and 236 mm, respectively. The estimates for ages 3 and 4 are in close agreement with the reported values of 197 and 215 mm. The reported value for age 5 is 221 mm which is about 15 mm less than the estimated value.

In this example for white crappie, even though there was an estimate of the age-length key, the more general approach of estimating the elements of the full parameter vector Θ had to be used, because of the apparent discrepancies in the data. Two important points emerge from this example: (a) the age-length key is valid only for the catch on which it is based, and (b) in reporting fish-length data it is important to back-calculate all the lengths to a standard time. The last point is extremely important if the catch over a year is to be summarized and used for production calculations.

CONCLUSION

We have used the length-frequency data in conjunction with the age-length key to obtain maximum likelihood estimates of the age structure in the population. The simulation results and examples presented lead us to believe that proper estimation of the age-length key will result in good estimates of the age structure. Since knowledge of the age structure is useful in estimating population parameters, such as growth and mortality, it is therefore important that proper sampling methods be used for estimating the age-length key. The model can be used as follows to obtain an "optimal" subsampling scheme for age determination.

1. Obtain a length-frequency distribution for a fish catch.

2. Use the techniques described to obtain (a) estimate of modes, (b) estimate of variances, and (c) estimate of proportions.

3. Use these results to determine the "optimal" sample size for age determination.

4. Obtain an age-length key.

5. Use the age-length key to re-estimate the proportions.

6. Re-evaluate step 3 and if the change in the optimal sample size is negligible, stop; otherwise, obtain additional samples for age-length key determination and repeat steps 3–6.

The algorithm described in steps 1–6 above has not been examined in this study. We hope to address this problem in more detail at a later date.

In addition to applying knowledge of the age structure to estimate functional parameters of fish populations, age distributions can be used in the context of evaluating the potential effects of power plant operations on fish populations. If an appropriate control system is available, then the age structures of fish in the control and stressed areas for the same time periods could be compared. Assume that the fish in a stressed area cannot migrate to and from the control area and that the two systems were similar before the intervention (preoperational period) by a power plant. Differential changes in the values of the mean lengths (μ_i's) and the proportions (ρ_i's) between the two areas can be tested for. In other words, a statistically significant differential change in the population characteristics in the stressed area as compared to the control area may be reflected in the estimated values of the parameters of the model (μ_i's and ρ_i's). For example, if the operation of a power plant increases the growth rate of certain age classes, then the estimated mean lengths for these age classes will be greater than the mean lengths in the control area during the same period. If the mortality rate of an age class increases in the stressed area, then a smaller proportion of that age class will exist in the stressed area as compared with the control area.

In the preceding discussion we have assumed that the control and stressed systems were similar before power plant intervention. This assumption is not realistic for natural systems. A more realistic assumption would be that when both the systems vary naturally without intervention, there exists a constant structural relationship between the two areas. This constant relationship would be reflected by a constant value r_i for the ratio of the proportions, ρ_i's, for age i between the two areas. This ratio can be estimated initially from the preoperational data. Changes in the value of this ratio after the start of plant operation would serve as an index of impact (see Thomas, *in this volume*, for a further discussion of the use of ratios).

ACKNOWLEDGMENTS

The authors wish to express their appreciation to W. Van Winkle and J. J. Beauchamp for their constructive criticism of the manuscript.

LITERATURE CITED

Allen, K. R. 1966. Determination of age distribution from IBM 7079, 7094, age-length keys and length distributions, Fortran IV. Trans. Am. Fish. Soc. **95**:230–231.

Chadwick, E. M. P. 1976. Ecological fish production in a small Precambrian shield lake. Environ. Biol. Fish. **1**:13–60.

Cohen, A. C. 1966. Discussion of "Estimation of parameters for a mixture of normal distributions by Victor Hasselblad." Technometrics **8**:445–446.

———. 1967. Estimation in mixtures of two normal distributions. Technometrics **9**:15–28.

Cramer, H. 1974. Mathematical methods of statistics. Princeton Univ. Press, Princeton, New Jersey. 575 p.

Davis, J. 1959. Management of channel catfish in Kansas. Univ. Kansas, Mus. Nat. Hist., Misc. Publ. No. 21. 56 p.

Dick, N. P., and D. C. Bowden. 1973. Maximum likelihood estimation for mixtures of two normal distributions. Biometrics **29**:781–790.

Euston, E. T. 1976. Trophic spectrum, p. 4-18 to 4-42. *In* T. W. Robbins and D. Mathur [eds.] Postoperational report No. 5 on the ecology of Conowingo Pond for the period of July 1975—December 1975. Ichthyological Associates Inc., Drumore, Pennsylvania.

Finnell, J. C., and R. M. Jenkins. 1954. Growth of channel catfish in Oklahoma waters: 1954 revision. Okla. Fish. Res. Lab., Rep. No. 41, 37 p.

Harter, H. L. 1952. Expected values of normal order statistics. Biometrika 48:151—165.

Hasselblad, V. 1966. Estimation of parameters for a mixture of normal distributions. Technometrics 8:431—444.

Ketchen, K. S. 1950. Stratified subsampling for determining age distributions. Trans. Am. Fish. Soc. 79:205—212.

Kumar, K. D., and E. H. Nicklin. 1976. Comparison of some models for security price behavior. Proc. 7th Annu. Pittsburg Conf. Modeling Simulation 7:98—103.

Mathur, D., P. G. Heisey, and N. C. Magnusson. 1975. Age and growth, p. 4—33 to 4—48. *In* T. W. Robbins and D. Mathur [eds.] Postoperational report No. 3 on the ecology of Conowingo Pond for the period of July 1974—December 1974. Ichthyological Associates Inc., Drumore, Pennsylvania.

MacDonald, P. D. M. 1971. Comment on "An estimation procedure for mixture of distributions" by Choi and Bulgren. J. Roy. Stat. Soc. B. 33:326—329.

Nelder, J. A., and R. Mead. 1965. A simplex method for function minimization. Comput. J. 7:308—313.

Pearson, K. 1894. Contributions to the mathematical theory of evolution. Phil. Trans. Roy. Soc. 185:71—110.

Rao, C. R. 1952. Advanced statistical methods in biometric research. Wiley and Sons, New York. 390 p.

Ricker, W. E. 1975. Computation and interpretation of biological statistics of fish populations. Fish. Res. Board Can., Bull. 191. 382 p.

Robbins, T. W., and D. Mathur. 1974. Peach Bottom Atomic Power Station preoperational report on the ecology of Conowingo Pond for Units No. 2 and 3. Ichthyological Associates, Inc., Drumore, Pennsylvania. 349 p.

Robbins, T. W., and D. Mathur. 1976. Peach Bottom Atomic Power Station postoperational report No. 5 on the ecology of Conowingo Pond for Units No. 2 and 3. Ichthyological Associates, Inc., Drumore, Pennsylvania.

Seber, G. A. F. 1973. The estimation of animal abundance and related parameters. Hafner Bros, New York. 506 p.

Stevens, R. E. 1959. The white and channel catfishes of the Santee—Cooper Reservoir and Tailrace Sanctuary, p. 203—219. *In* Proc. Thirteenth Annu. Conf., S.E. Assoc. Game Fish Comm.

Prediction of Fish Biomass, Harvest, and Prey-Predator Relations in Reservoirs

Robert M. Jenkins

National Reservoir Research Program
U.S. Fish and Wildlife Service
Fayetteville, Arkansas

ABSTRACT

Regression analyses of the effect of total dissolved solids on fish standing crops in 166 reservoirs produced formulas with coefficients of determination of 0.63 to 0.81. These formulas provide indexes to average biotic conditions and help to identify stressed aquatic environments. Simple predictive formulas are also presented for clupeid crops in various reservoir types, as clupeids are the fishes most frequently impinged or entrained at southern power plants. A method of calculating the adequacy of the available prey crop in relation to the predator crop is advanced to further aid in identification of perturbed prey populations. Assessment of stress as reflected by changes in sport fishing success can also be approached by comparison of the predicted harvest potential with actual fish harvest data. Use of these predictive indexes is recommended until more elaborate models are developed to identify power plant effects.

Key words: clupeid, fish harvest, fish standing crop, predator crop, prey crop, regression analysis, reservoir conditions, total dissolved solids

INTRODUCTION

Studies of the effects of power plants on reservoir fish populations have greatly intensified in the past decade. As more and larger power plants appear on drawing boards and on reservoir shores, prediction of their effects becomes increasingly important. Data are being laboriously acquired from field studies to permit better understanding of reservoir ecosystems, and attempts to model perturbations of these systems are gathering momentum. However, at present, broad extrapolation of field and laboratory measurements, combined with educated guesses, are required to construct compartment-type models. Field data are not yet adequate to permit precise portrayals of whole ecosystems or even of individual fish populations.

In addition to power plants, stresses that may contribute to reservoir fish population perturbations include changing patterns in flow and water-level fluctuation, excessive nutrient loading, overexploitation, invasion of new

282

aquatic plants and fish species, shoreline modification, water withdrawal for irrigation or industrial purposes, and natural eutrophication. The effects of these natural and man-induced stresses must be sorted out before the direct effects of thermal stress, impingement, or entrainment can be assessed. Establishment of expected fish standing crops and harvest levels, based on key environmental variables, should be of immediate help in the sorting process.

Until more precise models are developed, simpler methods of estimating fish crop and production can be used to provide indexes to average reservoir conditions and to help identify perturbed environments. Empirically derived formulas resulting from analyses of data on selected environmental factors vs fish crop and harvest from a large number of waters are now available for immediate application (e.g., Jenkins 1970, Jenkins and Morais 1971, Ryder et al. 1974, Schneider 1975).

PREDICTION OF FISH STANDING CROP

Since 1963 the National Reservoir Research Program has been compiling and analyzing available pertinent information on the biological, physical, and chemical characteristics of United States reservoirs greater than 200 hectares (ha) in area. The primary aim has been to seek statistically significant relations between fish standing crop and harvest with such variables as reservoir area, age, mean and maximum depth, water level fluctuation, outlet depth, thermocline depth, growing season length, total dissolved solids, shore development, and storage ratio. Data on standing crop are based on recovery of fish after application of rotenone in measured coves or in open water areas enclosed by block-nets.

Of 166 reservoirs included in this standing crop analysis, all but six lie south of the Mason-Dixon line and east of 100°W longitude. To reduce variability and increase predictive value, the total standing crop sample was divided into 97 hydropower and 69 nonhydropower reservoirs. The hydropower subsample was subdivided into 52 mainstream (storage ratio less than 0.165) and 45 storage reservoirs. The nonhydropower subsample was divided according to the prevalent chemical type of inflowing streams as defined by Rainwater (1962). Correlation analyses revealed significant differences in fish crop between 43 reservoirs with a water chemistry type in which carbonate-bicarbonate ions are dominant (chemical type 1 or 3), and 26 in which sulfate-chloride ions are dominant (chemical type 2 or 4).

Correlation and stepwise multiple regression analyses were performed on an IBM 370/155 computer at the University of Arkansas Computing Center. Partial correlation analysis of fish crop vs the 11 environmental variables consistently showed that total dissolved solids (TDS) residue on evaporation at 180°C was the most important independent variable. The addition of other variables in multiple regressions did not significantly increase the accuracy of predictions.

Although the statistically significant, positive correlation of dissolved solids with standing crop does not explain the nature of casual connections, it can serve as a useful predictor of fish biomass. Long-term changes in stressed fish populations may be measured by this simple relation. As noted by Patten (1976, p. 10), "The growing realization that ecosystems may be stable in their coarser attributes ..., even though specific components or processes may be subjected to destabilizing perturbation, is leading the ecologist to further appreciation of holistic aspects of systems that transcend and in a sense are independent of fine scale characteristics."

Regression of standing crop on TDS yielded formulas with relatively high coefficients of determination (R^2 = 0.63 to 0.81) for all four reservoir types (Figs. 1–4). Within the hydropower mainstream subsample, TDS content ranged from 20 to 800 ppm; most values corresponded to those of large southern rivers (40 to 150 ppm). The Wheeler Lake, Alabama, crop estimate appears as an outlier (Fig. 1) and contributes to a rather broad 90% confi-

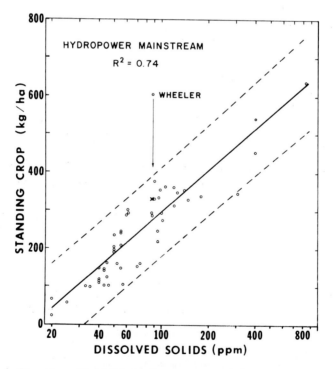

Fig. 1. Regression of total fish standing crop on total dissolved solids in 52 hydropower mainstream reservoirs (storage ratio: <0.165); dashed lines indicate 90% confidence limits. Standing crop = −431.4 + 363.2(log TDS); coefficient of determination R^2 = 0.74. Wheeler Lake, Alabama, appears as an outlier, on the basis of data collected in 1961 and 1968–72. More recent samples for the years 1973–75 yield a mean standing crop of 330 kg/ha, indicated by X. This value was not used in calculating the regression.

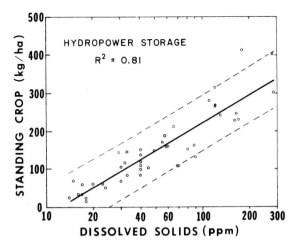

Fig. 2. Regression of total fish standing crop on total dissolved solids in 45 hydropower storage reservoirs (storage ratio: >0.165); dashed lines indicate 90% confidence limits. Standing crop = −268.6 + 245.0(log TDS); R^2 = 0.81.

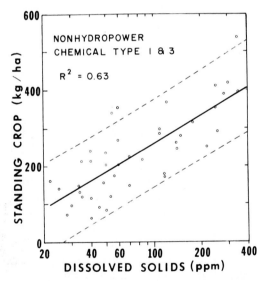

Fig. 3. Regression of total fish standing crop on total dissolved solids in 43 nonhydropower reservoirs of chemical type 1 or 3 (carbonate-bicarbonate ions dominant); dashed lines indicate 90% confidence limits. Standing crop = −236.7 + 247.8(log TDS); R^2 = 0.63.

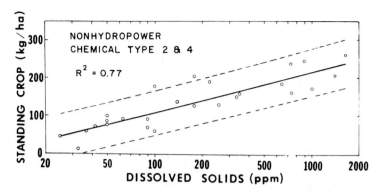

Fig. 4. Regression of total fish standing crop on total dissolved solids in 26 non-hydropower reservoirs of chemical type 2 or 4 (sulfate-chloride ions dominant); dashed lines indicate 90% confidence limits. Standing crop = $-104.4 + 106.8(\log TDS)$; $R^2 = 0.77$.

dence interval. Wheeler Lake is of particular interest because it is the site of the Browns Ferry Nuclear Plant. Awareness of the unusually high standing crop in this lake in 1969–72, prior to plant operation, is vital to the sound interpretation of post-operational data. However, more recent field estimates (1973–75), which were not included in this analysis, indicate that annual standing crops in Wheeler have declined to near predicted levels. The 1973–75 mean total crop was 330 kg/ha (G. E. Hall, Tennessee Valley Authority, Norris, *personal communication*), compared with a predicted 275 kg/ha (Fig. 1). Clupeids averaged 129 kg/ha for that period, compared with a predicted 130 kg/ha.

The subsample of hydropower storage reservoirs produced the highest R^2 value, 0.81, and a narrower confidence interval (Fig. 2). The range in TDS values (14 to 290 ppm) was smallest in this subsample. The nonhydropower chemical type 1 or 3 subsample evidenced the greatest scatter in crop values (Fig. 3) and the lowest R^2. The greatest variability in other environmental characteristics also occurred in this subsample which undoubtedly contributed to the lower R^2. The nonhydropower chemical type 2 or 4 subsample had the largest range in TDS content, and the calculated regression had the flattest slope (Fig. 4) — presumably a reflection of the depressing effect of high concentrations of sulfate-chloride ions on the fish standing crop.

Because clupeids (shads) are the fishes most frequently impinged at southern power plants (Griffith and Tomljanovich 1976) and make up 40 to 55% of the biomass in cove samples, regressions of the clupeid standing crop on TDS were also calculated for the four subsamples (Fig. 5). If mortalities induced by power plants are profoundly affecting the fish population in the receiving water, it should be first detected by changes in the clupeid crop. Coefficients of determination (0.39 to 0.70) were generally lower for clupeid crop regressions than for total crop regressions but still are considered useful for predictive purposes.

Regressions were also calculated for the non-clupeid portion of total standing crop on TDS (Fig. 6). The R^2 values ranged from 0.23 to 0.65, with the lowest value appearing in the nonhydropower, chemical type 1 or 3

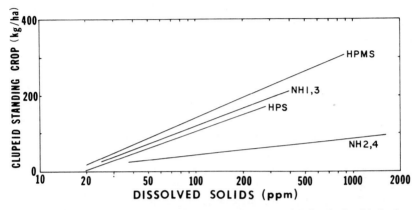

Fig. 5. Regression of standing crop of clupeid fishes on total dissolved solids for four reservoir types: 49 hydropower mainstream reservoirs (HPMS), clupeid crop = −198.6 + 169.2(log TDS), R^2 = 0.46; 40 nonhydropower reservoirs, chemical type 1 or 3 (NH1,3), clupeid crop = −192.1 + 155.7(log TDS), R^2 = 0.51; 42 hydropower storage reservoirs (HPS), clupeid crop = −181.7 + 143.9(log TDS), R^2 = 0.70; 23 nonhydropower reservoirs, chemical type 2 or 4 (NH2,4), clupeid crop = −39.5 + 42.0(log TDS), R^2 = 0.39.

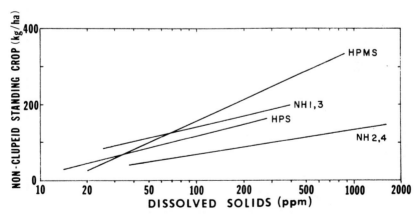

Fig. 6. Regressions of standing crop of non-clupeid fishes on total dissolved solids for four reservoir types: 49 hydropower mainstream reservoirs (HPMS), non-clupeid crop = −220 + 187.9(log TDS), R^2 = 0.52; 40 nonhydropower reservoirs, chemical type 1 or 3 (NH1,3), non-clupeid crop = −48.2 + 94.1(log TDS), R^2 = 0.23; 42 hydropower storage reservoirs (HPS), non-clupeid crop = −88.0 + 102.0(log TDS), R^2 = 0.65; 23 nonhydropower reservoirs chemical type 2 or 4 (NH2,4), non-clupeid crop = −57.3 + 62.8(log TDS), R^2 = 0.55.

subsample. For each of the four subsamples at given TDS levels, the summation of clupeid and non-clupeid crops estimated from the regression lines in Figs. 5 and 6, respectively, are nearly equal to the total standing crop estimated from the regression lines in Figs. 1–4.

Comparison of total crop regressions for the four reservoir types (Fig. 7) shows that at equal TDS levels above 50 ppm, hydropower mainstream reservoirs have the highest crops, followed by nonhydropower chemical type 1 or 3, hydropower storage, and nonhydropower chemical type 2 or 4. With the exception of chemical type 2 or 4, this difference is probably attributable to the phenomenon described by Aggus and Lewis (*in press*): the total quantity of nutrients passing through a reservoir appears to be of greater importance to fish production than the concentration of nutrients per unit volume. In a two-year analysis of 23 southern reservoirs, they found high positive correlations between total crop and quantity of water released. Mean water exchange rates were highest in the hydropower mainstream sample, followed by nonhydropower chemical type 1 or 3, and then hydropower storage. Thus, differences in the predicted crops at equal TDS levels may be a quantitative reflection of the effects of nutrient passage rate.

Fish standing crops in individual reservoirs are known to increase up to four times minimum levels in response to widely fluctuating environmental

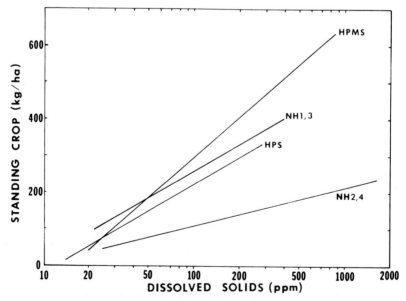

Fig. 7. Comparison of regressions of total standing crop on total dissolved solids for four reservoir types: hydropower mainstream (HPMS); nonhydropower, chemical type 1 or 3 (NH1,3); hydropower storage (HPS); and nonhydropower chemical type 2 or 4 (NH2,4). Regression formulas appear in legends of Figs. 1–4.

conditions and nutrient inflows. However, long-term sampling has indicated that biomass levels oscillate around a stable mean which can be predicted. For example, cove sampling conducted by the Arkansas Game and Fish Commission and the U.S. Fish and Wildlife Service (SCRI) on Bull Shoals Lake since impoundment provides a data set encompassing 25 years (1952–76). In that period, the total standing crop varied from 125 to 500 kg/ha and averaged 226 kg/ha. The predicted mean crop, based on average TDS content, is 224 kg/ha. The clupeid crop fluctuated from 40 to 300 kg/ha and averaged 107 kg/ha. The predicted clupeid crop is 110 kg/ha. Such close agreement between observed and predicted crops strongly suggests that the method can be used in evaluating stressed reservoir environments, at least until more sophisticated models with much higher data inputs are proven practical. Prerequisites for the method are standardized sampling methods and post-perturbation cove sample data over periods which include maximum fluctuation in natural environmental variables.

Patten (1976) found that a hypothetical elevation in water temperature of 3°C applied to his model of a Lake Texoma cove narrowed the range of variation of all model compartments as well as gross production and respiration. Total ecosystem biomass increased only 1.6%, but vertebrate biomass increased 33.3%. Detection of this magnitude of change (33%) in fish biomass in Bull Shoals Lake, by our simple prediction technique, would now require sampling for three successive years (Snedecor 1956, 90% confidence limit).

Sharp fluctuations in the fish crop usually occur in new reservoirs in response to rising water levels and nutrient input from freshly inundated vegetation and soil. After five to ten years of impoundment, environmental conditions tend to stabilize and fish biomass levels follow suit. In later years, total crops may increase if eutrophication occurs. Evaluation of man-induced stresses in new reservoirs would be difficult without previous knowledge of successional changes in the fish crop.

For example, Lake Keowee, a hydropower storage reservoir in South Carolina, is affected by a nuclear power plant and a headwater pumped storage operation. Estimates of the standing crop from the second through the ninth year of impoundment show a general decline in the total crop with age (Fig. 8). In the ninth year of impoundment, the field estimate of the fish crop equaled the predicted mean crop of 50 kg/ha. Without data on typical trends in the fish crop in new reservoirs and the expected crop level based on TDS levels, one might attribute the decline which occurred after nuclear plant operation began to the presence of the plant.

Threadfin shad were stocked in Lake Keowee in 1974 and crops have since averaged 4.7 kg/ha. The predicted crop when TDS = 20 ppm (Fig 5) is 5.5 kg/ha. On the basis of production and mortality hypotheses previously advanced (Jenkins 1974), the mean mid-winter crop of threadfin shad should be about 5 kg/ha or 36,600 kg for the entire lake. Comparison of this estimate with the biomass of threadfin impinged at the nuclear plant intake

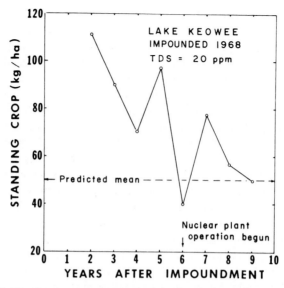

Fig. 8. Field estimates of total fish standing crop in Lake Keowee, South Carolina, through the ninth year of impoundment, compared with the predicted mean crop derived by the formula in the legend for Fig. 2. Unpublished field data were supplied by the South Carolina Wildlife and Marine Resources Department and Southeast Reservoir Investigations, U.S. Fish and Wildlife Service.

during the winter should provide insight into the relative impact of impingement losses on the total threadfin shad population.

PREY-PREDATOR RELATIONSHIPS

Inasmuch as power-plant-induced mortality primarily affects prey fishes, at least in reservoirs, analysis of the adequacy of the prey crop to sustain normal growth and well-being of the existing predator crop should provide a valuable measure of plant impact. A method has been developed for estimating prey-predator relationships in reservoirs on the basis of August cove samples (Jenkins and Morais, *in press*). Estimates were made of sizes of prey species which predators of a given size can swallow, and a computer program was developed to calculate biomass of prey available to the biomass of predators by 25.4-mm length classes. The calculations are cumulative to account for the ability of larger predators to ingest prey from the smallest to the largest they can swallow. A minimum desirable ratio of prey to predator biomass (PREY/PRED) of 1 to 1 was established for August samples. Plotting of the PREY/PRED values by 25.4-mm groups affords a simplified portrayal of a complex relationship and permits visual assessment of the relative adequacy of the prey base.

For example, plotting of PREY/PRED calculations for thermally stressed Lake Keowee (Fig. 9) indicates that prey biomass was inadequate for predators in 1973 and 1974, but more than adequate in 1975 and 1976. Nuclear plant operations began in 1973. Threadfin shad were stocked in 1974. Total available prey crops varied from 20 to 25 kg/ha in 1974–76, but predator crops declined from 19 kg/ha in 1974 to about 6 kg/ha in 1976. Biologists with Southeast Reservoir Investigations postulate that the apparent decline in predator crop reflects a shift in distribution away from the cove habitats sampled due to increased epilimnial water temperatures. Regardless of the cause of the decrease in predator crop, significant decreases in available prey have not occurred.

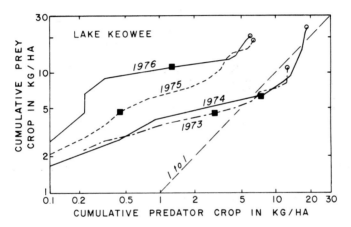

Fig. 9. Logarithmic plot of cumulative ratio of available prey crop versus predator crop by 25.4-mm length intervals (PREY/PRED), Lake Keowee, South Carolina, 1973–76, based on August standing crops. The 1:1 slope represents the minimum desirable PREY/PRED ratio. Solid squares indicate the PREY/PRED ratio for predators (largemouth bass equivalent) shorter than 165 mm. Open circles indicate the final PREY/PRED ratio. Maximum length of predators was 305 mm in 1973, 535 mm in 1974, 510 mm in 1975, and 355 mm in 1976.

SPORT FISH HARVEST AND ANGLER USE PREDICTIONS

The most important aspect of power-plant-induced mortality to fishery management agencies is the effect on sport fish harvest and angler use. Therefore, readily measured indexes to angler success and to quality and quantity of fishes harvested, applicable to a wide range of reservoir conditions, should be of immediate value in assessing perturbed fishing conditions. Annual estimates of use and harvest have been obtained from a nationwide distribution of U.S. reservoirs which were relatively unaffected by power plants. Regressions were computed relating sport fish harvest to surface area, dis-

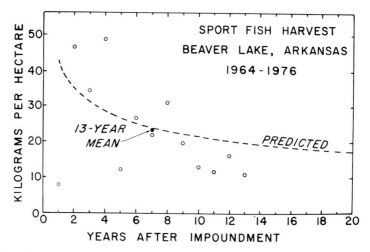

Fig. 10. Field estimates of sport fish harvest, Beaver Lake, Arkansas, from impound-ment in 1964 through 1976, and the 13-year mean, compared with predicted harvest based on a regression derived from data on 46 reservoirs: sport fish harvest = $-0.3994 - 0.1519$ log(surface area, in hectares) + 0.2027 log(total dissolved solids, in ppm) + 0.9796 log(growing season, in days) $-$ 0.3055 log(reservoir age in years), R^2 = 0.69.

solved solids content, growing season length, and reservoir age (Jenkins and Morais 1971). Using data from 46 reservoirs of less than 28,330 ha, with TDS less than 600 ppm and a growing season longer than 140 days, the authors derived a formula in 1972 for predicting sport fish harvest with a coefficient of determination of 0.69.

Applicability of the formula is illustrated by a comparison of 13 years of sport fish harvest data from Beaver Lake, Arkansas, and predicted values based on the formula (dashed curve, Fig. 10). Although there is considerable scattering of annual harvest values, the 13-year mean is very near the regres-sion prediction. Major effects of increased mortality, longer growing seasons, or increased TDS (Tilley 1975), any one of which may be associated with power plant operation, could be detected by long-term measurement of sport fish harvest. Additional formulas for estimating rates of angler success and use and for assessing the harvest of black basses, sunfishes, and commercial fishes are available (National Reservoir Research Program 1974).

CONCLUSION

Long-term monitoring of reservoir fish populations will be required to identify changes which can be positively attributed to power plant effects. Economically feasible monitoring necessitates selection of a comparatively small number of response indicators and adherence to standard procedures that will yield testable results. Ongoing monitoring of physical and chemical changes in reservoirs now provides data on independent variables required for using the simple indexes of fish crop and harvest which we have advanced. I recommend the use of these indexes as current best estimators and urge that research be accelerated to greatly refine the ability of fishery managers to predict optimum fish production potential in reservoirs.

LITERATURE CITED

Aggus, L. R., and S. A. Lewis. *In press*. Environmental conditions and standing crops of fishes in the predator-stocking-evaluation reservoirs. Proc. Annu. Conf. Southeast. Assoc. Game Fish Comm.

Griffith, J. S., and D. A. Tomljanovich. 1976. Susceptibility of threadfin shad to impingement. Proc. Annu. Conf. Southeast. Assoc. Game Fish Comm. **29**.

Jenkins, R. M. 1970. The influence of engineering design and operation and other environmental factors on reservoir fishery resources. Water Resour. Bull. **6**(1):110–119.

————. 1974. Reservoir management prognosis: migraines or miracles. Proc. Annu. Conf. Southeast. Assoc. Game Fish Comm. **27**:374–385.

Jenkins, R. M., and D. I. Morais. 1971. Reservoir sport fishing effort and harvest in relation to environmental variables, p. 371–384. *In* G. E. Hall [ed.] Reservoir fisheries and limnology. Amer. Fish. Soc., Spec. Publ. No. 8. 511 p.

Jenkins, R. M., and D. I. Morais. *In press*. Prey-predator relations in the predator-stocking-evaluation reservoirs. Proc. Annu. Conf. Southeast. Assoc. Game Fish Comm.

National Reservoir Research Program. 1974. A compilation of multiple regression formulas for use in estimating fish standing crop and angler harvest and effort in U.S. reservoirs. U.S. Fish and Wildlife Service, Fayetteville, Arkansas. 11 p. (*Mimeo*)

Patten, B. C. 1976. Ecosystem modeling and reservoir management. Okla. Acad. Sci. **5**:1–10.

Rainwater, F. H. 1962. Composition of rivers of the conterminous United States. U.S. Geological Survey. Atlas HA-61.

Ryder, R. A., S. R. Kerr, K. H. Loftus, and H. A. Regier. 1974. The morphoedaphic index, a fish yield estimator — review and evaluation. J. Fish. Res. Board Can. **31**:663–688.

Schneider, C. 1975. Typology and fisheries potential of Michigan lakes. Mich. Acad. **8**(1):59–84.

Snedecor, G. W. 1956. Statistical methods (5th ed.). Iowa State University Press, Ames. p. 501–3.

Tilley, L. J. 1975. Changes in water chemistry and primary productivity of a reactor cooling reservoir (Par Pond), p. 394–407. *In* E. G. Howell et al. [eds.] Mineral Cycling in Southeastern Ecosystems, Energy Res. and Devel. Admin. Symp. Ser., CONF-740513.

Part V

Use of Population Models

Effects of Power Station Mortality on Fish Population Stability in Relationship to Life History Strategy

Thomas J. Horst, Consultant

Environmental Division
Stone & Webster Engineering Corporation
Boston, Massachusetts

ABSTRACT

The use of mathematical models for the assessment of power station mortality on fish populations is discussed. The density-independent Leslie model is used to assess this impact by eigenvalue analysis and by simulation. Eigenvalue analysis is used to investigate the growth rate, age distribution, and stability of the fish population. The discrete-time simulation produces a trajectory of the population size through time.

The fish populations, modeled at equilibrium with a stable age distribution, are Atlantic silverside (*Menidia menidia*), Atlantic menhaden (*Brevoortia tyrannus*), cunner (*Tautogolabrus adspersus*), and winter flounder (*Pseudopleuronectes americanus*). The effect of power station mortality is modeled as a reduced survivorship in the first year of life. The sensitivity of the four fish populations to power station mortality is considered.

The order from most to least affected is silverside, menhaden, winter flounder, and cunner. These results are, in general, consistent with the life history theory of r- and k-selection. The distribution of these species is considered as it affects their susceptibility to power station operation. In general, localized species will be more susceptible than dispersed populations. Therefore, in selecting species for impact assessment, consideration must be given both to susceptibility and to sensitivity of the population to such perturbations.

Key words: cunner, eigenvalues, fish populations, impact assessment, Leslie model, life history strategy, menhaden, modeling, population stability, silverside, simulation, winter flounder

INTRODUCTION

This symposium considers the question, "How do mortality rates imposed by power stations on young fish affect adult fish populations?" A number of research areas involved in answering this question are being considered. This session considers mathematical models as a technique for predicting effects on fish populations of power station operation. This paper represents the development of ideas proposed by Horst (1975) and a synthesis of several years' experience in using population models to predict the effects of impacts of power station operation.

The potential sources of power station mortality to fish include entrainment of eggs and larvae in the circulating water, entrapment in the intake structure, and impingement on the fine mesh screens. Mortality may also result from the circulating water discharge into the water body. Any of these sources of mortality can be considered in mathematical models as incremental mortality to the population.

IMPACT ASSESSMENT AND MATHEMATICAL MODELING

The science or art of biological impact assessment is clearly in its infancy. Since the passage of the historic National Environmental Policy Act in 1969, scientists have responded in varying ways to the requirement to provide predictions of the environmental consequences associated with industrialization and the development of resources. Consideration of ecological values in the decision to develop resources depends in part upon the ability to quantify biological impacts in terms which are compatible with the overall cost-benefit analysis of the project.

I believe there is justifiable hesitation among ecologists in industry, consulting firms, regulatory agencies, and universities to undertake the challenge of impact analysis. However, it is clear that if ecologists do not assess ecological impacts, less qualified people will (Van Winkle et al. 1976).

The question becomes "What can the ecologist contribute to impact assessment?" This question does not emphasize how much descriptive data can be gathered, but what predictions can be made. Verification of these predictions will provide the insight to improve future predictions of impact and to guide the collection of data. By making more explicit the system descriptions and predictions, we will be in a better position to test predictions and to identify impacts when they occur.

These are difficult demands which are not easily achieved. However, failure to address the challenge implies a lack of understanding of ecological systems, which is not true. Ecologists must make predictions, properly qualified, and stop hiding behind the demand for more and more data. We must recognize that data will enhance our ability to predict and detect impacts, if the data are collected in a thoughtful experimental design which tests an explicit hypothesis. I, therefore, argue for explicit definitions of the potentially-affected biological system, for identification of the mechanisms of potential effect, and for the prediction of impact. The use of models, whether they are conceptual, graphical, or mathematical, is one of the best methods for achieving this explicitness.

LIFE HISTORY STRATEGY

Mathematical models for fish populations have been used for the assessment of the most probable impact associated with power plant operation. I will consider whether life history strategy provides an indication of the probable response of a population to increased mortality due to power station operation.

Organisms evolve a life history strategy in response to selective pressures from their environment. The life history strategy represents the partitioning in time of energy into maintenance, growth, and reproduction. The parameters of interest include, but are not limited to, length of life, age of first reproduction, the number of ages at which reproduction occurs, and the number of offspring produced.

The ecological theory of r- and k-selection, which was first described by MacArthur and Wilson (1967), describes how various life history strategies evolve. Populations that typically live well below the carrying capacity are opportunistic and are efficient reproducers. They represent the r-selection strategy. The alternate strategy (k-selection) is represented by populations which live near the carrying capacity and have high competitive ability. The r-selected species tend to be short-lived, to mature early, to have few reproductive periods and large numbers of offspring, and to occur in fluctuating environments. The k-selected species represent the alternate extreme to each of these characteristics.

Pianka (1970) emphasized that r- and k-selection represent extremes in a continuum. Populations can potentially exist at all points within this continuum. Pianka (1970) has tabulated characteristics of r- and k-selected populations, and fish generally are closer to the r-selected strategy. For example, they exhibit a Deevey (1947) type III survivorship curve, which has extremely low survivorship in early life, followed by increased survival later in life.

This paper investigates the effect of life history on the population growth rate when additional mortality occurs in early life. I consider departures from and recoveries to an equilibrium state, resulting from power station mortality. The models utilized for the population description are density independent. Therefore, they do not compensate for additional mortality rates due to power station operation. Since many populations exhibit some degree of compensatory regulation of size, this simplification may limit the degree of applicability of my results in such situations.

POPULATION MODEL

Fish population dynamics are modeled using the Leslie (1945) model. This discrete time population model is $N_{t+1} = A \times N_t$, that is,

$$
\begin{bmatrix} n_0 \\ n_1 \\ n_2 \\ \cdot \\ \cdot \\ \cdot \\ n_M \end{bmatrix}_{t+1}
=
\begin{bmatrix}
S_0 f_1 & S_1 f_2 & \cdots & S_{M-1} f_M & S_M f_{M+1} \\
S_0 & 0 & \ldots 0 & & 0 \\
0 & S_1 & \ldots 0 & & 0 \\
\cdot & & & & \cdot \\
\cdot & & & & \\
\cdot & & & & \\
0 & 0 & \ldots S_{M-1} & & 0
\end{bmatrix}
\times
\begin{bmatrix} n_0 \\ n_1 \\ n_2 \\ \cdot \\ \cdot \\ \cdot \\ n_M \end{bmatrix}_{t}
\cdot \qquad (1)
$$

The $M + 1$ column vector $\mathbf{N}_t$ is the population structure at time t. The elements correspond to the number of individuals at each of $M + 1$ ages. The $M + 1$ by $M + 1$ population projection matrix A has elements in the first row corresponding to the product of the age-specific survival from age i to $i + 1$ and the age-specific fecundity per fish at the beginning of age $i + 1$ for each of $M + 1$ ages and has elements in the subdiagonal corresponding to the age-specific survival from age i to $i + 1$ (where $i = 0$ to $M - 1$).

The relevant mathematical properties of the population projection matrix have been examined by several authors, but Sykes (1969) provides an especially clear discussion. This matrix, when primitive, has one positive real eigenvalue, and the remaining eigenvalues are negative or complex.

The biological interpretation of this positive real eigenvalue is the finite population growth rate, R. The positive eigenvector Z associated with this maximal eigenvalue has the biological interpretation of a positive multiple of the stable age distribution vector, $\mathbf{D}$; $\mathbf{D}$ has elements, d_j, such that

$$
1 = \sum_{j=0}^{M} d_j , \qquad (2)
$$

where d_j is the fraction of the population in age class j. $\mathbf{D}$ is independent of the initial population size $\mathbf{N}_0$ and depends only on A.

Therefore, if sufficient information for a species exists to estimate the age-specific fecundity and survivorship parameters in the population matrix A, the population growth rate and stable age distribution can be determined by eigenvalue analysis.

POPULATION PROJECTION MATRICES

Four species of marine fish representing different life history strategies are subjected to eigenvalue analysis and discrete time simulation to assess the effect of power station mortality.

The silverside (*Menidia menidia*) is a short-lived, shore-zone marine fish which occurs in relatively localized populations. Bayliff (1950) determined that the silverside generally reproduces only once — the year after it is born. Therefore the Leslie model (Eq. 1) reduces to the scalar form for silverside:

$$n_0(t + 1) = (S_0 f_1) n_0(t) .$$ (3)

Bayliff reports an average fecundity of 300 viable eggs per female. The number of males and females appears to be equal. Therefore, about 150 eggs are produced per adult per year (f_0). The survival ($S_0 = \dfrac{1}{f_0} = 0.00667$) was calculated to achieve an equilibrium condition ($R = 1.0$). The scalar equivalent to the eigenvalue is plotted in Fig. 1A.

The second fish species considered is the Atlantic menhaden (*Brevoortia tyrannus*). The population matrix for this species was developed by Vaughan et al. (1976) and is presented in Table 1. The eigenvalues of this matrix are plotted in Fig. 1B. The maximal eigenvalue is 1.0, indicating that the population is in equilibrium.

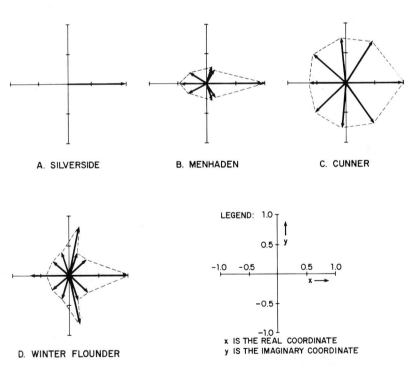

A. SILVERSIDE   B. MENHADEN   C. CUNNER

D. WINTER FLOUNDER

LEGEND:

x IS THE REAL COORDINATE
y IS THE IMAGINARY COORDINATE

Fig. 1. Eigenvalues for the equilibrium population matrix for silverside, menhaden, cunner, and winter flounder.

Table 1. Elements of the population projection matrix
for the atlantic menhaden

Age (i)	Eggs per adult (f_i)	Survivorship (S_i)
0	0	0.000077
1	7,497	0.2015
2	19,597	0.2015
3	29,567	0.2015
4	37,647	0.2015
5	43,537	0.2015
6	47,617	0.2015
7	51,355	0

Source: Vaughan et al. 1976.

The third species considered is the cunner (*Tautogolabrus adspersus*), a marine, nearshore fish. The population matrix for this species was developed by Horst (1977) and is presented in Table 2. The eigenvalues of this population projection matrix are plotted in Fig. 1C. The maximal eigenvalue is 1.0, indicating that the population is in equilibrium.

The fourth fish species considered is the winter flounder (*Pseudopleuronectes americanus*). The population matrix for this species was developed from data presented by Hess et al. (1976). The age-specific fecundity was multiplied by the proportion of females (0.7) to obtain the f_i elements. The S_0 element was calculated by the method of Vaughan and Saila (1976) to achieve an equilibrium population. The resulting population projection matrix is presented in Table 3, and the eigenvalues are plotted in Fig. 1D. The maximal eigenvalue has a value of 1.0, indicating that the population is in equilibrium.

Table 2. Elements of the population projection matrix
for the cunner

Age (i)	Eggs per adult (f_i)	Survivorship (S_i)
0	0	0.0002101
1	85	0.6252
2	698	0.6252
3	2,029	0.6252
4	3,967	0.6252
5	6,294	0.6252
6	8,796	0.6252
7	11,303	0

Source: Horst 1977.

Table 3. Elements of the population projection matrix
for the winter flounder

Age (i)	Eggs per adult (f_i)	Survivorship (S_i)
0	0	0.000057
1	0	0.1454
2	0	0.33
3	183,000	0.33
4	310,100	0.33
5	445,900	0.33
6	569,100	0.33
7	679,000	0.33
8	774,900	0.33
9	851,900	0.33
10	910,700	0.33
11	956,900	0.33
12	996,800	0

EFFECTS OF POWER PLANT IMPACT

The eigenvalues of the population projection matrices provide information on the ability of the population to withstand perturbations, such as power station mortality. The maximal eigenvalue is an estimate of the population growth rate, R, given the stable age distribution vector, **D**. If the population is perturbed from this equilibrium, the eigenvalues provide an understanding of how the population will respond.

If there is a maximal eigenvalue, its value is equivalent to the finite rate of population growth, R, and the other eigenvalues provide an indication of the population stability. It has been empirically determined that, the more closely the eigenvalues trace a circle of radius equal to the modulus of the maximal eigenvalue, the less R changes, and thus the less the population size changes, in response to perturbations. Therefore, an index, I, of relative stability of a population is the average deviation of the moduli of the eigenvalues from the modulus of the maximal eigenvalue. In other words,

$$I = \frac{1}{M+1} \sum_{i=0}^{M} (|\lambda_{max}| - |\lambda_i|)$$

or   (4)

$$I = |\lambda_{max}| - \frac{1}{M+1} \sum_{i=0}^{M} |\lambda_i|,$$

where $M + 1$ is the number of eigenvalues (the number of age classes in the population starting with zero), $|\lambda_{max}|$ is the modulus of the maximal eigenvalue ($\lambda_{max} = R$), and $|\lambda_i|$ is the modulus of the ith eigenvalue.

The modulus (or absolute value) of an eigenvalue (λ_i) is calculated as

$$|\lambda_i| = \sqrt{x_i^2 + y_i^2}, \tag{5}$$

where x_i and y_i are the x-axis and y-axis coordinates, respectively, of the eigenvalue plotted in the complex plane (see Fig. 1).

This index, I, which applies only when there is a maximal eigenvalue, can only take on values greater than 0.0 and less than 1.0 when the population is at equilibrium ($R = 1.0$). The closer this index is to zero, the more stable the population. The indices for menhaden (Fig. 1B), cunner (Fig. 1C), and winter flounder (Fig. 1D) are 0.60, 0.21, and 0.46, respectively. Therefore, the order of decreasing stability predicted by this index is cunner, winter flounder, and menhaden.

The effect of decreased survival of age-zero fish on population growth rate and population size was investigated for the four populations described above. The equilibrium S_0 values for each population were reduced to various levels to reflect a range in magnitude of power station impacts. The eigenvalues of these "perturbed" matrices were then determined. The effect of changing S_0 on the maximal eigenvalue or population growth rate (R) (Fig. 2) is, in general, consistent with the predictions based on the theory of r- and k-selection and with the predictions based on the relative stability index, I. The silverside, which is the shortest lived of the four species, is most affected. The cunner, which has the highest survival rates, is least affected by decreases in S_0.

This analysis considers only proportional reductions in the survival probability for age class 0. The actual number of organisms of each species which must suffer mortality to achieve these reduced survival rates will vary with the population size. For example, the winter flounder has localized populations which exhibit seasonal inshore-offshore migrations. These localized populations are relatively isolated, with little exchange with adjacent populations (Howe and Coates 1975). Therefore, throughout the geographic range of the winter flounder, only a localized population will probably be impacted from point-source perturbations. In contrast, the Atlantic menhaden stock is composed of essentially one interbreeding population throughout its geographic range (Dryfoos et al. 1973). Thus, a localized mortality, such as entrainment mortality at a particular power plant, will be "diluted" more widely for menhaden than for winter flounder. In other words, if a power plant entrains an equal number of winter flounder larvae and menhaden larvae, the flounder population will be affected more than the menhaden population.

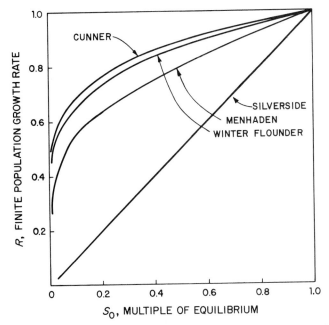

Fig. 2. Effect of reductions in survivorship of age class 0 (S_0) on the finite population growth rate (R), determined by eigenvalue analysis, for silverside, menhaden, cunner, and winter flounder.

Consider a hypothetical case where a power plant entrains one million winter flounder larvae and one million menhaden larvae. An estimate of the equilibrium population size of one localized winter flounder population was made by Hess et al. (1976). The population produces about 2×10^9 eggs. Given that the survival of eggs to flounder larvae is 0.1 (Hess et al. 1976), the flounder larvae production is 2×10^8. By comparison, the information presented by Schaaf and Huntsman (1969) and Jensen (1975) for menhaden results in an equilibrium production of about 4×10^{14} eggs per year. For an egg survival rate of 0.1, menhaden larvae production is 4×10^{13}.

The proportion of the larvae in the winter flounder and menhaden populations lost to entrainment can now be calculated, assuming all entrained larvae have recently hatched. The mortality rate due to entrainment for winter flounder is

$$M_F = \frac{1 \times 10^6}{2 \times 10^8} = 0.005 . \tag{6}$$

The mortality rate due to entrainment for menhaden is

$$M_m = \frac{1 \times 10^6}{4 \times 10^{13}} = 0.000000025 \ . \tag{7}$$

Thus, the proportional reduction in S_0 will be considerably greater for winter flounder than menhaden.

POPULATION SIMULATION

Another approach to assessing the effect of power station mortality on each of these four fish populations is a time simulation of the population using the Leslie model (Eq. 1). The initial population vector (N_0) was calculated from the stable age distribution (D) such that the equilibrium population contained one million individuals (counting age class 0) in the absence of any perturbation.

The effect of power station operation was introduced after 10 years of simulation as an additional mortality rate of 0.1 to age-zero fish (i.e., 0.9 S_0). This reduction appears reasonable, but may overestimate the effect of power stations in some situations. Hess et al. (1976) estimated a 1% incremental mortality of winter flounder larvae at the Millstone Nuclear Power Station. Since winter flounder is a species characterized by localized, relatively small populations, the proportion of the population affected is probably greater than for more dispersed populations. Following the initial 10 years, entrainment mortality was imposed for each of 5, 10, 20, and 40 years, after which the simulation was continued until the population reached a new equilibrium size.

The results of the simulations are presented in Figs. 3–6 for winter flounder, cunner, menhaden, and silverside, respectively, and the relative effect of power station mortality can be seen by comparing these four figures. The silverside had the greatest change in population size, which could be anticipated since it is the species that most resembles the "r-selected" extreme. The menhaden, cunner, and winter flounder matrices each had a maximal eigenvalue. Therefore, the relative effect of power station mortality could be predicted from the eigenvalue analysis and stability index, as well as from the life-history strategy. Again, the effect predicted by the stability index, I, in order of decreasing stability was cunner, winter flounder, and menhaden. The simulation results show the same ranking: the cunner is least affected and the menhaden is most affected (Figs. 3–5).

Therefore, the predictions from the theory of r- and k-selection are consistent with the results of the eigenvalue analysis and discrete-time simulation. These three approaches considered together provide an a priori basis for choosing species most likely to be affected by power station operation. However, in choosing the species to monitor, consideration must also be given to the proportion of individuals affected by the power station. To maximize the

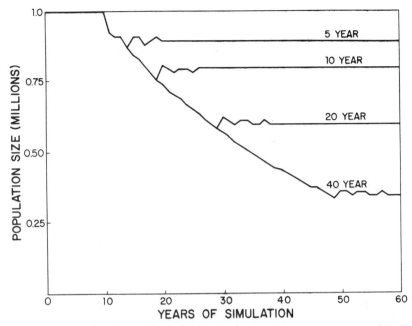

Fig. 3. Simulation of winter flounder population size when affected by 5, 10, 20, and 40 years of power station operation.

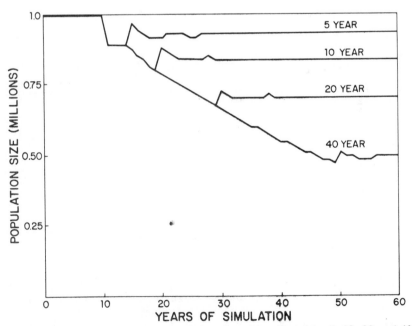

Fig. 4. Simulation of cunner population size when affected by 5, 10, 20, and 40 years of power station operation.

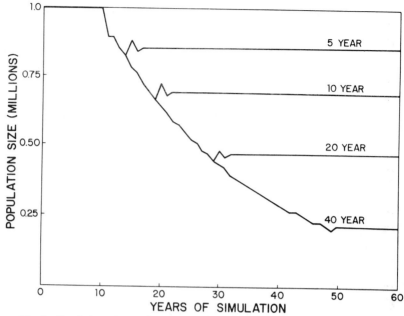

Fig. 5. Simulation of menhaden population size when affected by 5, 10, 20, and 40 years of power station operation.

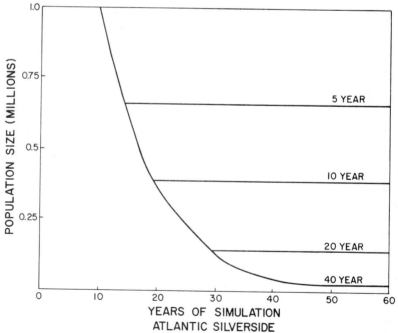

Fig. 6. Simulation of silverside population size when affected by 5, 10, 20, and 40 years of power station operation.

chance of detecting impacts, monitoring studies should focus on populations which have the greatest proportion of individuals affected and which are most susceptible to perturbations.

ACKNOWLEDGMENTS

I thank Stone & Webster for the financial support of this study, David Nemtzow for his assistance in the computer analysis, and Webb Van Winkle whose work has made this symposium possible.

LITERATURE CITED

Bayliff, W. H. 1950. The life history of the silverside, *Menidia menidia*. Ches. Biol. Lab. Publ. No. 97. 27 p.

Deevey, E. S. 1947. Life tables for natural populations of animals. Quart. Rev. Biol. **22**:283—314.

Dryfoos, R. L., R. P. Cheek, and R. L. Kroger. 1973. Preliminary analysis of Atlantic menhaden *Brevoortia tyrannus*, migrations, population structure, survival and exploitation rates, and availability as indicated from tag returns. Fish. Bull. **71**:719—734.

Emlen, J. M. 1973. Ecology: an evolutionary approach. Addison Wesley, Reading, Massachusetts. 493 p.

Hess, K. W., M. P. Sissenwine, and S. B. Saila. 1975. Simulating the impact of entrainment of winter flounder larvae, p. 1—30. *In* S. B. Saila [ed.] Fisheries and energy production. D.C. Health and Company, Lexington, Massachusetts. 300 p.

Horst, T. J. 1975. The assessment of impact due to entrainment of ichthyoplankton, p. 107—118. *In* S. B. Saila [ed.] Fisheries and energy production. D.C. Health and Company, Lexington, Massachusetts. 300 p.

————. 1977. Use of the Leslie matrix for assessing environmental impact with an example for a fish population. Trans. Am. Fish. Soc. **106** (*in press*).

Howe, A. B., and P. G. Coates. 1975. Winter flounder movements, growth and mortality in Massachusetts. Trans. Am. Fish. Soc. **105**(1):13—29.

Jensen, A. L. 1975. Computer simulation of effects on Atlantic menhaden yield of changes in growth, mortality, and reproduction. Ches. Sci. **16**(2):139—142.

Leslie, P. H. 1945. On the use of matrices in certain population mathematics. Biometrika **33**:183—212.

MacArthur, R. H., and E. O. Wilson. 1967. The theory of island biogeography. Princeton University Press, Princeton, New Jersey. 203 p.

Pianka, E. R. 1970. On r- and k-selection. Am. Nat. **104**:592—597.

Schaaf, W. E., and G. R. Huntsman. 1972. Effects of fishing on the Atlantic menhaden stock: 1955–1969. Trans. Am. Fish. Soc. **101**:290–297.

Sykes, Z. M. 1969. On discrete population theory. Biometrics **25**:285–293.

Usher, M. B. 1972. Developments in the Leslie model, p. 29–60. *In* J. N. R. Jeffers [ed.] Mathematical models in ecology. Blackwell Scientific Publ., Oxford. 398 p.

Van Winkle, W., S. W. Christensen, and J. S. Mattice. 1976. Two roles of ecologists in defining and determining the acceptability of environmental impacts. Int. J. Envir. Stud. **9**:247–254.

Vaughan, D. S., N. Buske, and S. B. Saila. 1976. Interim report on evaluating the effect of power plant entrainment on populations near Millstone Point, Connecticut. URI-NUS Co. Report No. 4. Graduate School of Oceanography, University of Rhode Island.

Vaughan, D. S., and S. B. Saila. 1976. A method for determining mortality rates using the Leslie matrix. Tran. Am. Fish. Soc. **105**(3):380–383.

Sensitivity Analysis Applied to a Matrix Model of the Hudson River Striped Bass Population

S. B. Saila and E. Lorda

Graduate School of Oceanography
University of Rhode Island
Kingston, Rhode Island

ABSTRACT

Given a set of age-specific fecundities and survivals for the Hudson River striped bass resulting in an equilibrium population, the effects of simulated increases in mortality of 2, 5, 10, and 20% on five young-of-the-year life stages, taken alone and in various combinations, are examined in terms of several population parameters defined for this purpose. The results indicate similar responses of the simulated population to additional mortality applied to any of the early life history stages taken individually. Additional simulated mortality applied to the same life history stages in various combinations resulted in greater reductions in the adult population. The relative insensitivity of the dominant latent root of the population matrix in comparison to four derived parameters is demonstrated for the simulated perturbations.

Key words: entrainment, impingement, Leslie matrix model, matrix projection, sensitivity analysis, striped bass

INTRODUCTION

The striped bass *Morone saxatilis* (Walbaum) is an important sport fish on both the Atlantic and Pacific coasts of the United States as well as in some inland waters of the southern states. It is also the object of a significant commercial fishery on the Atlantic Coast. The potential impact of electric power plants on the Atlantic striped bass stocks has received considerable attention in recent years. Both empirical studies of various aspects of striped bass life history and population dynamics, as well as model studies, have been made. Models have been developed to predict the impact of power plants on young-of-the-year (y-o-y) striped bass and on the entire population structure. The withdrawal of large amounts of water by power plants results in the entrainment of striped bass eggs and larvae through the cooling system and the impingement of some later life stages on the debris screens in front of the cooling water intakes.

Some indication of the research which has been performed on the striped bass is available from three recent bibliographies: Pfuderer et al. (1975), Rogers and Westin (1975), and Horseman and Kernehan (1976). The latter bibliography lists 1505 entries, many of which are relatively recent. No effort will be made in this study to review available literature. Only those references which have been utilized directly for data sources will be cited. Swartzman et al. (1977) provide a detailed review and comparison of the various striped bass models used in assessing the effects of power-plant-induced mortality.

Our objective in this study is to demonstrate the value of the Leslie matrix as a tool to critically examine the dynamics of the Hudson River striped bass population. Of special significance in our work is an examination of the available information on the survival rates of various y-o-y life stages and a sensitivity analysis of the effects of changes in these survival rates, due to entrainment and impingement, on the short-term dynamics of the population.

MATERIALS AND METHODS

The Leslie matrix model was introduced by Leslie (1945, 1948). It has been extensively used in population mathematics (Keyfitz, 1968) but overlooked by ecologists until recently. A major contribution to the generalization of this model was the theoretical work done by Lefkovitch (1965, 1967). Usher (1972, 1976) has done an up-to-date review of developments in the Leslie matrix.

Briefly, the Leslie matrix model is a formal way of simulating discrete time changes in a population age vector. Some obvious advantages of this discrete model include its computational straightforwardness and consideration of age structure without assuming a stable age distribution. The stable age distribution and the intrinsic rate of increase or decrease for a population can be elucidated by examining the dominant latent root (eigenvalue) of the matrix and the associated eigenvector. A less-known advantage of the model is that matrix projection techniques permit a detailed analysis of short-term population dynamics. The background and methodology for this further analysis will be described and illustrated in detail.

An apparent disadvantage of the Leslie matrix model is that, without modification, it implies a linear (density-independent) stock-recruitment relationship, since it does not incorporate compensatory phenomena. This limitation is not considered prohibitive for the purposes at hand for reasons which follow. Both the Beverton-Holt (1957) and Ricker (1958) stock-recruitment models are based on two premises. The first is that recruitment is zero when stock is zero and that recruitment is approximately proportional to stock at low levels of stock size. The second premise is that at high levels of stock, recruitment is stable (Beverton-Holt model) or inversely related to stock size (Ricker model). In each model the ascending left limb covers the range of stock sizes of primary environmental concern and may be approximated by a straight line.

The basic Leslie matrix model has a close correspondence to an exponential growth equation, which limits the utility of the Leslie model for purposes of long-term predictions. However, the major interest in most environmental studies is the population responses on a relatively short time scale (e.g., 20 years), such as in this study. Thus the density-independent Leslie model seems appropriate if it is not extended too far in time. Mendelsohn (1976) has recently summarized some of the limitations of the Leslie model and the relationship of the dominant latent root of the Leslie matrix to optimal harvesting policies of a population.

Several criteria to assess and compare population behavior in detail are proposed in the following material.

Since the stable age-distribution vector K associated with the dominant latent root λ of a Leslie matrix, M, is meaningful only as a set of proportions among age classes, stable distributions must be normalized for comparative purposes. If K is normalized to unity in its last element, the resulting new vector still has the same set of proportions among age classes. This new vector is denoted U, the stable unit age-distribution vector. The sum U of the elements of the vector U is called the stable population unit, and it is a useful index for some comparative purposes. The true size or total number of individuals in any population with a stable distribution and characterized by the same Leslie matrix must be an exact multiple of the parameter U defined above.

The dominant latent root, λ, of the Leslie matrix and the stable unit, U, can be combined to form a new concept of growth. This concept is not to be confused with the intrinsic natural rate of growth or increase r (derived from $\lambda = e^r$); λ is a specific growth rate, that is, the number of individuals in age class i at time $t + 1$ per individual in age class i at time t. It represents the proportionality constant by which K and U change per time step, and $\lambda = 1.0$ implies an equilibrium population. Hence, since U equals the total number of individuals in a stable unit population at time t, the product λU will represent the size of the population at time $t + 1$, and the change or growth G may be defined as

$$G = \lambda U - U = (\lambda - 1)U . \tag{1}$$

Thus, $G = 0$ implies no growth, $G > 0$ implies an increasing population, and $G < 0$ implies a decreasing population.

Since this growth G is measured over one unit of time, it seems reasonable to consider G as the stable population growth rate. It is obvious that G

will remain constant through time for any given population having a stable distribution. The stable growth rate G does not refer to the growth of a population of a given size, but specifically to the growth of the stable unit U of a population.

Similarly, the product $(\lambda - 1)U$ is a new age-distribution vector G; the sum of the elements of the vector G equal the stable growth rate G. Because the vector G is still proportional to either U or K, G can be properly defined as the stable vector of age-specific growth rates.

In addition to λ, the two population parameters U and G, along with associated stable vectors U and G, represent more efficient criteria for population comparisons than λ and the stable distribution vector K alone. The following numerical examples support this assertion:

$$M_1 = \begin{bmatrix} 0 & 10.8 & 12 \\ 0.33 & 0 & 0 \\ 0 & 0.6 & 0 \end{bmatrix} \quad U_1 = \begin{bmatrix} 23.5315 \\ 3.6157 \\ 1.0000 \end{bmatrix} \quad G_1 = \begin{bmatrix} 27.5177 \\ 4.2282 \\ 1.1694 \end{bmatrix} \quad (2)$$

$$\lambda_1 = 2.1694 \qquad U_1 = \overline{28.1472} \qquad G_1 = \overline{32.9153}$$

$$M_2 = \begin{bmatrix} 0 & 7.2 & 12 \\ 0.33 & 0 & 0 \\ 0 & 0.4 & 0 \end{bmatrix} \quad U_2 = \begin{bmatrix} 24.6228 \\ 4.5298 \\ 1.0000 \end{bmatrix} \quad G_2 = \begin{bmatrix} 19.9913 \\ 3.6777 \\ 0.8119 \end{bmatrix} \quad (3)$$

$$\lambda_2 = 1.8119 \qquad U_2 = \overline{30.1526} \qquad G_2 = \overline{24.4809}$$

$$M_3 = \begin{bmatrix} 0 & 9 & 12 \\ 0.33 & 0 & 0 \\ 0 & 0.6 & 0 \end{bmatrix} \quad U_3 = \begin{bmatrix} 20.8732 \\ 3.4053 \\ 1.0000 \end{bmatrix} \quad G_3 = \begin{bmatrix} 21.7749 \\ 3.5524 \\ 1.0432 \end{bmatrix} \quad (4)$$

$$\lambda_3 = 2.0432 \qquad U_3 = \overline{25.2785} \qquad G_3 = \overline{26.3705}$$

$$M_4 = \begin{bmatrix} 0 & 9 & 12 \\ 0.33 & 0.4 & 0 \\ 0 & 0.4 & 0 \end{bmatrix} \quad U_4 = \begin{bmatrix} 25.5485 \\ 4.3580 \\ 1.0000 \end{bmatrix} \quad G_4 = \begin{bmatrix} 24.3784 \\ 4.1584 \\ 0.9542 \end{bmatrix} \quad (5)$$

$$\lambda_4 = 1.9542 \qquad U_4 = \overline{30.9065} \qquad G_4 = \overline{29.4910}$$

By considering only the dominant latent root λ, the conclusion is that population 3 increases faster than population 4, since $\lambda_3 > \lambda_4$. However, there are other aspects of growth to be considered. For example, the stable population units U_3 and U_4 are not equal, and the fact that $G_3 < G_4$ implies a time lag of more than one time step for population 3 to become larger than population 4. The actual time lag can be estimated by projecting the two populations over time.

The projections are based on the following:

$$U(t + n) = M^n U(t) , \qquad (6)$$

where n is the number of equal time intervals or reproductive cycles in the projection, $U(t)$ is the stable unit age-distribution vector at time t, and $U(t + n)$ is the stable unit age-distribution vector after n reproductive cycles. In addition,

$$\lambda^n U(t) = M^n U(t) , \qquad (7)$$

where $\log_e \lambda = r$ and $\lambda^n = e^{nr}$, as shown by Leslie (1945). Again, λ is the dominant latent root of the matrix M, and $U(t)$ is the stable unit population distribution at time t or eigenvector associated with λ and normalized to unity in its last element.

From Eqs. (6) and (7) it follows that

$$U(t + n) = \lambda^n U(t) . \qquad (8)$$

The values of U at various times for populations 3 and 4 are

	Pop. 3	*Pop. 4*	
Value of U at time t =	25.28	30.91	
Value of U at time $t + 1$ =	51.65	60.40	
Value of U at time $t + 2$ =	105.53	118.03	(9)
Value of U at time $t + 3$ =	215.62	230.65	
Value of U at time $t + 4$ =	440.55	450.74	
Value of U at time $t + 5$ =	900.13	880.84	

Thus, five time steps are required before population 3 becomes larger than population 4. For populations 1 and 2, $\lambda_1 > \lambda_2$ and $U_1 < U_2$, but only one time step is required before population 1 becomes larger than population 2, because $G_1 > G_2$ and $(G_1 - G_2) > (U_2 - U_1)$.

Suppose now that the four populations are harvested on the same basis, that is, the net increase of individuals in each age class is removed after each reproductive cycle. The yield would be exactly proportional to G for each population for each reproductive cycle, and if populations 3 and 4 had the same size (i.e., equal number of stable population units) at the beginning, then population 4 would be consistently more productive than population 3. This result might seem paradoxical after showing that population 3 actually increases faster than population 4.

In general, it may be expected that the highest yields will correspond to the populations with the largest rates of increase (i.e., largest λ). However, this expected result does not hold for populations 3 and 4 because the birth rate per individual for age class 1 is higher in population 4 than in population 3 (i.e., $9/0.4 = 22.5$ in population 4; $9/0.6 = 15.0$ in population 3). Populations 1 and 2 have identical birth rates for both age classes 1 and 2, and thus, population 1, with the greater rate of increase $(\lambda_1 > \lambda_2)$, has the greater stable population growth rate $(G_1 > G_2)$.

In the case of complex populations with many age classes and with differences in both birth and death rates, it is not feasible to assess which is the most productive population unless the parameter G is available. If the vector **G** is also available, it will provide a breakdown of production by age class, which is relevant from the point of view of harvesting and management.

Dissimilarity between populations has been examined so far in light of different rates of increase and/or different stable unit age distributions. In addition, even if two population matrices have identical λ's and **U**'s, they may still represent populations that behave quite differently. Williamson (1967) illustrated this "ultimate" dissimilarity with a simple numerical example. This different behavior of two apparently identical populations is due to differences between the two populations in the "resilience" of the population structure which results in different rates of recovery to their stable age distributions following perturbation. This structural property of a population was also discussed by Lefkovitch (1967). If $\lambda^{(1)}$ is the maximal latent root of a Leslie matrix (i.e., the latent root with the largest modulus or absolute value) and $\lambda^{(2)}$ is the latent root with the second largest modulus, then the time to stabilization is directly proportional to the ratio $\lambda^{(2)}/\lambda^{(1)}$. However, what is even more important from a practical point of view is how the population size changes during the stabilization period. Analysis of the trajectory of the population size after a perturbation has occurred constitutes the most relevant aspect of our study.

We will next define the mean reproductive value, $\overline{V}$, of a population. Keyfitz (1968) has defined the reproductive value, V, of a human population with a stable age distribution as the total number of progeny that all the individuals in that population are expected to produce during their entire life time. This number may be calculated assuming a constant and known schedule of births and deaths. If this total is then divided by the number of individuals in the population, the result is the average number of births per individual. We call this average the mean reproductive value, $\overline{V}$, of a population.

Using either **U** or **U** normalized to the vector **N**, the elements of which add up to $N = 1.0$, it is possible to compute the mean reproductive value of the population. Although the process is somewhat tedious it can be easily programmed and is clear in meaning and derivation. An example follows. Consider the following population matrix M and stable unit age-distribution **U** and **N**:

$$M = \begin{bmatrix} 0 & 9 & 12 \\ \frac{1}{3} & 0 & 0 \\ 0 & \frac{1}{2} & 0 \end{bmatrix} \quad U = \begin{bmatrix} 24 \\ 4 \\ 1 \end{bmatrix} \quad N = \begin{bmatrix} 0.827586 \\ 0.137931 \\ 0.034483 \end{bmatrix} \tag{10}$$

$$U = 29 \qquad N = 1.000000$$

Suppose that there is one individual in each age class. The number of births each individual will generate during its entire life can be calculated as follows:

Individual starting in age class	While in age class 0	While in age class 1	While in age class 2	Total	
0	1 × 0 = 0	⅓ × 9 = 3	(⅓)(½) × 12 = 2	5	
1		1 × 9 = 9	½ × 12 = 6	15	(11)
2			1 × 12 = 12	12	

Since the stable unit age-distribution vector, **U**, is not $\begin{bmatrix} 1 \\ 1 \\ 1 \end{bmatrix}$ but instead is $\begin{bmatrix} 24 \\ 4 \\ 1 \end{bmatrix}$,

the total reproductive potential will be

$$\begin{aligned} 5 \times 24 &= 120 \\ 15 \times 4 &= 60 \\ 12 \times 1 &= 12 \end{aligned} \tag{12}$$

$$U = 29 \qquad 192 = \text{total reproductive potential}$$

The ratio, $192/29 = 6.6207$, is the mean reproductive value, $\overline{V}$, of this unit population. Both the mean reproductive value and the contributions of each age class to it can also be computed by using the vector **N** as follows:

$$\begin{aligned} 5 \times 0.827586 &= 4.1379 \equiv 62.50\% \\ 15 \times 0.137931 &= 2.0690 \equiv 31.25\% \\ 12 \times 0.034483 &= 0.4138 \equiv 6.25\% \end{aligned} \tag{13}$$

$$6.6207 \qquad 100.00\%$$

In summary, we have described three parameters in addition to λ, the dominant latent root. They are, U, the stable population unit; G, the stable growth rate, and $\overline{V}$, the mean reproductive value of a population. These parameters will be utilized in the analyses described in the next section.

Our Leslie matrix model used to simulate the dynamics of the Hudson River striped bass population assumes that the population has a constant sex ratio (1:1 for this study) and that there are always sufficient males to fertilize the eggs spawned by the females. The model time step is a year and spawning occurs at the start of each year. This implies that all the surviving mature females from a given age class spawn as they enter the next age class. It is also assumed in our model that all the individuals die after the 20th year. Although this last assumption is obviously not true, in our case its acceptance only implies that the negligible contribution of eggs from females beyond the 20th age class is not taken into consideration.

The life-table data used for the analysis of the dynamics of the Hudson River striped bass population are shown in Table 1. This life table is based primarily on material from LMS (1975), McFadden (1977), and USNRC (1975). Some comments on the estimates of fecundities and survival probabilities chosen as input parameters for our model seem warranted. The fecundity rates were directly derived from Table 7.8-8 in McFadden (1977). As an example, the fecundity estimate per individual entering age class 4 (Table 1) is obtained as

$$f_4 = 400{,}000 \times 0.06 \times 0.50 = 12{,}000 \text{ eggs} ,$$

where 400,000 is the estimated (McFadden 1977) number of eggs spawned by a mature female entering age class 4; 0.06 is the estimated (McFadden, 1977) fraction of sexually mature females entering that age class; and 0.50 is the fraction of individuals entering age class 4 that are assumed to be females.

Survival probabilities were either derived or taken directly from the three above mentioned sources. The survival probability through age class 0 was taken from McFadden (1977, Tables 7.7-2 and 12.2-1), where an annual mortality rate of 0.999965 was estimated. Although in various models six life history stages are generally recognized for striped bass within age class 0 (Swartzman et al. 1977), we decided to consider the first two stages (i.e., eggs and yolk-sac larvae) as only one and to keep the other four stages separate. The survival probability of 0.058373 for life stage 1 (eggs and yolk-sac larvae) was directly derived from the 24.73% daily mortality rate given in Table 7.7-2 of McFadden (1977). The survival probability of 0.20 for stage 3 (Juvenile 1) was taken from LMS (1975), since this value is in good agreement with the 5.06% daily mortality rate given in Table 7.7-2 of McFadden (1977). The survival probabilities for stages 4 and 5 (Juvenile 2 and 3) were taken from Table 12.2-1 in McFadden (1977). Finally, the survival probability for stage 2 (post-yolk-sac larvae) was derived from the survival probability of 0.000035 for the entire age class 0, given the survival probabilities for life stages 1, 3, 4, and 5.

Survival probabilities for age classes 1 and 2, which are not yet subjected to fishing mortality, were taken directly from the best estimates given in USNRC (1975). The total survival probability for each of the remaining age classes 3 to 20 was assumed to be the same. A value of 0.487857, which is within the range of values (0.4–0.6) commonly used by both LMS (1975) and USNRC (1975), was used.

This set of age-specific fecundity and survival probabilities yields an equilibrium striped bass population as shown in Table 1. Since a stable age distribution is only a set of proportions, the age distribution for Table 1 has been normalized to unity in the last age class (AG20), that is, the vector $\mathbf{U}$ as previously defined. The number of individuals entering each of the other age classes is back-calculated in Table 1 starting from the single individual entering AG20. If the total probability of survival for age class i is denoted as p_i, then the number of individuals entering age class i, AG_i, in terms of the number of individuals entering age class $i + 1$, AG_{i+1} is calculated as $AG_i = AG_{i+1}/p_i$. It is obvious that the equilibrium condition will be met only when the sum of eggs spawned by all the females in the reproductive age classes equals the number of eggs entering age class 0. This equality is shown in Table 1 proving the equilibrium state of the model striped bass population. It should also be noted that the total number of eggs produced by females of ages 5, 6, 7, and 8 account for 77.76% of the total egg production. In contrast, the eggs produced by the last four age classes account for less than 0.01% of the total egg production.

The population response to perturbations imposed on the five entrainable and impingeable life stages within age class 0 shown in Table 1 will be examined by means of the sensitivity analysis that follows.

SENSITIVITY ANALYSIS

The mean reproductive value of a population is related to the sensitivity of that population to changes in its schedule of births and deaths. For two similar populations subjected to an equal increase in the mortality rate in the same age class, the population having the lower $\overline{V}$ value will experience a larger reduction in population size. For a single population with two age classes subjected one at a time to an equal increase in mortality rate, the larger decrease in population size will occur when the age class affected is the one that contributes the larger percent to the total reproductive value of the population. Therefore, if the life table of a population is available and if it can be reasonably assumed that the current age distribution is near its theoretical stable distribution, it will be possible to determine what kind of response one may expect from a given perturbation.

In order to quantify the effect of a given perturbation, matrix projection techniques can be used. The procedure utilized in our analysis is as follows:

Let M be the Leslie matrix of a population and let each reproductive cycle last one year, as for many fish species. If the population has an initial stable distribution $\mathbf{K}_0$ which is normalized to $\mathbf{U}_0$, then $M\mathbf{U}_0 = \mathbf{U}_1$; $\lambda\mathbf{U}_0 = \mathbf{U}_1$; and the sums U_0 and U_1 of the elements in the vectors $\mathbf{U}_0$ and $\mathbf{U}_1$ will

Table 1. Hudson River striped bass equilibrium life-table parameters[a]

Age class[b] and life stage	Duration (days)	Probability of survival through the life stage or age class (p_i)	Number of individuals entering the life stage or age class (AG_i)	Eggs per individual[c] (F_i)	Total eggs	Percent of total eggs
AG0						
(1) Eggs and yolk-sac	10	0.05837300	23,699.9549E6[d]	0	0	0.000
(2) Post yolk-sac	24	0.03673973	1,383.4375E6	0	0	0.000
(3) Juvenile 1	30	0.20000000	50.8271E6	0	0	0.000
(4) Juvenile 2	145	0.51000000	10.1654E6	0	0	0.000
(5) Juvenile 3	156	0.16000000	5.1843E6	0	0	0.000
Total	365	0.00003500				
AG1	365	0.40000000	829.4984E3	0	0	0.000
AG2	365	0.60000000	331.7994E3	0	0	0.000
AG3	365	0.48785700	199.0796E3	0	0	0.000
AG4	365	0.48785700	97.1224E3	12,000	1,165,468,800	4.918
AG5	365	0.48785700	47.3818E3	86,250	4,086,680,250	17.243
AG6	365	0.48785700	23.1156E3	230,750	5,333,924,700	22.506
AG7	365	0.48785700	11.2771E3	427,500	4,820,960,250	20.342
AG8	365	0.48785700	5.5016E3	675,000	3,713,580,000	15.669
AG9	365	0.48785700	2.6839E3	750,000	2,012,925,000	8.493
AG10	365	0.48785700	1.3094E3	900,000	1,178,460,000	4.972
AG11	365	0.48785700	0.6388E3	1,100,000	702,680,000	2.965
AG12	365	0.48785700	0.3116E3	1,100,000	342,760,000	1.446
AG13	365	0.48785700	0.1520	1,100,000	167,200,000	0.705

Table 1 (continued)

Age class[b] and life stage	Duration (days)	Probability of survival through the life stage or age class (p_i)	Number of individuals entering the life stage or age class (AG_i)	Eggs per individual[c] (F_i)	Total eggs	Percent of total eggs
AG14	365	0.48785700	74.1728	1,100,000	81,590,080	0.344
AG15	365	0.48785700	36.1857	1,300,000	47,041,410	0.198
AG16	365	0.48785700	17.6535	1,350,000	23,832,225	0.101
AG17	365	0.48785700	8.6124	1,400,000	12,057,360	0.051
AG18	365	0.48785700	4.2016	1,450,000	6,092,320	0.026
AG19	365	0.48785700	2.0498	1,500,000	3,074,700	0.013
AG20	365	0.48785700	1.0000	1,550,000	1,550,000	0.007
Totals					23,699,877,095[e]	99.999

[a]Based on data and estimates from LMS (1975), McFadden (1977), and USNRC (1975).

[b]AG0 = age class 0, AG1 = age class 1, . . . , AG20 = age class 20.

[c]Calculated by adjusting the number of eggs per female for the observed age-class specific fraction of females sexually mature [from Table 7.8-8, McFadden (1977)] and assuming a sex ratio of 1:1.

[d]To be read as $23,699.9549 \times 10^6$.

[e]The difference between this total and the number of eggs entering AG0 is less than 0.00033%.

represent the initial population size and the population size after one year, respectively. Therefore,

$$\frac{U_1}{U_0} = \lambda .$$
(14)

Similarly,

$$M\mathbf{U}_1 = \mathbf{U}_2 \; ; M^2 \mathbf{U}_0 = \mathbf{U}_2 \; ; \lambda^2 \mathbf{U}_0 = \mathbf{U}_2 \; ; \frac{U_2}{U_1} = \frac{U_1}{U_0} = \lambda .$$
(15)

In general,

$$M^n \mathbf{U}_0 = \lambda^n \mathbf{U}_0 = \mathbf{U}_n \; ; \frac{U_1}{U_0} = \frac{U_2}{U_1} = ... = \frac{U_n}{U_{n-1}} = \lambda .$$
(16)

By making $U_0 = 100$, then the population size after n years as a percent of the initial population size is

$$100\lambda^n = U_n .$$
(17)

Let us now assume that the initial population represented by M is perturbed so that the schedule of births and deaths is changed, and let M_p be the new Leslie matrix of the perturbed population (i.e., a new matrix including the change which occurred in M). Then, still using the $\mathbf{U}_0$ associated with M before the perturbation,

$$M_p \mathbf{U}_0 = \mathbf{U}_{p,1} ,$$
(18)

where $\mathbf{U}_{p,1} \neq \mathbf{U}_1$, since $\mathbf{U}_0$ is not the stable age distribution associated with the new matrix M_p.

We will define a pseudo-root of the matrix M_p as the ratio

$$\hat{\lambda}_{p,1} = U_{p,1}/U_0 ,$$
(19)

where, of course, $\hat{\lambda}_{p,1} \neq \lambda$. This pseudo-root is actually an estimate of the new true root, λ_p, of M_p. If the number of cycles n is large enough,

$$M_p{}^n \mathbf{U}_0 = \mathbf{U}_{p,n} \; ; \frac{U_{p,n}}{U_{p,n-1}} = \hat{\lambda}_{p,n} \cong \lambda_p ,$$
(20)

where λ_p is the true root.

When $\hat{\lambda}_{p,n} \cong \lambda_p$, n represents the "stabilization" period or time units that it takes for the perturbed population to converge to its new stable distribution $\mathbf{U}_p$. In other words, the sequence of ratios of pseudo-roots $\hat{\lambda}_{p,1}$, $\hat{\lambda}_{p,2},...,$ $\hat{\lambda}_{p,n}$ converges asymptotically to the true root λ_p. The geometric mean ratio, $\bar{\lambda}$, of this sequence may be thought of as the geometric mean yearly ratio of the perturbed population size to the initial population size; that is,

$$\bar{\lambda} = (\hat{\lambda}_{p,1} \cdot \hat{\lambda}_{p,2} \cdot ... \cdot \hat{\lambda}_{p,n})^{1/n} = \left[\prod_{i=1}^{n} \hat{\lambda}_{p,i} \right]^{1/n} = (U_{p,n}/U_0)^{1/n}. \quad (21)$$

This geometric mean ratio provides a valuable index to assess the extent of the change in population size during the critical stabilization period, since it takes into account the relative size of the perturbed population after n time steps, $U_{p,n}/U_0$, as well as the number of time steps, n, required for the perturbed population to converge to its new stable age distribution, $\mathbf{U}_p$. Therefore, while λ_p is the constant specific growth rate of the perturbed population after stabilization, $\bar{\lambda}$ is the exact average of the variable specific growth rates during the stabilization period. Note that stabilization refers to the age distribution and does not necessarily imply equilibrium.

These two specific growth rates and the specific growth rate associated with the initial population, λ, can be used to define an empirical index of perturbation, P, as

$$P = 100 \left[\left(\frac{\bar{\lambda} \cdot \lambda_p}{\lambda^2} \right)^{1/2} - 1.00 \right]. \quad (22)$$

The quantity $(\bar{\lambda} \cdot \lambda_p/\lambda^2)^{1/2}$ is the geometric mean of the ratios $\bar{\lambda}/\lambda$ and λ_p/λ, and thus reflects both the sequence of transient growth rates during the stabilization period as well as the stable growth rate of the perturbed population after stabilization, in each case relative to the growth rate of the unperturbed population. Subtracting 1.0 from this geometric mean of the two ratios, although not necessary, translates the scale for P so that for $\lambda = 1.0$, P will be positive if the perturbation results in an increase in population size and negative if the perturbation results in a decrease in population size.

In addition, the population size n years after the perturbation, as a percent of the population size before perturbation, is given by

$$100(\bar{\lambda})^n = 100U_{p,n}/U_0, \quad (23)$$

where n is the stabilization time.

Since in reality no population actually reaches stabilization after a perturbation (since other perturbations will probably occur before stabilization), for most practical purposes the interesting period to look at is a short one following occurrence of the perturbation. In this study, it was found that the stabilization time ranged between 36 and 59 years for the type of perturbations simulated, and it was decided that the most realistic approach would be to look at the time series of population size during only the first 20 years. An outline of the procedure is as follows.

Number of years of perturbation	Population size as a percent of initial population
0	100
1	$100\hat{\lambda}_{p,1}$
2	$100\hat{\lambda}_{p,1} \cdot \hat{\lambda}_{p,2}$
.	.
.	.
.	.
	20
20	$100 \prod_{i=1} \hat{\lambda}_{p,i}$ (24)

This procedure is easily programmable on a digital computer, and it has the advantage that it is possible to truncate the initial vector U_0 at any age class or even at any life stage within age class 0, so that a particular segment of the population can be followed through the stabilization period. Truncation of the initial vector is valid because each age class in a population converges to its equilibrium size at the same rate. The only difference between any two age classes is the amplitude of the oscillations (i.e., the true size changes), but not the stabilization time.

Next, it will be shown how to compute the elements of the Leslie matrices M and M_p from the striped bass life table (see Table 1).

Lefkovitch (1965) demonstrated that the Leslie matrix of a population classified into age groups was only a particular case of a more general matrix of a population classified into stages according to morphological differences. Although it is true that the stages explicitly considered in this generalized matrix do not need to have equal durations, all the stages represented in the model must occur concurrently at some point within each model time step. In the striped bass population the five life stages considered in age class 0 (Table 1) do not all occur simultaneously within the model time step of a year, and thus it is not possible to utilize a Leslie matrix with all five life stages.

Examples of the proper procedure to simulate this population are shown in Tables 2 and 3. The method used to adjust the coefficients of the Leslie matrix for each simulation should be apparent from the footnotes in these two tables.

The validity of this procedure is proven by the fact that both matrices yield exactly the same dominant latent root λ, which is as it should be since both matrices represent the same equilibrium population and set of life cycle

Table 2. Elements of the Leslie matrix and of the stable
age-distribution vector U for the Hudson River striped bass
population at spawning time

Age class	Probability of survival (Subdiagonal) (p_i)	Fecundity[a] (Top row) (F_i)	U Number of individuals
0 (eggs)	0.000035[b]	0	23,699.9549E6[c]
1	0.400000	0	829.4984E3
2	0.600000	0	331.7994E3
3	0.487857	0	199.0796E3
4	0.487857	12,000	97.1224E3
5	0.487857	86,250	47.3818E3
6	0.487857	230,750	23.1156E3
7	0.487857	427,500	11.2771E3
8	0.487857	675,000	5.5016E3
9	0.487857	750,000	2.6839E3
10	0.487857	900,000	1.3094E3
11	0.487857	1,100,000	0.6388E3
12	0.487857	1,100,000	0.3116E3
13	0.487857	1,100,000	0.1520E3
14	0.487857	1,100,000	74.1728
15	0.487857	1,300,000	36.1857
16	0.487857	1,350,000	17.6535
17	0.487857	1,400,000	8.6124
18	0.487857	1,450,000	4.2016
19	0.487857	1,500,000	2.0498
20		1,550,000	1.000
	$\lambda = 1.000000$		

[a]Fecundity rates taken directly from Table 1.

[b]0.000035 = (0.058373)(0.03673973)(0.20)(0.51)(0.16). (These survival probabilities for the five life stages in age class 0 are from Table 1. Survival probabilities for age classes 1 to 20 are directly taken from the same table).

[c]To be read as $23,699.9549 \times 10^6$.

Table 3. Elements of the Leslie matrix and of the stable
age-distribution vector U for the Hudson River striped bass
population 35 days after spawning

Age class	Probability of survival (Subdiagonal) (p_i)	Fecundity[a] (Top row) (F_i)	U Number of individuals
0 (Juvenile 1)	0.016320[b]	0	50,827.1092E3[c]
1	0.400000	0	829.4984E3
2	0.600000	0	331.7994E3
3	0.487857	0	199.0796E3
4	0.487857	25.7353	97.1224E3
5	0.487857	184.9725	47.3818E3
6	0.487857	494.8684	23.1156E3
7	0.487857	916.8200	11.2771E3
8	0.487857	1447.6106	5.5016E3
9	0.487857	1608.4562	2.6839E3
10	0.487857	1930.1474	1.3094E3
11	0.487857	2359.0691	0.6388E3
12	0.487857	2359.0691	0.3116E3
13	0.487857	2359.0691	0.1520E3
14	0.487857	2359.0691	74.1728
15	0.487857	2787.9907	36.1857
16	0.487857	2895.2211	17.6535
17	0.487857	3002.4516	8.6124
18	0.487857	3109.6820	4.2016
19	0.487857	3216.9124	2.0498
20		3324.1428	1.0000
	$\lambda = 1.000000$		

[a]Fecundities from Table 2 are reduced by a factor of $(0.058373)(0.03673973) = 0.00214460826$, which is the probability of survival through the first two life stages of age class 0 (see Table 1).

[b]$0.01632 = (0.20)(0.51)(0.16)$. (These survival probabilities for the last three life stages in age class 0 are from Table 1. Survival probabilities for year classes 1 and 20 directly taken from the same table.)

[c]To be read as $50,827.1092 \times 10^3$.

parameters. The only difference in the equilibrium age-distribution vectors, U, in Tables 2 and 3 is the number of individuals in age class 0, which in the case of Table 2 represents the equilibrium number of eggs (23.7×10^9) and in the case of Table 3 represents the equilibrium number of Juvenile 1's (50.8×10^6).

Three more matrices were constructed in a similar manner for the other three life stages within age class 0. It makes no difference in terms of the validity of the model whether, for instance, the mature females in the population are "spawning" Juvenile 1's, larvae, or eggs. As long as the survival and fecundity rates are properly adjusted, the same dominant latent root will be obtained.

These five matrices were utilized in the next step of our study to analyze the effect of decreasing the survival rate in each of the five y-o-y life stages by some specified percent. If a decrease of 5% were to be simulated in the survival rate of the Juvenile 1 life stage for instance, the new M_p matrix would be derived from M (Table 3) by simply multiplying the survival rate for age class 0 (i.e., 0.016320) by the factor $1.0 - 0.05 = 0.95$.

In order to simulate a decrease of 5% in the survival rates for each of (eggs/yolk-sac larvae), post-yolk-sac larvae, and Juvenile 1's, the matrix M_p would be derived from M (Table 3) as follows:

The survival rate 0.016320 becomes

$$(0.016320)(0.95) = 0.0159936 , \qquad (25)$$

and all the fecundities are reduced by a factor of

$$(0.95)(0.95) = 0.9025 . \qquad (26)$$

The factor $(0.95)^2$ represents the reduction in the survival rate of the two life stages preceding the Juvenile 1 stage, and consequently, this factor must be applied to the fecundity rates which control the size of the cohort actually reaching the Juvenile 1 life stage.

Any other survival reduction, either in a single stage or in any combination of stages, can be simulated in a similar manner, proving the outstanding flexibility and realism of the Leslie matrix as a model for this type of simulation.

RESULTS AND DISCUSSION

The life-table data shown in Table 1 were utilized to construct the initial 21×21 matrix (Table 2). The results of the analysis of this initial matrix produced a dominant latent root $\lambda = 1.000000$, indicating a stable population with zero growth, as shown in the first row of Table 4. The eigenvector

Table 4. Results of simulating annual entrainment and impingement losses of 2, 5, 10, and 20% for each of the five life stages of age class 0[a]

Percent decrease and stage number[b]	λ_p	$\bar{V}$	$U \times 10^{-3}$	$G \times 10^{-3}$	n	$\bar{\lambda}$	P	Population size after 20 years[c]
Initial	1.000000	8.050380	1550.020	0.000				100.000
2% (1)	0.997496	7.964439	1481.450	-3.710	36	0.997301	-0.260	94.427
5% (1)	0.993659	7.834522	1381.980	-8.763	42	0.993238	-0.655	86.466
10% (1)	0.987040	7.615167	1225.100	-15.878	48	0.986294	-1.333	74.224
20% (1)	0.972845	7.164734	943.868	-25.630	52	0.971433	-2.786	53.347
2% (2)	0.997496	7.956174	1481.450	-3.710	36	0.997301	-0.260	94.427
5% (2)	0.993659	7.826581	1381.980	-8.763	42	0.993238	-0.655	86.466
10% (2)	0.987040	7.607757	1225.100	-15.878	48	0.986294	-1.333	74.224
20% (2)	0.972845	7.158355	943.868	-25.630	52	0.971433	-2.786	53.347
2% (3)	0.997496	7.732720	1481.450	-3.710	36	0.997301	-0.260	94.427
5% (3)	0.993659	7.611729	1381.980	-8.763	42	0.993238	-0.655	86.466
10% (3)	0.987040	7.407047	1225.100	-15.878	48	0.986294	-1.333	74.224
20% (3)	0.972845	6.985158	943.868	-25.630	52	0.971433	-2.786	53.347
2% (4)	0.997496	6.925055	1481.450	-3.710	36	0.997301	-0.260	94.427
5% (4)	0.993659	6.832824	1381.980	-8.763	42	0.993238	-0.655	86.466
10% (4)	0.987040	6.675697	1225.100	-15.878	48	0.986294	-1.333	74.224
20% (4)	0.972845	6.347425	943.868	-25.630	52	0.971433	-2.786	53.347
2% (5)	0.997496	6.153201	1481.450	-3.710	36	0.997301	-0.260	94.427
5% (5)	0.993659	6.085002	1381.980	-8.763	42	0.993238	-0.655	86.466
10% (5)	0.987040	5.968003	1225.100	-15.878	48	0.986294	-1.333	74.224
20% (5)	0.972845	5.720217	943.868	-25.630	52	0.971433	-2.768	53.347

[a]Definitions of the symbols used in this table are as follows: λ_p = dominant latent root; U = stable population unit for age classes 1–20; G = stable growth rate for age classes 1–20; $\bar{V}$ = mean reproductive value; n = stabilization period in years (see Eq. 20); $\bar{\lambda}$ = the geometric mean yearly ratio of the perturbed adult population size to the initial adult population size during the entire stabilization period, and P is the perturbation index for the adult population (see Eq. 22).

[b]The stage numbers correspond to the life stage numbers in Table 1.

[c]The adult population size after 20 years of perturbation as a percent of the initial adult population size (see Eq. 24).

associated with λ was normalized to unity in its last element (Table 2) and found to be equal to the expected stable age-distribution in the life table (Table 1).

The perturbation analysis of this equilibrium population was intended to simulate annual entrainment and/or impingement losses of 2, 5, 10, and 20% for each of the five y-o-y life stages taken alone and in various combinations. The range of values used (i.e., 2–20%) is considered adequate for the purposes at hand because it spans a range greater than has been reported for the available model studies to date.

Table 4 represents the results of simulated annual entrainment and impingement losses of 2, 5, 10, and 20% for each of the five y-o-y life stages in Table 1 taken one at a time. In addition to the dominant latent root, λ_p, the average reproductive value for the entire population, $\overline{V}$, the stable population unit for the adult population (i.e., age classes 1–20), U, the stable growth rate for the adult population, G, the length of the stabilization period in years, n, the geometric mean yearly ratio of the perturbed adult population size to the initial adult population size during the n years of stabilization, $\overline{\lambda}$, the perturbation index, P, and the adult population size after the first 20 years of perturbation as a percent of the initial adult population size (Eq. 24) are given.

It is clear from an examination of these simulation results for each life history stage (Table 4) that the dominant latent root, λ_p, and the geometric mean yearly ratio, $\overline{\lambda}$, change very little (less than 3%) for any of the five sets of simulations. However, the mean reproductive value for the population, $\overline{V}$, the stable adult population unit, U, the stable growth rate for the adult population, G, the perturbation index, P, and the relative population size after 20 years change considerably. It is evident from this table that a given percent decrease applied to a single life stage in age class 0 results in the same reduction in the adult population after 20 years regardless of which life stage suffers the increased mortality. It is also evident that entrainment and/or impingement losses up to 20% in any one life stage alone are not adequate to reduce the adult population size in 20 years to values less than 50% of the initial value.

Table 5 provides a more realistic estimate of the population effects of entrainment and/or impingement mortalities. In this table, which is organized in a manner identical to Table 4, power plant mortality is assumed to occur to a progressively greater number of the y-o-y life stages. That is, the first set of simulation results [labeled (1–2)] involved the indicated percent decreases in survival applied to each of life stages 1 and 2, whereas the last set of results [labeled (1–5)] involved the same percent decreases in survival applied to each of the five y-o-y life stages.

It is immediately evident that, for a given percent decrease in survival, the greater the number of life stages to which this percent decrease is applied, the greater the effect on all parameters (Table 5). The dominant latent root, λ_p,

Table 5. Results of simulating annual entrainment and impingement losses of 2, 5, 10, and 20% for successive life stages of age class 0[a]

Percent decrease and stage number[b]	λ_p	$\bar{V}$	$U \times 10^{-3}$	$G \times 10^{-3}$	n	$\bar{\lambda}$	P	Population size after 20 years[c]
Initial	1.000000	8.050380	1550.020	0.000				100.000
2% (1–2)	0.995000	7.871472	1416.000	−7.079	41	0.994720	−0.514	89.378
5% (1–2)	0.987378	7.618371	1232.680	−15.559	48	0.986779	−1.292	75.240
10% (1–2)	0.974378	7.203295	970.085	−24.904	50	0.973154	−2.626	55.878
20% (1–2)	0.946784	6.398463	579.581	−30.843	57	0.944766	−5.423	29.485
2% (1–3)	0.992514	7.567121	1353.540	−10.132	45	0.992161	−0.766	84.606
5% (1–3)	0.981155	7.208781	1100.000	−20.729	49	0.980350	−1.925	65.514
10% (1–3)	0.961861	6.637586	769.580	−29.351	54	0.960421	−3.886	42.183
20% (1–3)	0.921774	5.589314	358.946	−28.078	58	0.919181	−7.952	16.509
2% (1–4)	0.990037	6.694063	1293.920	−12.891	45	0.989587	−1.019	80.097
5% (1–4)	0.974991	6.275712	982.036	−24.559	51	0.974015	−2.549	57.081
10% (1–4)	0.949633	5.627282	611.666	−30.807	54	0.947846	−5.126	31.935
20% (1–4)	0.897775	4.498887	224.254	−22.924	59	0.894725	−10.375	9.367
2% (1–5)	0.987570	5.870168	1237.000	−15.376	46	0.987036	−1.269	67.856
5% (1–5)	0.968885	5.408443	877.103	−27.291	53	0.967757	−3.168	49.767
10% (1–5)	0.937641	4.712956	487.079	−30.374	55	0.935570	−6.339	24.246
20% (1–5)	0.874748	3.565228	141.362	−17.706	57	0.871165	−12.704	5.392

[a]The symbols are the same as for Table 4.

[b]The stage numbers correspond to the life stage numbers in Table 1. For example, the notation (1–4) indicates that the given annual percent decrease occurred in each of life stages 1, 2, 3, and 4.

[c]The adult population size after 20 years of perturbation as a percent of the initial adult population size (see Eq. 24).

and the geometric mean yearly ratio, $\overline{\lambda}$, are reduced by about 15% in the case of a 20% decrease in survival of stages 1 through 5. However, the other four derived parameters described for Table 4 change even more markedly. Examination of Table 5 clearly suggests that entrainment and impingement losses of all five life stages in excess of 2% cause serious population effects leading to reductions of the adult population of more than 30% after 20 years. It is also clear that losses up to 5% in life stages 1 to 4 can be tolerated, using a threshold of 50% reduction of the adult population. Up to 10% decreases in stages 1 and 2 are tolerated by this criterion.

In summary, the results of these simulations suggest the following:

1. The dominant latent root, λ, of the population matrix is relatively insensitive to increased mortality in age class 0.

2. The derived population parameters and measures of perturbation, consisting of the mean reproductive value, $\overline{V}$, the stable population unit, U, the stable growth rate, G, the perturbation index, P, and the relative adult population size after a finite period (such as the 20-year period used in this study) are more descriptive of population effects resulting from perturbations than is the dominant latent root λ_p.

3. If only one of the five life stages in age class 0 is subjected to increased mortality, the population seems capable of tolerating losses up to 20% before being reduced to about 50% of its initial size in 20 years.

4. If each of the five life stages in age class 0 is subjected to increased mortality, the percent loss per life stage must be less than 5% in order to not have the size of the adult population reduced by 50% or more in 20 years.

5. Any reduction in the fishing mortality in one or several of the age classes 3 to 20 will permit a higher tolerance for additional mortality in the y-o-y life stages.

6. The derived index of perturbation, P, together with the technique described to follow any segment of the population through the stabilization period, can be very useful in assessing the effect of power-plant-induced mortality on fish populations.

LITERATURE CITED

Beverton, R. J. H., and S. J. Holt. 1957. On the dynamics of exploited fish populations. Fish. Invest. Ser. 2, 19. 533 p.

Horseman, L. O., and R. J. Kernehan. 1976. An indexed bibliography of the striped bass, *Morone saxatilis*, 1970–1976. Ichthyological Associates, Inc., Bull. No. 13. 118 p.

Keyfitz, N. 1968. Introduction to the mathematics of populations. Addison-Wesley, Reading, Massachusetts. 450 p.

LMS (Lawler, Matusky & Skelly Engineers). 1975. Report on development of a real-time, two-dimensional model of the Hudson River striped bass population. LMS Project No. 115–49. 71 p.

Lefkovitch, L. P. 1965. The study of population growth in organisms grouped by stages. Biometrics **21**:1–18.

———. 1967. A theoretical evaluation of population growth after removing individuals from some age groups. Bull. Entomol. Res. **57**:437–445.

Leslie, P. H. 1945. On the use of matrices in certain population mathematics. Biometrika **33**:183–212.

———. 1948. Some further notes on the use of matrices in population mathematics. Biometrika **35**:213–245.

McFadden, J. T. [ed.] 1977. Influence of Indian Point Unit 2 and other steam electric generating plants on the Hudson River estuary, with emphasis on striped bass and other fish populations. Submitted to Consolidated Edison Co. of New York, Inc.

Mendelssohn, R. 1976. Optimization problems associated with a Leslie matrix. Am. Nat. **110**:339–349.

Pfuderer, H. A., S. S. Talmage, B. N. Collier, W. Van Winkle, Jr., and C. P. Goodyear. 1975. Striped bass – a selected, annotated bibliography. ORNL-EIS-75-73, ESD 615. Oak Ridge National Laboratory, Oak Ridge, Tennessee. 158 p.

Ricker, W. E. 1953. Handbook of computations for biological statistics of fish populations. Fish. Res. Board Can., Bull. 119. 300p.

Rogers, B. A., and D. T. Westin. 1975. A bibliography on the biology of the striped bass, *Morone saxatilis* (Walbaum). Mar. Tech. Rept. No. 37, Univ. of Rhode Island. 134 p.

Swartzman, G., R. Deriso, and C. Cowan. 1977. Comparison of simulation models used in assessing the effects of power-plant-induced mortality on fish populations. (*In this volume*).

Usher, M. B. 1972. Developments in the Leslie matrix model, p. 29–60. *In* J. N. R. Jeffers [ed.] Models in ecology. Proc. of British Ecological Society Symposium, Vol. 12, Blackwell Sci. Pub. Ltd., London.

———. 1976. Extensions to models, used in renewable resource management, which incorporate the arbitrary structure. J. Environ. Manage. **4**:123–140.

USNRC (U.S. Nuclear Regulatory Commission). 1975. Final environmental statement related to operation of Indian Point Nuclear Generating Station, Unit No. 3, USNRC Docket No. 50-286, Vols. I and II, February 1975.

Williamson, M. H. 1967. Introducing students to the concepts of population dynamics, p. 169–175. *In* J. M. Lambert [ed.] The teaching of ecology. Proc. of British Ecological Society Symposium, Vol. 12, Blackwell Sci. Pub. Ltd., London.

Comparison of Simulation Models Used in Assessing the Effects of Power-Plant-Induced Mortality on Fish Populations [*]

Gordon Swartzman, Rick Deriso,
and Chris Cowan

Center for Quantitative Science
University of Washington
Seattle, Washington

ABSTRACT

This paper compares eight models predicting the impact of power plant operation upon economically important fish species. The review compares the biological rationale behind model equations; parameter values are tabulated and compared. This paper focuses on differences between the models and the effect of these differences on model predictions. In order to expedite the evaluation of the models, generalized model simulators were developed for the young-of-the-year and life-cycle submodels. Criteria used to evaluate model predictions are percent reduction in young-of-the-year and annual loss in yield to the fishery due to plant operation. Major differences affecting model predictions of these impact criteria include the definition of life stages within age class 0, density-dependent or density-independent mortality for larvae and juveniles, density-dependent or density-independent fishing mortality, and the method for computing recruitment of young-of-the-year fish to age class 1. Major differences in parameter values include entrainment factors, total egg production, equilibrium population size, and survival probabilities for the life cycle models. Recommendations are made regarding our preference between model approaches or parameter values when these differ markedly. A suggestion is made for improved model documentation in the future and for increased cross fertilization of modeling ideas and background data and information.

Key words: density dependent, density independent, life cycle model, model predictive ability, model simulations, parameter values, percentage reduction, regression equation, scaling factor, striped bass, young-of-the-year model

*Work supported under Contract AT(49-24)-0222 with the Nuclear Regulatory Commission, Washington, D.C.

INTRODUCTION

This paper discusses a review and comparison of eight models developed for the expressed purpose of evaluating the impact of power plant operation on populations of commercially and recreationally important fish species spawning upriver or in the neighborhood of a power plant. Most of these models are complex computer simulations using data from ongoing sampling programs. All of them simulate the entrainment of fish eggs and larvae through the cooling systems of the power plants. Most of the models reviewed are part of a family of models, evolving through time and generally increasing in complexity as new data and information become available and as the issues become more well defined. Many of the models have been used as part of the licensing procedures for the Nuclear Regulatory Commission (NRC).

Our objectives in this review have been

1. To evaluate the relative predictive ability of various models;

2. To review and evaluate the various models with regard to the biological realism of model assumptions;

3. To pinpoint model similarities and differences and to ascertain how important these differences are to model prediction. This objective is important since several of the models give strikingly disparate predictions of plant impact;

4. To look for ideas for possible general approaches to modeling power plant impact.

MODELS REVIEWED

The following terminology is used in this report in referring to the various modeling groups and models: LMS refers to Lawler, Matusky & Skelly Engineers; ORNL refers to Oak Ridge National Laboratory; UEC refers to United Engineers & Constructors; JHU refers to The Johns Hopkins University Applied Physics Laboratory; MPPSP refers to the Maryland Power Plant Siting Program; and URI refers to the University of Rhode Island.

Hudson River Striped Bass Models (Bowline, Indian Point, and Roseton Power Stations)

1. The 1972 LMS model, developed for the applicant, involving no longitudinal segmentation or vertical stratification of the Hudson River (Lawler 1972a) — referred to as the LMS Completely Mixed Model.

2. The 1972 LMS model, developed for the applicant, involving longitudinal segmentation but no vertical stratification of the Hudson River (Lawler 1972b) — referred to as the LMS 1-D Model, the LMS 1-D(67) Model with 1967 data, and the LMS 1-D(73) Model with 1973 data (Lawler 1974).

3. The 1973 ORNL model, developed for NRC, involving longitudinal segmentation but no vertical stratification of the Hudson River (USNRC 1975, Eraslan et al. 1976, Van Winkle et al. 1974) — referred to as the ORNL 1-D Model.

4. The 1975 LMS model, developed for the applicant, involving both longitudinal segmentation and vertical stratification of the Hudson River (LMS 1975) — referred to as the LMS 2-D Model.

Chesapeake and Delaware (C&D) Striped Bass Models
(Summit Nuclear Power Station)

5. The United Engineers & Constructors model, developed for the applicant (UEC 1975) — referred to as the Delmarva Model.

6. The ORNL model, developed for NRC (Christensen et al. 1975) — referred to as the ORNL Summit Model.

7. The JHU model, developed for MPPSP (Warsh 1975, Portner 1975) — referred to as the JHU Model.

Winter Flounder Model (Millstone Nuclear Power Station)

8. The URI model, developed for the applicant (Sissenwine et al. 1974) — referred to as the Winter Flounder Model.

Although other impact "models" exist besides those reviewed here, they either have been superseded by the ones we will review or are simple enough that we have put them into the category of a calculation (for example, a local entrainment calculation).

When discussing a property of the models (e.g., the life cycle submodels) that applies to all three of the LMS models or both of the ORNL models, the designations LMS and ORNL, respectively, are used.

In general, the models are conveniently partitioned into two submodels. The first simulates the annual effect of plant entrainment (and in some cases, plant impingement) on recruitment of young-of-the-year into the adult population and is called the young-of-the-year (y-o-y) model. The second submodel simulates the subsequent, long-term effect of reduced recruitment on the adult population and is called the life cycle model. The young-of-the-year models generally are based on a hydrodynamic model which determines water flow and transport of eggs and larvae. None of the life cycle models explicitly considers spatial phenomena.

APPROACH USED IN THIS REVIEW

Questions relevant to our goals were formulated to guide our review (Table 1). For each of the y-o-y and life cycle models the following procedure was adopted.

Table 1. Questions formulated to guide the review of models

Presentation and comparison of model assumptions and systems studied

In what ways are the systems studied similar and in what ways are they different? What are the model assumptions in the physical and biological domains? What are the major similarities and differences in the models? What statistical assumptions are made in using data? How relevant are the data to the study site?

Simulation study and sensitivity analysis

How different are model predictions for models developed for the same system? How sensitive is model output to areas where the models are different? How sensitive are the models to changes in parameter values? What are sensitivity criteria?

Model evaluation

How do these models compare with classical fisheries models and other more mechanistic fish models? Are any of these models good predictive tools and does increased model complexity increase predictivity? Is there a best model? How can we judge the relative quality of models? How close to field data are simulations? What is the relative cost of the various modeling approaches?

General conclusions and recommendations

Based on this review, what recommendation can we make about using models for decision making or for predicting the impact of a plant on an ecological system? What work on further model comparison, field research, or model development might improve the usefulness of present generation models as decision aids or as predictors of plant impact upon ecological systems?

Compare the Model Equation Forms and Underlying Model Assumptions

Graphs of model equations were lifted from model reports where available. Otherwise, they were computer generated or hand drawn. When these equations differed between models, primary literature sources were consulted to help judge the adequacy and scientific rationale of model assumptions.

Tabulate and Compare Parameter Values and Investigate Data Sources (When Available) Used in Obtaining Them

In most models parameter values are based on data or at least have a rationale for their selection. Parameter differences were investigated to see whether the differences could be traced to the use of different data sets. Statistical assumptions inherent in deriving some of the parameter values from data sources were also examined. Relevance of data sources was considered.

Compare Model Simulations

Many of the models had differing assumptions and/or parameter values, and no prior investigation was available to test the effect of these differences on model predictions. Thus, we decided to develop general simulators (models that simulate equations in the reviewed models) for the y-o-y and life cycle models which could utilize equations or parameter values from many of the models. These simulators, discussed in the body of this paper, have proven invaluable in our analysis. The main reasons we chose to adapt the models to a generalized simulator framework, rather than lift the code entirely, were (1) relative inflexibility of the code of any one model to include the formulation of another model, (2) expense of running several of the models at the time step and spatial resolution utilized by model developers, and (3) lack of portability or documentation for several of the model codes.

Decide upon Model Prediction Criteria

Since the y-o-y models all predict the impact of power plant entrainment on y-o-y recruited into the adult population, we chose the percentage reduction in y-o-y with plant operation as our measure of impact, thus following the lead of most of the models. Percentage reduction (PR) is given by

$$PR = \frac{x - y}{x} \times 100 \, ,$$

where x = y-o-y population without plant operation and y = y-o-y population with plant operation.

Ultimately we are concerned in these models with the long-term impact of plant operation on the fish population itself and on the fishery in terms of loss of fish yield. A number of criteria are suggested in the various life cycle models for translating PR into such a quantity.

Perform a Sensitivity Analysis

Since many of the models had different parameter values, even when the equation forms were the same, it was of interest to investigate the effect of changing parameter values on the various impact criteria. Many of these sensitivity studies were done by the original modeling groups, and the results from their studies were also reviewed and compared.

Recommend Model Improvements and Evaluate a "Best" Approach

Various models had various strengths. In some cases we leaned towards one and sometimes towards another as a "best" approach. In many cases the question was one of simplicity of representation vs accuracy of model results. Simulation study is helping us to decide which is to be preferred.

SYSTEMS MODELED

The three systems studied in these impact models are the Hudson River (Bowline, Indian Point, and Roseton), Chesapeake and Delaware Canal (Summit), and the Niantic River—Long Island Sound (Millstone). All three areas are estuarine in nature, all are mixed by tidal fluctuations, and all are major spawning areas for commercially and recreationally important, East Coast fish species.

YOUNG-OF-THE-YEAR MODELS

General Structure

In estimating how many eggs, larvae, and juveniles are entrained, all models divide the y-o-y into life stages.

As they mature, larvae and juveniles are increasingly able to avoid entrainment. This phenomenon is modeled either by making individuals above a certain age unentrainable or by slowly increasing resistance to entrainment. Most of the models divide the estuary spatially into segments and use a hydrodynamic model to move the organisms between the segments. Model validation consists of comparing observed and simulated spatial and temporal distributions of eggs, larvae, and juveniles. More rigorous validation, involving independent data sets, seems appropriate for all models.

Entrainment mortality in all models is density independent, while natural mortality is either density dependent or density independent. Entrainment is inflicted upon some fraction of the population of each life stage in the neighborhood of the plant. Estimates of the size of these populations near the plant either are obtained from the hydrodynamic spatial model (which may include such factors as migration, in addition to transport) or are assumed to be a constant fraction of the total population in the system. The y-o-y models predict either the number of y-o-y fish surviving to age-class 1 (i.e., yearlings) with and without power plant mortality or PR itself. One or the other of these outputs from the y-o-y models is then used (in most cases) in conjunction with a Leslie matrix model (the life cycle model) to predict long-term fishery yield or adult population reduction.

Our review treated each of these aspects of the y-o-y models in some detail. However, neither space nor time permits a full description in this paper. Thus, we have focused on the highlights of our work, emphasizing the biological factors.

Life Stages and Life Stage Parameters

As mentioned earlier, each model divides the y-o-y into life stages. Table 2 compares the life stages and the parameter values for the duration and the survival fraction for the different models. The choice of life stages and life stage parameters vary considerably among the models. Although seemingly innocuous, the choice of life stages and parameter values bears upon the estimates of PR obtained from a model. For instance, choosing a longer duration for an entrainable life stage lengthens the time that life stage is subject to entrainment, thereby potentially raising the entrainment loss. The duration of the Juvenile I life stage used in the ORNL 1-D model (40 days) as compared to that used in the LMS models (30 days) is an example.

Equations for Survival and Transfer

Equations for survival and transfer between life stages have used three approaches. These are (1) not dividing the y-o-y into life stages, (2) treating each life stage as a group of cohorts, where a cohort represents individuals entering that life stage at the same time step, and (3) treating individuals in each life stage as indistinguishable by cohort but uniformly distributed in age within the life stage.

Approach (1) is used in the Winter Flounder Model. Here the survival from spawned eggs in year t, $N_0(t)$, to one-year-old fish in year $t + 1$, $N_1(t + 1)$, is modeled using the Ricker spawner-recruit relationship (Ricker 1954):

$$N_1(t + 1) = N_0(t) \exp[a - bN_0(t)] \qquad (1)$$

Table 2. Comparison of values used for the life stage duration and equilibrium survival fraction[a]

Model	Eggs	Yolk-sac larvae	Post-yolk-sac larvae	Juvenile			Source
				I	II	III	
LMS Completely Mixed	1.5 (0.01–0.1)	21 (0.005–0.1)	b	30 (0.2)	137.45[c] (0.4–0.6)	159 (0.184–0.4)	Table 3, Lawler (1972a)
LMS 1-D 1967	1.5 (0.1)	28 (0.15–0.5)	b	30 (0.2)	123[d] (0.5)	158 (0.1898)	Table 4, Lawler (1972b)
LMS 1-D 1973	1.5 (0.1)	28 (0.15)	b	30 (0.2)	123[d] (0.53)	158 (0.186)	TI (1973)
LMS 2-D	1.5,2.25,3 (0.1)	6 (0.15)	22 (0.15)	30 (0.2)	123[d] (0.53)	158 (0.186)	Table 8, LMS (1975)
ORNL 1-D[e]	2[f]	6	22	40	123	172	Table B–23, USNRC (1975)
JHU[g]	3	10	b	67	275	275	Clark (1972) and Table I–2, Portner (1975)
Delmarva[h]	2–3 (0.1)	10–15 (0.1)	78–72 (0.1)	i	275	275	Mansueti (1961) and Portner (1975)

Footnotes for Table 2

[a]Entries without parentheses are durations (i.e., number of days required to pass through the given life stage). Entries in parentheses are survival fractions (i.e., probability of survival through the given life stage).

[b]Yolk-sac larvae and post-yolk-sac larvae are combined into one life stage.

[c]Minimum = 116 days, maximum = 153.5 days.

[d]Minimum = 98.5 days, maximum = 147.5 days.

[e]The formulation for mortality and transfer to the next life stage in this model is sufficiently different from that used in the other models that there does not appear to be a set of input parameters equivalent in concept to the survival fractions of the other models.

[f]Duration of the egg life stage is a function of water temperature.

[g]Estimates of PR from this model are independent of survival. Life-stage survivals are varied, but the total survival fraction through the entrainable life stages is held constant at 0.001.

[h]These durations are varied in sensitivity studies.

[i]Post-yolk-sac larvae and Juvenile I are combined into one life stage.

The parameters a and b are estimated, in theory, from data on eggs and age-1 adults, but in lieu of adequate data they are estimated from information derived from properties of the Ricker curve and of the adult population (Sissenwine et al. 1974). Recruitment to year class 1 is a function of the number of eggs produced during the previous year $[N_0(t)$ in Eq. 1] multiplied by $[1.0 -$ (the density-independent, fractional reduction in recruitment during year class 0 resulting from entrainment)]. This procedure assumes implicitly that all compensation occurs within age class 0 after entrainment mortality has already taken place. This assumption is questionable; general opinion is that compensation is most likely to occur during the early life stages (i.e., yolk-sac and post-yolk-sac larvae), which are entrainable.

This first approach has the benefit of simplicity and relatively low data requirements. There is, however, scanty supporting evidence for the estimated values of a or b or even for the applicability of the Ricker formulation to this species.

Approach (2) is used in the LMS, JHU, and Delmarva Models. The time step ranges from daily (Delmarva) to a tenth of a day (LMS 1-D). The differential equation for the rate of loss due to mortality is given by

$$\frac{dn_i^{(k)}}{dt} = -m_i {}^* n_i^{(k)} . \tag{2}$$

where $n_i^{(k)}$ = the number of individuals in the ith life stage which are in the kth cohort (e.g., k days old). The age-class-specific mortality parameter m_i may be either a constant (Delmarva, JHU) or a function of the density of fish in that life stage (LMS 1-D, 2-D). One important feature about the cohort model is that fish mature into a new life stage when they have lived for a time equal to the sum of the durations of the preceding life stages. The cohort approach also means that there is no ambiguity about the age distribution of individuals in a life stage; they are all aged to within one time step. The cohort method accurately represents the mechanics of aging, assuming all fish within a cohort mature at the same rate, but it is very expensive in computer storage and time.

With approach (3) each life stage is viewed as containing an amorphous mass of individuals who are not distinguishable by age. In the case of the ORNL 1-D model, which illustrates one form of approach (3), the fish are assumed to be uniformly distributed by age within a life stage, no matter what pattern of recruitment into that life stage has actually occurred in the model. This assumption of uniform age distribution within a life stage seems like a conceptual disadvantage of the ORNL formulation, since in reality the assumption is rarely satisfied. Furthermore, the assumption is not required in order to use approach (3). A second conceptual disadvantage of the ORNL formulation is that three independent parameters (apparent survival probability, growth rate coefficient, and mortality rate coefficient) are used to represent two biological phenomena (time required to pass through a life stage and mortality rate within a life stage). Eraslan et al. (1976) point out that their particular formulation is not intended to imply that the three parameters for each life stage reflect three independent biological phenomena; there is an implicit overlap or redundancy among the three parameters. Their argument is that this redundancy allows flexibility in terms of adjusting parameter values to fit the simulated temporal distributions of the various life stages to the observed temporal distributions. Their point that data are not available with which to accurately estimate life stage durations and mortality rates is certainly valid.

Approach (3), whether implemented as in the ORNL 1-D model or in some other form, has some advantages, the main benefit being greatly reduced computer cost. Another benefit is that the formulation is not more precise than the data base. Since it is impossible to age larvae to the fraction of a day, it seems that keeping track of cohorts at small time intervals is inconsistent relative to the information potentially obtainable to validate the model.

Eraslan et al. (1976) assert that the inaccuracy resulting from assuming a uniform age distribution within each life stage can be effectively reduced by increasing the number of life stages to make the life-stage durations closer to the time step. Although this change could be made, it was never done in any of the reported results, and it would involve a significant increase in computer cost. In the limit, as life-stage duration approaches the time step, approach (3) reduces to the cohort approach.

In future work we would like to develop a compromise approach between the cohort and dynamic pool models that will have reasonable accuracy without computing expense. Initial development in this area has involved basing growth rates within an age class on an assumed negative exponential age distribution instead of on a uniform age distribution. This assumption seems more realistic because cohorts near the end of a life stage have been subjected to mortality for a longer time than cohorts just entering the life stage.

Compensatory Mortality

It is difficult to realistically quantify the survival history for y-o-y fish. Even if for two models the numbers of young fish entrained are equivalent, considerable variation in predicted PR has been reported. The major reason for this variation seems to be the degree of compensatory mortality incorporated in the model. If mortality of a fish is independent of (dependent on) the density of other fish, then mortality is termed density independent (density dependent or compensatory). As an example of the sensitivity of PR to density-dependent mortality, LMS (Lawler 1973, Case #17) reports a decrease of PR from 13.16 to 3.36 within a single juvenile life stage due solely to density-dependent mortality within this juvenile life stage.

In the development of a fish population model, it is important to consider the notion of compensatory mortality somewhere in the life history of the fish. The persistance of populations over time, in the face of environmental variation, is direct evidence supporting this phenomenon (Lawler 1972a). However, the life stage or life stages in which compensation occurs and the manner in which compensatory mortality acts are poorly understood. It is assumed that some compensation occurs in the y-o-y population (Lawler 1972a, 1972b; USNRC 1975), although there are currently two hypotheses for the manner in which compensatory mortality acts.

HYPOTHESIS 1 (used in ORNL 1-D model): For densities of y-o-y fish below an age-specific, critical density, survival is density independent; above this critical density, survival is density dependent in that it decreases as density increases.

HYPOTHESIS 2 (used in LMS 1-D and 2-D models): Survival is always density dependent, with a plateau of approximate, density-independent survival centered about an age-specific, equilibrium density. Survival is higher at low densities than at high densities.

Both of these hypotheses account for compensatory mortality and long-term persistence of populations. In Fig. 1, graphs for both hypotheses are presented in terms of survival vs density. The general shape of a spawner-recruit curve can be inferred for each hypothesis and is illustrated in Fig. 2. Further study of the range of possible shapes of the spawner recruit curve for each hypothesis is needed using a range of reasonable input values for the parameters in these two functions. The biological arguments presented by ORNL (USNRC 1975, pp. V-127 to V-144; Van Winkle et al. 1976) for hypothesis 1 rather than hypothesis 2, however, seem more plausible to us.

A literature review of spawner-recruit data for anadromous fish populations did not indicate any studies on striped bass with data over enough years and of decent quality to infer a spawner-recruit relationship. On the other hand, quite good information on spawner-recruit relations for other anadromous fish is available.

In a summary of studies, Larkin and McDonald (1968) state: "Shepard and Withler (1958) . . . describe for Skeena [River in British Columbia] sockeye [salmon] of ages 4 and 5 [major spawning class] a reproductive curve [spawner recruit curve with recruits being mature adults] with an almost linear ascending limb and a very precipitous descending limb generally after the Ricker type (Ricker 1954)." This study and preliminary results from other studies (Killick and Clemens 1963, Henry 1954, Anon. 1962) tend to support a reproductive curve of the Ricker type with an almost linear ascending left limb. Both hypotheses 1 and 2 appear to result in a reproductive

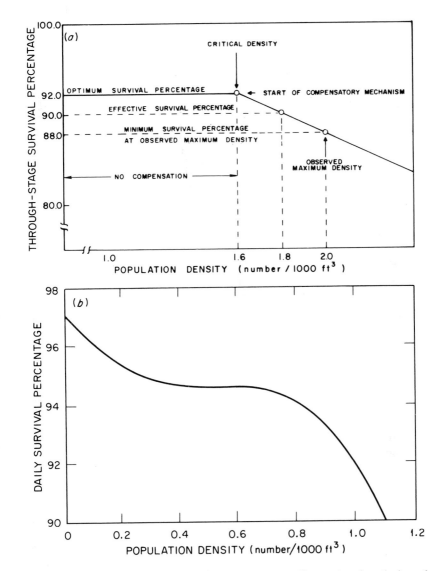

Fig. 1. Curves of survival percentage for the Juvenile I life stage based on the hypothesis of compensatory mortality used in the (a) ORNL 1-D model (USNRC 1975) and (b) LMS models (Lawler 1972b).

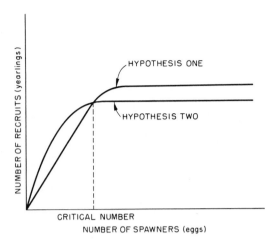

Fig. 2. Comparison of typical spawner-recruit curves resulting from density-dependent mortality based on hypothesis 1 (ORNL) and hypothesis 2 (LMS).

curve more similar to the Beverton–Holt type than the Ricker type, although the ascending left limb for hypothesis 1 will never have any curvature, while the ascending left limb for hypothesis 2 will always have some curvature.

Regardless of which hypothesis is more reasonable for striped bass, a major question about critical population values for hypothesis 1 still remains unanswered. If the critical population values (Fig. 1a) are larger than values observed in the y-o-y striped bass, then operationally, the survival equations would be density independent. Indeed, this was precisely the assumption made for the majority of the runs in initially applying the ORNL 1-D Model to the Hudson River striped bass population (USNRC 1975), and this assumption results in relatively large PR values. In some of the initial runs and in a more recent application of the model, lower critical population values were used (USNRC 1975, p. V-149; Eraslan et al. 1976). Critical density values [e.g., the equilibrium values of LMS 1-D(67)] in the middle range of life-stage population densities simulated by the model should result in substantially reduced PR. If Lawler (1972a) is correct in his statement, "It is recognized that the current condition of striped bass in the River [Hudson] may more likely be one of population growth," then high critical population values would be supported. However, more work is needed before a definitive answer can be reached.

For the presently used models we prefer hypothesis 1 over hypothesis 2 and suggest that PR predictions be presented for various critical population values. However, further work is needed using a spawner-recruit curves.

Some of the models (JHU, LMS, Completely Mixed, ORNL Summit) assume density-independent mortality rates. For these models and for the

ORNL 1-D model, when the critical population values for compensation are higher than the densities attained in running the model, there are some analytical advantages. For example, the PR predictions for passive organisms (moved entirely according to hydrodynamic equations) are completely independent of survival rates and total egg production, and are relatively insensitive to active fish migration.

We have recently developed a model simulator for comparing the y-o-y models in greater detail. This simulator can be used to investigate the sensitivity of the PR predictions to migration preference in a density-independent formulation or to look at the sensitivity of PR to changing the critical density above which density-dependent mortality will operate.

Comparison of Impact Predictions

Table 3 compares values of PR for the various models for minimum, best estimate, and maximum values of entrainment factors (f-factors; the f-factor is a measure of the fraction of fish in the neighborhood of the power plant which are actually entrained) and for varying degrees of compensation (in the LMS models). Results from the LMS and ORNL models are strikingly different. These differences are primarily due to compensation, transport defect factor and dispersion coefficients, the choice of life stages, and entrainment factors. The PR values for the three Summit models are all in quite close agreement. Compared to the PR values from models applied to the Hudson River striped bass, the Summit PR values are relatively low. This difference is due to closed-cycle cooling at the proposed Summit power plant and once-through cooling at the power plants on the Hudson River (Bowline, Danskammer, Indian Point, Lovett, and Roseton).

GENERAL LIFE CYCLE MODEL COMPARISON

General Structure

Six of the eight models reviewed include a life cycle model to translate the effect of power plant mortality into a long-term impact on the adult population and thence the fishery. These life cycle models all take the form of a modified Leslie matrix (Leslie 1945). The Leslie matrix, which consists of age-specific fecundity parameters in the first row, age-specific probabilities of survival along the subdiagonal, and zeros for all the other elements, transforms (via matrix multiplication) the age-specific population vector at time t into the age-specific population vector at time $t + 1$.

Table 3. Comparison of predictions of percentage reduction (PR) for various models

Model	Compensation[a]	Entrainment factors[b]	PR	Plants operating[c]	Source
LMS 1-D 1967	High	Best estimate	2.5	Indian Point Units 1 & 2	Table 24, Lawler (1972b)
	High	Maximum	4.0		Table 26, Lawler (1974)
LMS 1-D 1973	High	Best estimate	2.77	Indian Point Units 1, 2, & 3 and Cornwall	Table 26, Lawler (1974)
	Low	Best estimate	4.88		
LMS 2-D	High	Best estimate	1.257	Indian Point Units 1, 2, & 3	Table 36, LMS (1975)
	Low	Best estimate	3.138		
	Low	Minimum	2.44		
ORNL 1-D	None	Minimum	18.0	Bowline Unit 2, Indian Point Units 1, 2, & 3, Roseton Units 1 & 2	Table B−34, USNRC (1975)
	None	Best estimate	34.0		
	None	Maximum	42		
ORNL Summit			4.5	Summit	p. 2−1, Christensen et al. (1975)
JHU			1.0−5.0	Summit	p.17, Portner (1975)
Delmarva			0.71−5.53	Summit	Table I−3, Portner (1975)

[a]With reference to the LMS models: high compensation implies $K_0 = 0.5K_E$; low compensation implies $K_0 = 0.8K_E$.

[b]Values for the entrainment factors (f-factors) are given in the original sources.

[c]For each model the plants tabulated in this column were assumed to be operating in the model run with power plant mortality and were assumed to be not operating in the corresponding model run without power plant mortality.

In Leslie's original formulation (Leslie 1945) only the female population of a species was treated. This convention is followed in the ORNL models, while the LMS and Winter Flounder models represent the age structure of the entire adult population and compute fecundities by multiplying total adults in each age class by the female fraction in that age class.

If the population size and age distribution do not change from year to year, then the population is said to have attained an equilibrium or steady state condition. All the models reviewed here assume that without plant operation the population maintains an equilibrium condition. While some parameter values are estimated from data, the remainder are assumed or computed so that an equilibrium condition arises. One place the models differ is in which parameters are estimated from data, which are assumed, and which are calculated to give an equilibrium condition. As we show later the choice of parameter estimation method can result in widely varying parameter values, resulting in different predictions of the long-term effect of power plant mortality.

Several of the models depart from the assumption of constant (i.e., density-independent) adult mortality by making fishing mortality depend upon the size of the female population legally available to the fishery. Such an assumption tempers the long-term effect on adult populations of density-independent losses, such as entrainment and impingement losses, by reducing fishing mortality, and thus fishery yield.

Comparison of Life Cycle Model Equations

The survival-fecundity equations applicable to all the models reviewed can be represented as follows:

$$N_i(t) = N_{i-1}(t-1)S_{i-1}(t-1) \text{ for } i = 1, ..., a , \tag{3}$$

$$N_0(t) = \sum_{i=1}^{a} N_{i-1}(t-1)S_{i-1}(t-1)F_i , \tag{4}$$

where $N_i(t)$ = female population in age class i at time t, $S_i(t)$ = fraction of females surviving from age class i to $i+1$ from time t to $t+1$, F_i = fecundity (female eggs/female) of an individual in age class i, and a = total number of age classes considered in the model.

If the total population (and not just the female population) is modeled, then $N_i(t)$ represents the total population in age class i, F_i becomes total number of eggs per female and the fecundity equation becomes

$$N_0(t) = \sum_{i=1}^{a} N_{i-1}(t-1)S_{i-1}(t-1)FF_iF_i , \tag{4'}$$

where FF_i is the fraction of females in age class i.

For models having density-independent fishing mortality (LMS 1-D and Winter Flounder), survival fractions are obtained from the integral of a mortality rate accrued over the year.

$$S_i(t) = \exp(-M_i \Delta t) , \qquad (5)$$

where M_i is the mortality rate due to both fishing and natural losses and Δt is the time step (one year). Fishing mortality is included as part of total mortality M. Parameter values for fishing mortality are given explicitly in the documentation for the Winter Flounder model and can be estimated from information given in the documentation for the LMS 1-D and mixed models.

In those models with density-dependent fishing mortality (ORNL and LMS 2-D), the survival fraction S_i in age class i is given by

$$S_i(t) = NS_i FS_i(t) , \qquad (5')$$

where NS_i = conditional probability of survival from death due to natural causes for age class i, and FS_i = conditional probability of survival from death due to fishing for age class i.

The density-dependent fishery mortality model assumes that survival probability drops with increased biomass densities (Fig. 3).

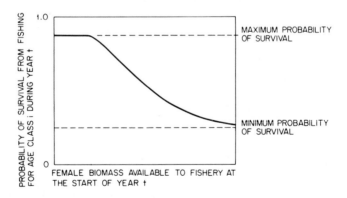

Fig. 3. The density-dependent function for fishing mortality used in the ORNL and LMS 2-D models (USNRC 1975, Fig. B-42).

Measures of Power Plant Impact

The measures of power plant impact involve comparison of population size or of fishery yield with and without power plant operation after n years of operation.

These measures include:

1. Relative number or PR in number of y-o-y (ORNL, LMS).

2. Percent reduction in total adult population (LMS, Winter Flounder).

3. Relative number of female fish available to the fishery (ORNL).

4. Relative yield or loss of yield to the fishery (ORNL, JHU).

5. Fractional change in probability of surviving fishing (ORNL).

In all the above cases "relative X" is defined as

$$\frac{X \text{ with plant operating}}{X \text{ without plant operating}},$$

and "percent reduction in X" is defined as

$$\frac{X \text{ without plant operating} - X \text{ with plant operating}}{X \text{ without plant operating}} \times 100.$$

Our impression is that the ultimate measures of long-term plant impact are the effects on fishery yield and on recruitment into the adult population (i.e., measures 1 and 4), since these effects represent future potential yield to the fishery and are also a measure of population vigor.

The Summit models (JHU and ORNL Summit) measure impact as reduction in fishery yield and try to estimate the fraction of the Atlantic coast fishery originating in the C&D canal. This approach involves estimation of such parameters as the fraction of the Delaware River and Chesapeake Bay adult stock arising in the C&D canal, the fraction of the Middle Atlantic and North Atlantic fishery spawning within the Chesapeake-Delaware drainage system, and estimates of commercial and sports catches for the various fisheries potentially impacted.

The JHU and ORNL Summit models each compute annual loss in fishery yield due to plant operation by multiplying the estimated entrainment fraction by the fractional contribution of the impacted fish stock to the fishery of interest. The ORNL Summit model adds to this approach by multiplying the above product by a scaling factor. This scaling factor is obtained by running the ORNL life cycle model with a given loss in yearlings, obtaining the resulting long-term annual loss in yield, and calculating the ratio: annual loss in yield/annual loss in yearlings. The JHU model does not utilize a life cycle model and therefore implicitly assumes a scaling factor of 1.0.

Comparison of Parameter Values

The major differences in formulation between the life cycle models have already been mentioned. Here we will focus on data sources supporting the various formulations and on striking differences in parameter values.

LMS assumes in all their life cycle models that the female fraction is 0.5 at age 1 and increases linearly to 0.7 at age 11. Evidence cited includes a sampling of spawning adults in the Roanoke River by Trent and Hassler (1968) and in the Hudson River by Texas Instruments (1973).

The Roanoke River study did show the female fraction as increasing with age. This increase is probably due to differential fishing mortality during the spawning migration up the Roanoke River on two- and three-year-old striped bass, since in North Carolina males become sexually mature at age 2 and 3 whereas females do not become sexually mature until age 3, 4, and 5. The Hudson River stock is not subject primarily to a river fishery, and thus the earlier spawning of males would not be expected to affect sex ratios as strongly. The TI data used by LMS involved one-year and eleven-year old fish only. The sample of eleven-year olds was, in our opinion, too small to justify use of female fractions increasing with age.

Our recommendation is to use a constant sex ratio, until studies definitely confirm age-varying sex ratios for striped bass. The assumption of unequal sex ratios implies that there are sex-selective sources of mortality, since one sex must die more than another to have sex ratio change with age.

Both the Winter Flounder and ORNL models estimate fecundity using a regression equation of eggs per female on weight, which in turn is regressed to length, which is regressed to age. Thus, fecundities depend only on the age, and a data set for age-specific fecundities could just as easily have been read in directly, since fecundities do not change during a model run. The only advantage of the regression relationship is that it supplies a framework for integrating new fecundity, length, and weight data simply by changing regression coefficients, although there may be no biological rationale for such a parameter change. The parameter values for the density-dependent, fishing mortality function used in the ORNL and LMS 2-D models are identical, except for the parameter RATIO which is 2.37 for LMS and 2.0, 3.0, or 5.0 for ORNL. Estimates of this parameter were based on commercial fisheries landings from the Hudson River, New York, Middle Atlantic, and North Atlantic (Koo 1970). Based on Hudson River and New York landings data for over 30 years, values of RATIO less than 3.0 seem low.

Several other parameters in this model (i.e., PMAX, PMIN and D) are based on recaptures of tagged fish by commercial and sports fishermen. These studies are fraught with experimental uncertainties, such as increased catchability of tagged fish and nonreturn of tagged fish. Also tag-recapture data are only peripherally related to the problem of estimating maximum and minimum levels of fishery exploitation. Worse yet, model predictions of yield are most sensitive to just these parameters which are known with so little certainty (USNRC 1975, Appendix B). Since the density-dependent fishing function makes yield so dependent upon parameters estimated from scanty data, it seems reasonable to question why the function should be used in the first place. Unfortunately, adequate data are not available to estimate parameter values for density-dependent functions for natural mortality or fecundity either.

Another phenomenon that makes the use of a deterministic Leslie model seem questionable is the great variability in year class strength, with occasional dominant year classes influencing the fishery for several years (Raney 1952, Koo 1970). This variability is not predicted by a deterministic model in which fecundity depends only on the adult population (not that peak years could be predicted by any other model at present).

We think the use of a density-dependent function for fishing mortality should depend upon a thorough study of the fishery. Some fisheries are not species selective and would not be expected to be density dependent. However, there are other fisheries, for example a single-species sports fishery, that should be density dependent since lowered fish stock would discourage fishing effort. The striped bass sports fishery is, from what we have read, probably of the latter variety, while the striped bass commercial fishery is probably more density independent. Thus, the relative importance of the density-dependent and density-independent parts of a fishing mortality function for striped bass might depend on the ratio of sport to commercial catch. Estimates of this ratio (Portner 1974, Christensen et al. 1975) range from 1:1 for Chesapeake Bay to 9:1 for the North Atlantic.

A major difference between the life cycle models is in how equilibrium conditions and survival estimates are obtained. The ORNL model bases estimates for $NS_1 - NS_{15}$ (survival from natural mortality) on a tag-recapture study (Sommani 1972) of the California striped bass population, which included data for three-year-old and older fish. Estimates of fishing mortality at equilibrium are calculated assuming that the female biomass available to the fishery at equilibrium is at a specified point in the density-dependent region of the fishing mortality function (Fig. 3). Natural survivals for age classes 1 and 2 were made somewhat smaller than Sommani's estimates (0.8) to account for hypothesized higher natural mortality in younger age classes. Specifying the number of eggs at equilibrium is then sufficient to generate a unique, equilibrium age distribution and a unique survival fraction of eggs to one-year-old adults (S_0). In short, S_0 is calculated after all other parameters are estimated.

The LMS 1-D model takes S_0 from the young-of-the-year submodel, assumes values for $NS_1 - NS_3$, takes average egg production and fecundities from Texas Instruments (1973) data, and then estimates $S_4 - S_{15}$ assuming they are all equal. In applying the 2-D model, LMS uses the same type of estimates for fecundity, egg production, S_0, and $NS_1 - NS_3$ as in the 1-D model (LMS 1975). Then, assuming an initial adult population, they compute fishing mortalities. Finally, they compute estimates of natural survival, using a Newton-Raphson technique, such that the model egg production equals the estimated egg production based on observed egg standing crops. The new adult population is then compared with the initial population, and if they differ, the new population is used to compute a new fishing mortality, and so on until the process converges on a unique equilibrium population.

Comparison of Impact Predictions

Survival for age classes 1—3 is quite a bit lower in the LMS 1-D model than the ORNL model. Having low survival in younger age classes, coupled with density-dependent fishing mortality for older age classes, can result in quite a different value for relative yield compared to model runs having higher survival values for the younger age classes. For example, at the end of 40 years with PR = 10% for each year, the relative yields from runs with low (0.4), average (0.6), and high (0.8) survival probabilities for age classes 1—3 were 0.78, 0.83, and 0.86, respectively (USNRC 1975, Appendix G). LMS assumed survival probabilities for age classes 1—3 of 0.16 (Lawler 1972b) and 0.30 (Lawler 1974, LMS 1975). Thus, the LMS 1-D model, without compensation in the y-o-y submodel, would predict even greater reduction in relative yield than the ORNL model. However, this difference does not show in the results presented by LMS, because they did not present relative yield as an output from their model. Sommani (1972, p. 91) used an annual probability of survival of 0.83 for age classes 1 and 2 and 0.64 to 0.71 for age class 3 in his study of the population dynamics of the California striped bass population. The difference in survival probabilities for the first three age classes is the major reason why the ORNL equilibrium adult population size is two orders of magnitude greater than the LMS value.

Table 4 shows a comparison of the long-term impacts predicted by the various life cycle models. The various Hudson River predictions are not comparable, since the values for PR used by LMS are in the 1 to 5% range, while the ORNL values are in the 10 to 50% range.

The JHU and ORNL Summit predictions of annual reduction in yield are quite similar (Table 4), being based on similar data sets to estimate the total catch of the Atlantic coast fishery and the relative proportion of the catch coming from stock spawned in the C&D canal. The ORNL Summit model uses a life cycle submodel to compute a scaling factor for yield, which converts PR values into values for annual depletion of yield after long-term plant operation.

Our comparison of life cycle models has raised a number of questions, such as the sensitivity of model results to parameter estimation techniques or to density-dependent vs density-independent fishing mortality, that cannot be answered by information given in the model reports. We decided to adapt the ORNL life cycle model to a more generalized framework that allows either density-dependent or density-independent fishing mortality. Also, our generalized life cycle model allows fecundities to be read in directly as data or to

Table 4. Comparison of life cycle model impact predictions

Model	PR[a]	y-o-y compensation	PR in total adults (Number of years)			PR in 1-year-old fish (Number of years)			Source
			5	10		5	10		
LMS 1-D(67)	2.07	High	2.52	3.93		2.71	4.01		Table 1,
	3.42	Low	4.93	9.74		5.68	7.48		Lawler (1973)
	3.13	None	4.82	11.39		5.55	12.00		
			7	10	40	7	10	40	
LMS 2-D	1.21	High	1.29	1.64	2.18	1.33	1.68	2.18	Tables 36 and 37,
	1.26	High	1.34	1.70	2.26	1.38	1.75	2.26	Lawler (1975)
	2.44	Low	2.64	3.70	6.82	2.81	3.91	6.99	
	3.14	Low	3.46	4.86	8.95	3.61	5.03	8.99	
	4.47	Low	4.93	6.88	12.42	5.13	7.11	12.46	

Model	PR	y-o-y compensation	Relative yield (Number of years)				PR in 1-year-old fish (Number of years)				Source
			5	10	20	40	5	10	20	40	
ORNL	10	None	0.96	0.90	0.85	0.83	10	14	17	18	Figs. B−47
	25	None	0.88	0.75	0.64	0.60	25	33	38	42	and B−48,
	50	None	0.78	0.52	0.35	0.26	50	62	70	75	USNRC (1975)

Table 4 (continued)

Model	PR	y-o-y compensation	PR in annual yield	Source
ORNL	0.5	None	0.03	Tables 8.1, 8.2, and 8.3; Christensen et al. (1975)
Summit	2.75	None	0.77	
	5.0	None	3.7	
JHU	2.5	None	0.45	Portner (1975)
	5.0	None	1.7	

Model	PR	y-o-y compensation	PR in total adults	Source
			35 years	
Winter	1.0	Best estimate	6.0	Sissenwine et al. (1974)
Flounder	1.0	None	9.0	

[a]PR = percentage reduction.

be computed indirectly by regression, and it allows either the female or the total adult population to be represented in the model.

We have used this model so far to corroborate equilibrium population distributions for the ORNL model, the LMS Completely Mixed and 1-D models, and the Winter Flounder model. We are also running the models with equivalent scenarios (e.g., 20 years of plant operation and 20 years of subsequent "recovery") and equivalent PR values in order to generate comparable predictions of yield and other measures of plant impact.

Future plans for our simulator include sensitivity analyses and testing of different hypotheses, for example, partly density-dependent and partly density-independent fishing mortality, depending upon the type of fishery. Another projected use of the simulator is to analyze the survival-fecundity matrices arising from the density-dependent fishing function using eigen-analysis. The purpose of this analysis will be to see how rapidly an asymptotic steady-state condition occurs with different scenarios of plant operation (assuming the system will behave in the future according to the age-specific fecundity and survival parameters in the matrix at present). This analysis involves assuming that the survival fractions based on density-dependent fishing mortality are actually constant coefficients in the Leslie matrix.

CONCLUSIONS

Our continuing review of impact models is focusing on major differences between the models with respect to biological assumptions, equations, and parameter values. Our focus is on locating differences between the models which result in different values of PR, long-term fishery yield, or first-year adult population size. In cases where differences in model output are "large," we try to indicate why. Also, in some cases we suggest our preference between existing approaches.

The major differences in biological assumptions in the striped bass y-o-y models are the choice of life stage durations and the inclusion of compensatory mortality at both high and low fish densities. While literature evidence seems to support compensation occurring in many species, the evidence also indicates that compensatory mortality is more likely to be a phenomenon that occurs at high population densities, as in the ORNL model. Spawner-recruit curves for salmon support a linear ascending left limb as obtained with the ORNL model. Our recommendation is to use compensation functions of the form used by ORNL and to focus on determining the critical densities above which mortality rates begin to increase. It may be possible to estimate these critical densities using fisheries data and other sampling data, coupled with a stock-recruitment or stock-progeny analysis. Perhaps the scientific community can become more involved in hypothesizing and testing possible biological mechanisms for density-dependent mortality in commercially and recreationally important fish populations, so advances can be made both in pure and applied science. Certainly, compensation is the one most important factor in accounting for the differences between the LMS and ORNL model predictions (see Table 4).

In the striped bass life cycle models, major differences in predictions of yield and population reduction result from using density-dependent vs density-independent fishing mortality and from using different values for the probability of natural survival of fish in age classes 1–3.

There are unresolved conceptual enigmas in the aging and mortality equations in the ORNL 1-D model, in the application of the density-dependent function for fishing mortality based on adult female stock in the LMS 2-D model, and in inclusion of entrainment loss as a loss of spawned eggs in the Ricker relationship used in the Winter Flounder model.

Our one major complaint has been the opacity and incompleteness of the documentation for many of the models. Computer codes examined were, in general, not designed for portability. We would like to see impact models subscribe to set standards of documentation and to have certain sensitivity studies run compulsorily, so that straightforward model comparisons could be accomplished relatively rapidly.

While this review is far more comprehensive than others (Lewis 1975, Wallace 1975), it only touches the tip of the iceberg. The job has been far more demanding and time consuming than we originally expected. Our hope is that a period of evaluation of the present models can ensue, in which we cooperatively decide what elements a model should have and what tests and sensitivity analyses would be useful. Such a review could address questions as: What constitutes an adequate data set? How might a model be used to complement a monitoring program? How much and what kind of model documentation is needed? How can we bring investigators to work on research suggested by our models, so that answers to fundamental questions, such as the form of the compensatory mortality function, can be based on a sounder scientific footing? This conference, to my mind, represents a beginning in this effort. To proceed further we need a cooperative spirit, portability of information, and intellectual honesty and objectivity.

ACKNOWLEDGMENTS

Several people besides the authors have also been indispensible to this work. Stan Clark has worked with us deciphering the life cycle model code we adapted for our life cycle simulator and coding and documenting the y-o-y simulator in his inimitable, structured programming style. Kandy Kroll has always been ready with a smile to type a rough draft or hunt down a reference, and Bernadette Hunter has always been ready to tackle a table poorly organized and indecipherable. Mary Rokes worked under some pressure to prepare the final manuscript. We really appreciate her effort.

We also appreciate the cooperation from investigators who developed and applied the models we are reviewing, especially Webb Van Winkle and Sig Christensen of the ORNL staff. Arsev Eraslan of the University of Tennessee,

Saul Saila and Kurt Hess (formerly) of the University of Rhode Island, Tom Englert, John Lawler, and Fehmi Aydin of Lawler, Matusky & Skelly Engineers, and Ken Warsh and Ed Portner of The Johns Hopkins University, Applied Physics Laboratory, have all been very helpful in giving us their time and information. We also appreciate the care taken by Webb Van Winkle and Tom Englert in reading and editing this manuscript.

LITERATURE CITED

Anon. 1962. The exploitation, scientific investigation, and management of salmon (genus *Oncorhynchus*) stocks on the Pacific coast of Canada in relation to the abstention provisions of the North Pacific Fisheries Convention. Bull. Int. North Pacif. Fish. Comm. No. 9. 112 p.

Christensen, S. W., W. Van Winkle, and P. C. Cota. 1975. Effect of Summit Power Station on striped bass populations. Testimony before the Atomic Safety and Licensing Board in the matter of Summit Power Station, Units 1 and 2, USAEC Docket Nos. 50–450 and 50–451, March 1975.

Clark, J. R. 1972. Effects of Indian Point Units 1 and 2 on Hudson River aquatic life. Testimony before the Atomic Safety and Licensing Board in the matter of Indian Point Unit No. 2, USAEC Docket No. 50–247, October 30, 1972.

Eraslan, A. H., W. Van Winkle, R. D. Sharp, S. W. Christensen, C. P. Goodyear, R. M. Rush, and W. Fulkerson. 1976. A computer simulation model for the striped bass young-of-the-year population in the Hudson River. ORNL/NUREG-8, ESD-766, Oak Ridge National Laboratory, Oak Ridge, Tennessee. 208 p.

Henry, D. A. 1954. Age and growth study of Tillamook Bay chum salmon (*Oncorhynchusketa*). Contr. Fish Comm., Ore., No. 19. 28 p.

Killick, S. R., and W. A. Clemens. 1963. The age, sex ratio and size of Fraser River sockeye salmon 1915–1960. Bull. Int. Pacif. Salm. Fish. Comm. No. 14. 140 p.

Koo, T. S. Y. 1970. The striped bass fishery in the Atlantic states. Chesapeake Sci. **11**:73–93.

Larkin, P. A., and J. G. McDonald. 1968. Factors in the population biology of the sockeye salmon in the Skeena River. J. Anim. Ecol. **37**:229–258.

Lawler, J. P. 1972a. The effect of entrainment at Indian Point on the population of the Hudson River striped bass. Testimony before the Atomic Safety and Licensing Board in the matter of Indian Point Unit No. 2, USAEC Docket No. 50–247, April 5, 1972.

————. 1972b. Effect of entrainment and impingement at Indian Point on the population of the Hudson River striped bass. Testimony before the Atomic Safety and Licensing Board in the matter of Indian Point Unit No. 2, USAEC Docket No. 50–247, October 30, 1972.

————. 1973. Response to questions on the sensitivity of the model presented in the testimony of October 30, 1972, on the effect of entrainment and impingement at Indian Point on the population of the Hudson River striped bass. Testimony before the Atomic Safety and Licensing Board in the matter of Indian Point Unit No. 2, USAEC Docket No. 50–247, February 5, 1973.

————. 1974. Effect of entrainment and impingement at Cornwall on the Hudson River striped bass population. Testimony before the Federal Power Commission in the matter of Cornwall, USFPC Project No. 2338, October 1974.

LMS (Lawler, Matusky & Skelly Engineers). 1975. Report on development of a real-time, two-dimensional model of the Hudson River striped bass population. LMS Project No. 115–149. 71 p.

Leslie, P. H. 1945. The use of matrices in certain population mathematics. Biometrika **33**:183–212.

Lewis, B. G. 1975. Some aspects of aquatic ecosystem models, and the role of such models in the choice of a cooling system for steam electric power plants. The Social Consequences of Waste Heat Discharge Alternatives. Progress Report FY 1975. Argonne National Laboratory, Argonne, Illinois.

Mansueti, R. J. 1961. Age, growth and movements of the striped bass, Roccus saxatilis, taken in size-selective fishing gear in Maryland. Chesapeake Sci. **2**:9–36.

Portner, E. M. 1975. Testimony on striped bass entrainment by Summit Power Station. Testimony before the Atomic Safety and Licensing Board in the matter of Summit Power Station, Units 1 and 2, USAEC Docket Nos. 50–450 and 50–451, March 14, 1975.

Raney, E. C. 1952. The life history of the striped bass. Bull. Bingham Oceanogr. Collect. **14**:5–97.

Ricker, W. E. 1954. Stock and recruitment. J. Fish. Res. Board Can. **11**:559–623.

————. 1958. Handbook of computations for biological statistics of fish populations. Fish. Res. Board. Can., Bull. No. 119. 300 p.

Shepard, M. P., and F. C. Withler. 1958. Spawning stock size and resultant production for Skeena sockeye. J. Fish. Res. Board Can. **15**:1007–1025.

Sissenwine, M. P., D. W. Hess, and S. B. Saila. 1974. Interim report on evaluating the effect of power plant entrainment on populations near Millstone Point, Connecticut. S. B. Saila, Project Director, Marine Experiment Station, Graduate School of Oceanography, University of Rhode Island. MES-NUS Co. Rep. No. 3.

Sommani, P. 1972. A study on the population dynamics of striped bass in the San Francisco Bay Estuary. Ph.D. Thesis, Univ. Washington, Seattle, Washington. 133 p.

TI (Texas Instruments, Inc.). 1973. First Annual Report. Hudson River Ecological Study in the Area of Indian Point. Prepared for Consolidated Edison Co., April 1973.

Trent, L., and W. W. Hassler. 1968. Gill net selection, migration, size and age composition, sex ratio, harvest efficiency, and management of striped bass in the Roanoke River, North Carolina. Chesapeake Sci. 9(4):217–232.

UEC (United Engineers & Constructors). 1975. Applicant's supplemental testimony (mathematical model) on entrainment of striped bass by Summit Power Station. Testimony before the Atomic Safety and Licensing Board in the matter of Summit Power Station, Units 1 and 2, USAEC Docket Nos. 50–450 and 50–451, March 21, 1975.

USNRC (U.S. Nuclear Regulatory Commission). 1975. Final environmental statement related to operation of Indian Point Nuclear Generating Station, Unit No. 3, Docket No. 50–286, Vols. I and II, February 1975.

Van Winkle, W., B. W. Rust, C. P. Goodyear, S. R. Blum, and P. Thall. 1974. A striped bass population model and computer programs. ORNL/TM-4578, ESD-643, Oak Ridge National Laboratory, Oak Ridge, Tennessee. 200 p.

Van Winkle, W., S. W. Christensen, and G. Kauffman. 1976. Critique and sensitivity analysis of the compensation function used in the LMS Hudson River striped bass models. ORNL/TM-5437, ESD-944, Oak Ridge National Laboratory, Oak Ridge, Tennessee. 100 p.

Wallace, D. N. 1975. A critical comparison of the biological assumptions of Hudson River striped bass models and field survey data. Trans. Am. Fish. Soc. 4:710–717.

Warsh, K. L. 1975. Hydrological-biological models of the impact of entrainment of spawn of the striped bass (*Morone saxatilis*) in proposed power plants at two areas in the Upper Chesapeake Bay. Report No. JHU, PPSE-T-1, Johns Hopkins University Applied Physics Laboratory. 94 p.

Part VI

Conclusions and Recommendations

Conclusions and Recommendations for Assessing the Population-Level Effects of Power Plant Exploitation: The Optimist, the Pessimist, and the Realist*

Webster Van Winkle

Environmental Sciences Division
Oak Ridge National Laboratory
Oak Ridge, Tennessee

ABSTRACT

Two questions are addressed: "To what extent can we presently assess the population-level effects of power plant exploitation?" and "What can we do to improve these assessments?" Each of these questions is addressed from the point of view of an optimist, a pessimist, and a realist. The realist adopts a flexible blend of the optimist's and pessimist's positions. He envisions a continuum of possible assessment situations. At one extreme, population-level assessments are not feasible because of the dominating influence of the pessimist's unavoidable limitations. At the other extreme, population-level assessments are both necessary and possible using the tools proposed by the optimist.

Key words: case histories, compensation, generation time, monitoring programs, optimist, pessimist, population models, realist, research proposals, stock-recruitment relationships, time frame for decision making, unavoidable limitations, natural variation

INTRODUCTION

The purpose of this conference was to evaluate our ability to assess the population-level effects of power plant exploitation. The two questions I will address are: "To what extent can we presently assess the population-level effects of power plant exploitation?" and "What can we do to improve these assessments?" I will approach each of these two questions from three points of view: that of the optimist, that of the pessimist, and finally, that of the

---

*Research sponsored by the Energy Research and Development Administration under contract with Union Carbide Corporation. ESD Publication No. 1055, Environmental Sciences Division, Oak Ridge National Laboratory.

realist. I am primarily interested in the realist, but it will prove valuable to arrive at the point of view of the realist by first considering the views of the optimist and the pessimist.

THE OPTIMIST

Let me start by identifying the optimist. He is a scientist concerned with generic or site- and species-specific research which he believes is essential for assessing the effects of power-plant-induced mortality on fish populations. His research or study proposal emphasizes the positive. In particular, he recommends what can be done to improve the situation in one or more of the areas 1–4 in Fig. 1. With respect to my two questions, the optimist does not attempt to give a general or overall answer to the first question (How well can we do at present?). Rather, he points to some narrow and specific limitation in one of these four areas — namely, the limitation he will address in his research. His proposed research answers my second question (What can we do to improve the situation?).

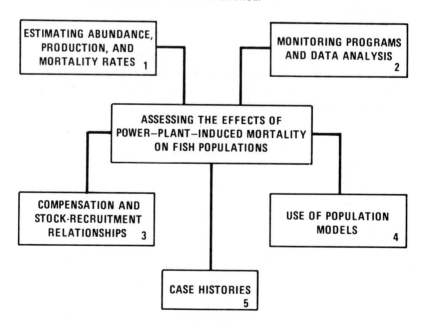

AREAS OF ACTIVITY GERMANE TO ADDRESSING
THE TOPIC OF THIS CONFERENCE.

| ESTIMATING ABUNDANCE, PRODUCTION, AND MORTALITY RATES 1 | MONITORING PROGRAMS AND DATA ANALYSIS 2 |

ASSESSING THE EFFECTS OF
POWER–PLANT–INDUCED MORTALITY
ON FISH POPULATIONS

| COMPENSATION AND STOCK-RECRUITMENT RELATIONSHIPS 3 | USE OF POPULATION MODELS 4 |

CASE HISTORIES
5

Fig. 1. Five areas of activity, each of which is germane to assessing the effects of power-plant-induced mortality on fish populations.

By way of illustrating the optimist's narrow or piecemeal approach to answering my two questions, I consider each of the four areas in Fig. 1. In each area I identify two research topics and briefly indicate why the optimist believes that research on this topic will improve the scientific basis for assessing the population-level effects of power plant exploitation. The individual papers in this book have been organized according to these areas, and although I have not attempted to exhaust all possibilities, I have identified at least one paper for each research topic. Further details and additional research topics may be found in the individual papers.

Estimating Abundance, Production, and Mortality Rates, Especially for Young Fish

Research in this area might emphasize one of the following:

1. Collection of fish abundance data. Rationale: An impingement exploitation rate cannot be estimated without an estimate of the initial size of the impingeable population (Campbell et al., *this volume*). Estimating production and mortality rates requires estimates of abundance at two or more times (Polgar, *this volume*).

2. Evaluation of collection-gear efficiency (Kjelson, *this volume*). Rationale: Estimates of the relative efficiency of collection gear are essential in order to compare data collected with different gear types. Estimates of the absolute efficiency are needed to estimate absolute abundance.

Monitoring Programs and Data Analysis

Research in area 2 of Fig. 1 might emphasize one of the following:

1. General design of monitoring programs (Thomas, *this volume*). Rationale: We have already learned several lessons from existing monitoring programs. These lessons now need to be reflected by modifying the design of future monitoring programs. In particular, for each assessment, specific and testable hypotheses must be formulated. Monitoring programs must be designed to test for impacts and not merely to describe the ecosystem.

2. Design of specific sampling programs (McCaughran, *this volume*). Rationale: Greater attention during the design phase of a sampling program must be given to the number of samples and to the distribution of sampling effort both in space and time. This increased attention is required in order to be able to detect a specified difference or to achieve a desired level of precision.

Compensation and Stock-Recruitment Relationships

Research in this area might include the following:

1. Field and laboratory studies (Chadwick et al., *this volume*). Rationale: Field and laboratory studies of the timing and mechanisms of compensation, especially during the crucial and poorly understood larval life stage, will increase our understanding of the role of density-dependent factors in regulating population size.

2. Model studies (Christensen et al., *this volume*; Goodyear, *this volume*). Rationale: Given the impossibility of a priori directly measuring compensatory capabilities and compensatory reserve, model studies can be used profitably to investigate the effects of the timing of compensation and the degree of compensation required to offset a specified incremental mortality.

Use of Population Models

Although this fourth area overlaps with the third area, two additional research topics are:

1. Density-independent and density-dependent Leslie population models (Horst, *this volume*; Saila and Lorda, *this volume*). Rationale: Applications of the Leslie model, both without and with compensation, can be of value in forecasting multi-year effects at the population level.

2. Evaluation of population models (Swartzman, *this volume*). Rationale: Thorough evaluation, including sensitivity analyses, of models used in assessing the population-level effects of incremental mortality of young fish is essential in order to establish the degree of uncertainty associated with forecasts from these models.

In summary, the optimist believes that results from research in each of these four areas will improve, albeit in an incremental fashion, the scientific basis for assessing population-level effects of power plant exploitation. Conceiving and performing this research are both a responsibility and a reward (not to mention a livelihood) for our optimist.

THE PESSIMIST

As with the optimist, let me start by identifying the pessimist. He is a scientist reviewing an optimist's research proposal or a manager deciding whether to fund an optimist's research proposal.

The pessimist's answer to my first question (How well can we do at present?) is that the "science" of assessing the effects of power-plant-induced mortality on fish populations is fraught with uncertainty and is of limited value in guiding decision makers. The pessimist's answer to my second question (What can we do to improve the situation?) is that there is little we can do to improve this unfortunate state of affairs because of unavoidable limitations imposed by both the fish populations and the decision-making process. The pessimist's unavoidable limitations include (1) an inability to quantify compensatory capabilities; (2) large, uncontrollable, and unpredictable natural variations in population size; (3) a time frame for decision making that is generally less than the generation time of the fish populations of interest; and (4) uncontrollable effects from other sources of impact. I will elaborate briefly on each of these four unavoidable limitations.

Inability to Quantify Compensatory Capabilities

With respect to this first unavoidable limitation, the pessimist asks pointedly, "Can we determine a priori whether a given level of incremental exploitation will or will not significantly reduce stock size, possibly to below the replacement line?" The pessimist's answer is "No, not with an acceptable degree of certainty even in the best of circumstances." Limiting himself to this volume, he points to the case histories for the salmon stocks in the Columbia River (Salo and Stober, *this volume*) and for the striped bass population in the Sacramento–San Joaquin Estuary (Chadwick et al., *this volume*).* The pessimist next asks "Is the best we can do to (a) establish extensive, long-term monitoring programs, (b) allow (with warnings) the level of total exploitation to increase until a posteriori we have hard data indicating over-exploitation, and then (c) attempt to reduce the level of total exploitation and/or establish a stocking program?" The pessimist's answer is "Yes, that is the best we can do," and again he points to the case histories for salmon (Salo and Stober, *this volume*) and striped bass (Chadwick et al, *this volume*).

To a major extent the fundamental problem is that compensatory capabilities and "compensatory reserve" are not amenable to direct measurement. In fact, whether or not these capabilities and a value for maximum sustainable yield can even be inferred with reasonable accuracy from a long time series of stock-recruitment data seems less certain now than it did ten

---

*As pointed out by McFadden (*this volume*) and earlier by ORNL (USNRC 1975, p. V-164+), there is a need to distinguish between exploitation impacts (e.g., fishing, entrainment, and impingement) and impacts involving habitat destruction (e.g., construction of a dam). The case history of the Columbia River salmon clearly involves both types of impacts.

years ago (Larkin 1977). The pessimist argues that the inability to directly quantify compensatory capabilities imposes a major limitation on the ability to forecast population effects of power plant exploitation.

Large, Uncontrollable, and Unpredictable
Natural Variations in Population Size

Large, uncontrollable, and unpredictable natural variations in population size, particularly due to the occurrence of dominant year classes, has been a long-standing source of frustration for quantitative fishery biologists. The pessimist's position is that although multiple regression and covariance analyses may help us account for such natural variations, these statistical techniques will not save us, since the variation of the remaining residuals generally is still large (Leggett, *this volume*). Especially with short time series (e.g., less than ten years), natural variations may mask a real reduction in population size due to exploitation. Alternatively, a short-term trend due solely to natural variations may suggest a real reduction in population size due to exploitation, when in fact there is none. Thus, the inability to account for or to predict the natural variation in a time series of annual population estimates as completely as desired imposes a second limitation on the ability to assess population effects of power plant exploitation in terms of both interpreting case histories and forecasting effects.

Time Frame for Decision Making vs Generation Time for Fish

The pessimist's third unavoidable limitation is that the generation times for the fish species of primary interest are as long as or longer than the time frame for decision making. For environmental decisions relating to siting, construction, and operation of a power plant, one to five years appears to be a typical time frame. By way of comparison and with reference to fish populations highlighted at this conference, the number of years required for the majority of females to become sexually mature is five years for the Connecticut River American shad population (Leggett, *this volume*), generally four years for the various salmon stocks in the Columbia River (Salo and Stober, *this volume*), and four to six years for striped bass (Chadwick et al., Christensen et al., Goodyear, and Saila and Lorda, *all in this volume*).

Two generic examples of how sound decision making can be hindered by these long generation times illustrate the pessimist's concern. First, stocking is commonly proposed as a mitigation measure to offset the effects of entrainment and impingement mortality. A reasonable a priori criterion (i.e., a criterion to be satisfied before a decision is made) for accepting stocking as a feasible mitigation measure is that stocked females actually reproduce. To adequately satisfy this criterion requires a minimum of one generation from the first attempts at stocking. The second example involves conditional decisions, by which I mean a decision to allow a plant to operate in a given manner for a specified number of years while operational data are collected. Then, at the end of the specified number of years, the initial decision is reevaluated in light of the new and old data combined. The pessimist contends that the interim period of operation must extend over preferably more than one generation — for example, more than four to six years for shad, salmon, and striped bass. The case histories presented by Leggett, Salo and Stober, and Chadwick et al. (*all in this volume*) are greatly strengthened by having a time series of population estimates that extend over several generations.

Uncontrollable Effects from Other Sources of Impact

The pessimist's fourth and final unavoidable limitation is that any assessment of the population-level effects of power plant exploitation will be hindered by the simultaneous occurrence of other impacts, which may be both uncontrollable and unmonitored. These other impacts may involve exploitation, habitat alteration, or both. According to the pessimist, these other impacts can greatly increase the uncertainty involved in any population-level assessment, and in fact, they can render useless an assessment that focuses on a single source of impact (Chadwick et al., *this volume*; Salo and Stober, *this volume*).

In summary, the pessimist's position is that we cannot do very well at present in assessing the population-level effects of power plant exploitation and that we cannot do much to improve the situation because of unavoidable limitations.

THE REALIST

My characterizations of the optimist and pessimist have been deliberately one-sided in order to highlight the strong points of their respective positions.

The realist recognizes that in any situation the pessimist's unavoidable limitations are real, will continue to plague us, and cannot be ignored by scientists or decision makers. These limitations, in turn, limit the rigor with which we can assess impacts at the population level in even the best of situations, and the optimist does a disservice to decision makers and the field of assessing population effects by ignoring these limitations. In fact, population-level analyses may be appropriate only when considerable background data have been collected (e.g., Chadwick et al., Christensen et al.,

Horst, Salo and Stober, and Swartzman, *all in this volume*). This situation, of course, emphasizes the need for systematic and long-term studies of major fish stocks by appropriate state, federal, or international agencies.

The realist also recognizes that when the necessary background data are available, the unavoidable limitations of the pessimist should not paralyze our efforts to perform population-level assessments and to improve the scientific basis for such assessments. As the years go by, we can improve this scientific basis (in an incremental fashion) in two general ways. First, we can perform studies on selected research topics as suggested by the optimist. Second, we can accumulate well-documented, long-term case histories covering a range of species and habitat types.

In summary, the realist adopts a flexible blend of the optimist's and the pessimist's positions. The realist envisions a continuum of possible assessment situations. At one extreme the realist judges that a population-level assessment is not feasible due to the dominating influence of the pessimist's unavoidable limitations. (Alternatively, the realist may judge that a population-level assessment is not necessary because there is sufficient evidence already available to conclude that exploitation or habitat destruction due to power plant operation will not be a problem for any fish population.) At the other extreme, the realist judges that a population-level assessment is both feasible and necessary. The unavoidable limitations of the pessimist are still a problem, but now they are counterbalanced by the availability of extensive background data resulting from a systematic and long-term study of a population that is of major interest to the sport and/or commercial fishermen. The realist concludes that this flexible blend of the optimist's and pessimist's positions provides a realistic framework for assessing population-level effects (including when to attempt such an assessment), for establishing research priorities, and for providing guidance in the decision-making process.

LITERATURE CITED

Larkin, P. A. 1977. An epitaph for the concept of maximum sustained yield. Trans. Am. Fish Soc. **106**:1–11.

USNRC (U.S. Nuclear Regulatory Commission). 1975. Final environmental statement related to operation of Indian Point Nuclear Generating Station, Unit No. 3, USNRC Docket No. 50-286, Vol. I, February 1975.

SUBJECT INDEX

Abundance estimates, 71, 110, 365. *See also* Sampling methods
    Accuracy of, 71
    Methods (Acoustical, Area-density, Direct counts, Mark-recapture), 71
    Minimum-variance unbiased estimator of average density, 110
    Precision of, 71
    Young-of-the-year fish, 46, 91

Abundance indices, 91

Abundance of adults, 91
    Angler surveys, 91
    Catch and exploitation estimates, 91
    Indices derived from sport fishing (Creel-census, Party-boat catch-per-unit-effort, Party-boat catch records, Postcard surveys), 91
    Petersen mark-recapture estimates, 91
    Sampling bias (Age, Sex), 91

Abundance of young striped bass, 91
    Egg and larva surveys, 91
    Egg and post-larva sampling (Pump, Plankton net), 91
    Fall and winter midwater trawl survey, 91
    Summer tow-net survey, 91
    38-mm index, 18

Age
    Exponential distribution, 110
    Stable distribution vector, 297, 311
    Structure, 256
    Uniform distribution, 110

Age-length key, 256

Allowable impact, 36

American shad (*Alosa sapidissima*), 3

Analysis of covariance, 243

Analysis of variance, 128, 229, 243
    Normal distribution, 229
    Number of replicates, 71, 229, 243
    Number of stations, 229
    Transformations, 229, 243
    Unequal variances, 229

Atlantic menhaden (*Brevoortia tyrannus*), 71, 128, 297

Atlantic silverside (*Menidia menidia*), 297

Bertalanffy growth curve, 256

Beverton-Holt, 185, 196

Biases, 71, 128

Black crappie (*Pomoxis nigromaculatus*), 46

Brown's Ferry Nuclear Plant, 282

Cannibalism, 18, 153, 196
    Rate per unit stock, 196

Case histories, 3, 18, 36, 91, 365

Catch curve, 256

Catch efficiencies, 71

Catch catfish (*Ictalurus punctatus*), 256

Clupeid (shad), 282

Coefficient of variation, 71, 243

Columbia River, 36

Compensation, 3, 18, 110, 365
   Compensatory capacity, 153
   Compensatory mechanisms (Fecundity, Growth, Mortality), 153
   Compensatory ratio, 186
   Compensatory reserve, 153, 186, 196
   Compensatory response, 185
   Critical ratio, 186
   Fishing, 153
   Functions, 196
   General concept, 153
   Lack of, in models, 297
   Natural, 153
   Reality of, 153
   Timing of, 153, 196

Confidence intervals (Empirical, Half-width, Normal approximation), 128

Connecticut River, 3

Conowingo Pond, 256

Control stations, 229, 243, 256

Critical period concept, 110

Cumulative effects, 36

Cunner (*Tautogolabrus adspersus*), 297

Delta (Inflow, Outflow), 18

Demand for more and more data, 297

Density-dependent (compensatory) mechanisms, 3, 18, 153, 185, 186, 196, 333. *See also* Compensation.
   Cannibalism, 196
   Competition, 196
   Growth, 185
   Inverse form, 153
   Mortality, 153, 196

Density-independent mechanisms, 3, 18, 153, 196, 297, 311, 333
   "Critical impact," 196
   Depletion, 196
   Diversions (Source of mortality), 18
   Index of survival from, 3

Eigenvalue (Dominant latent root, Maximal, Modulus of), 297, 311

Eigenvector, 297, 311

Entrainment, 18, 110, 127, 153, 184, 196, 311

Environmental Technical Specifications, 229

Equilibrium point, 3, 128, 153, 184, 186, 196, 297, 311, 333

Error
    Biologically acceptable range, 3
    In numerical predictions, 3

Expected order statistic, 256

Experimental design (Completely random, Factorial), 229

Exploitation rates, 153, 184, 186
    50%, 153
    Maximum sustained yield, 153
    Reversibility, 153

Exponential age distribution, 110, 333

Extinction, 153

Fecundity, 3, 128, 196
    Lifetime, 3
    Per unit stock, 196
    Potential, per recruit, 186

Fishing effort (Catchability coefficient, Effective, Weighted), 128

Fish lifts (Efficiency of, Flow regime), 3

Geometric mean ratio, 311

Generation time, 365

Growth rates, 3
    Finite, 297
    Intrinsic, 128, 311
    Mean daily, juvenile shad, 3
    Stable, 311
    Transient, 311

Gulland, J. A., 153

Hatchery, 36

Holyoke Dam, 3

Hudson River, 153, 184, 185, 196, 311

Hypothesis
    Alternative, 229
    Explicit, 297
    Null, 229

Impact assessment, 3, 36, 153, 196, 365
    Limitations, 36, 365
    Models, 196
    Schindler (1976), 3

Impact factor, 196

Impingement, 18, 153, 184, 196, 311

Index of relative stability, 297

Intervention analysis, 229

Intraspecific competition, 3, 110

Juveniles, 71
    Growth rates, 3

Keowee Reservoir, 46, 282

Length-frequency distribution, 256

Length of fish in age class (Mean, Variance, Weighted mean), 256

Leslie matrix, 128, 196, 297, 311, 333

Life cycle, 333
    Model, 333
    Parameters, 3

Life history strategy (r- and k-selection), 297

Life table data, 311
    American shad, 3
    Atlantic menhaden, 71, 128, 297
    Atlantic silverside, 297
    Cunner, 297
    Salmon, 36
    Striped bass, 18, 91, 110, 184, 186, 196, 311, 333
    Winter flounder, 297

Logistic model, 153. *See also* Transformation

Log normal distribution, 229

Logarithms of ratios, 243

Malthus, T., 153

Maryland Power Plant Siting Program, 110

Matrix
    Density-independent, 297, 311
    Dominant latent root of, 297, 311
    Equilibrium, 297, 311
    Leslie, 297, 311, 333
    Perturbed, 297, 311
    Population projection, 297, 311

Maximum likelihood estimates, 256

Mean age, 110

Minimum detectable difference, 229

Mitigation, 36
    Mitigative efforts, 36

Mixture of normals, 256

Models, 365
    Beverton-Holt, 185, 196
    "Impact," 196
    Leslie matrix, 128, 196, 297, 311, 365
    Logistic, 153
    Predictive ability, 333
    Ricker, 3, 36, 91, 153, 185, 196
    Schaefer, 153
    Simulation, 333
    Stock-recruitment, 196

Monitoring program, 91, 110, 229, 243, 256, 365

Monte Carlo simulation, 128

Mortality, 3, 91, 110, 365
    Capacity of fish population to withstand, 153
    Conditional rate of, 184, 186
    Incremental, 3
    Natural instantaneous rate (Deterministic, Nondeterministic) for age class 0, 128
    Total instantaneous rate, 128

Multiple regression, 3, 46, 282
    Step-wise, 3, 282

Natural variation, 365

Neomysis mercedis, 18

Noncentrality parameter, 229

Normal approximation, 128

Null hypothesis, *See* Hypothesis, null

Oconee Nuclear Station, 46

Optimist, 365

Parrish, B. B., 153

Patton, B. C., 282

Peach Bottom Atomic Power Station, 256

Percentage reduction, 196, 333

Percent change detectable, 229

Perturbations, 297, 311
    Empirical index of, 311
    Power plants, 297
    Susceptibility to, 297

Pessimist, 365

Population abundance, 18, 71. *See also* Stock size

Potomac River, 110

Power function, 229, 243

Precision of estimates, 71, 243

Preoperational-operational changes, 229, 243

Predation, 3, 282

Prey-predator relationship, 282

Probability model, 256. *See also* Model

Production
    Early life stages, 110, 365
    Returning adults per spawner, 36, 311

Pseudo-root, 311

Ratio of measurements, as an index of change, 243

Realist, 365

Recruitment, 3, 36, 153, 186, 196

Regression (Functional, Iterative procedure, Step-wise), 127. *See also* Multiple regression

Regression equation, 243, 333

Replacement
    Level, 3, 153
    Line, 36
    Reproduction, 153

Reproduction curve, 3, 36, 153, 186, 196, 333

Reproductive value, 311

Research proposals, 365

Reservoir, 46, 282
    Aging, 46, 282
    Chemical type, 282
    Hydropower/nonhydropower, 282
    Indexes to average conditions, 282

Ricker, 3, 36, 91, 153, 186, 196, 256

Rotenone, 46, 282

Sacramento–San Joaquin Estuary, 18, 91

Salmon, 3, 36

Samples
    Number of, 71, 229, 243
        Acceptable error, 71
        Approximate procedure for estimating, 71, 243
        Half-width of the confidence interval, 71, 128
        Level of precision, 71
    Subsampling, for age determination, 256
    Variability of, 71

Sampling gear, 71, 91
    Gear catch efficiency, 71

Sampling methods, 46, 71
    Acoustical, 110
    Blockage, 71
    Encircling, 71
    Enclosure, 71
    Oblique tow, 91, 110
    Seine, 71
    Trawl, 46, 71

Sampling stations (Control, Number of, Treated), 243

Scaling factor, 333

Sensitivity analysis, 196, 311, 333

Silverside. *See* Atlantic silverside

Simulation models. *See* Models

Simulation study, 196, 256, 297, 311, 333

Spawner-recruit curves, 3, 36, 153, 186, 196, 333
    Replacement line, 36
    Reproduction line, 153

Spawning escapement, 3, 36

Sport fishery, 18, 128, 282
    Angler use predictions, 282

Stable age distribution, **196**, 297, 311

Stable age structure, 110, 311

Stability, 196

Stabilization period, 311

Stations. *See* Sampling stations

Steelhead, 36

Stock
    Depletion percentage, 196
    Parent, 153
    Progeny, 153
    Stock-recruitment (stock-progeny) relationship, 3, 184, 196, 365
        Beverton-Holt type, 196
        Equilibrium point, 184, 196
        Ricker type, 153, 184, 186, 196
        Slope at equilibrium, 196
        Slope at the origin, 196
    Size, 3. *See also* Population abundance
    Value, 196

Stocking, 36, 365

Stopping rule, 243

Striped bass (*Morone saxatilis*), 18, 91, 110, 184, 186, 196, 311, 333

Sunfishes, 46
    Bluegill (*Lepomis macrochirus*), 46

Survival
    From density-independent factors, index, 3
    Of egg to adult (Probability), 3
    Of young to adult, 18
        Abundance index, 18
        Influencing factors (Diversions, Outflow, Power plants, Timing), 18
        Multiple correlation, 18
    Through life stage k, 110

Thermocline, influence on distribution, 46

Threadfin shad (*Dorosoma petenense*), 46, 282

Time frame for decision making, 365

Time series analysis, 243

Total dissolved solids (TDS), 282

Transformation
    Logarithmic, 243
    Logistic, 256

Unavoidable limitations, 365

Uniform age distribution, 110, 333

Vector, 256, 297, 311
    Of age-structure parameters, 256
    Stable age distribution, 311
    Stable age-specific growth rates, 311
    Stable unit age distribution, 311

Viability index, 185

Water temperature, 3, 46, 110
    "Mean responses," 46

Wheeler Lake, 282

White crappie (*Pomoxis annularis*), 256

Winter flounder (*Pseudopleuronectes americanus*), 297

Year-class, 3, 46, 110
    Strength, 3, 46
    Success, 110

Yellow perch (*Perca flavescens*), 46

Young-of-the-year fish, 46, 71, 196, 256, 297
    Abundance, 46, 91
    Vertical distribution, 46

Young-of-the-year model, 311, 333

Zion Nuclear Power Plant, 229